Bevel Gear

Jan Klingelnberg
Editor

Bevel Gear

Fundamentals and Applications

Editor
Jan Klingelnberg
Klingelnberg GmbH
Peterstraße 45, 42499
Hückeswagen
Germany

Editorial team:
Hartmuth Müller
Klingelnberg GmbH, Peterstraße 45, 42499 Hückeswagen, Germany

Joachim Thomas
ZG Hypoid GmbH, Georg-Kollmannsberger-Straße 3, D-85386 Eching-Dietersheim, Germany

Hans-Jürgen Trapp
Klingelnberg GmbH, Peterstraße 45, 42499 Hückeswagen, Germany

Claude Gosselin Ing. Ph.D., Involute Simulation Softwares Inc., Quebec, Canada

ISBN 978-3-662-43892-3 ISBN 978-3-662-43893-0 (eBook)
DOI 10.1007/978-3-662-43893-0

Library of Congress Control Number: 2015949639

Printed on acid-free paper

This Springer Vieweg imprint is published by Springer Nature
The registered company is Springer-Verlag GmbH Berlin Heidelberg

Preface

In all textbooks on gear technology, cylindrical gears usually come to attention first because of their wide use, whereas bevel gears are dealt with superficially. Therefore, readers with a deep interest in bevel gears are not satisfied.

Although the essential differences between bevel and cylindrical gears are always outlined, the true characteristics and features of bevel gears, and their "three-dimensional" nature which varies along the face width, are not sufficiently detailed.

In this book, a team of authors from academia and industry aims to provide a comprehensive textbook on bevel gears.

After covering the major fields of application for these machine elements, this book presents the geometrical attributes of bevel gears and the different cutting methods based on gear theory.

The three-dimensional aspect of the gear teeth is treated in detail, with chapters on tooth flank development, load capacity, and noise behaviour. Descriptions of production processes and necessary technologies provide a knowledge base for sound decision-making.

The aim of this textbook is to introduce the reader to all aspects of the complex world of bevel gears and to present in detail and in understandable form the results of rapid developments in recent years

I like to thank all the co-authors for their contributions and for sharing their knowledge gained through many years of professional experience.

Hückeswagen, Germany
2016

Jan Klingelnberg

Symbols and Units

Symbols	Definition	Units
A	Auxiliary factor for dynamic factor	–
AE	Distance between contact points A and E (complete path of contact)	mm
A_m	Area of the half of ellipse of contact pressure distribution over the middle contact line	mm^2
A_r	Area of the half of ellipse of contact pressure distribution over the root contact line	mm^2
A_t	Area of the half of ellipse of contact pressure distribution over the tip contact line	mm^2
a	Hypoid offset	mm
a_p	Hypoid offset in the pitch plane	mm
a_{rel}	Relative offset	–
a_v	Centre distance of virtual cylindrical gears	mm
B	Accuracy grade acc. ISO 17485	–
B	Auxiliary factor for dynamic factor	–
B_M	Thermal contact coefficient	$\frac{N}{mm^{1/2} m^{1/2} s^{1/2} K}$
b	Minor half-axis of the contact ellipse	mm
b	Face width	mm
b_{2eff}	Effective face width (wheel)	mm
b_e	Face width from calculation point to outside	mm
b_H	Semi-width of Hertzian contact band	mm
b_i	Face width from calculation point to inside	mm
b_v	Face width of virtual cylindrical gears	mm
b_{veff}	Effective face width of virtual cylindrical gears	mm
C_a	Tip relief	μm
C_{eff}	Effective tip relief	μm

Symbols	Definition	Units
C_{ZL},C_{ZR},C_{ZV},	Factors for determining lubricant film factors	–
$C_{1,2}$, C_{2H}	Experimental weighting factors for calculation of bulk temperature	–
c	Component of circumferential speed in plane of contact	m/s
c	Clearance	mm
c_{be2}	Face width factor	–
c_{ham}	Mean addendum factor of wheel	–
c_M	Specific heat per unit	N m/kg K
c_α	Component of circumferential speed in plane of contact in tooth depth direction	m/s
c_β	Component of circumferential speed in plane of contact in direction of face width	m/s
c_γ	Mesh stiffness	N/(mm ·µm)
D	Sum of damage	–
D_1, D_2, D_3	Integration constants of the sliding velocity over the path of contact	–
d_a	Tip diameter of virtual crossed axes helical gear	mm
d_{ae}	Outside diameter	mm
d_b	Base diameter of virtual crossed axes helical gear	mm
d_e	Outer pitch diameter	mm
d_m	Mean pitch diameter	mm
d_s	Pitch diameter of virtual crossed axes helical gear	mm
d_T	Tolerance diameter	mm
d_v	Reference diameter of virtual cylindrical gear	mm
d_{va}	Tip diameter of virtual cylindrical gear	mm
d_{van}	Tip diameter of virtual cylindrical gear in normal section	mm
d_{vb}	Base diameter of virtual cylindrical gear	mm
d_{vbn}	Base diameter of virtual cylindrical gear in normal section	mm
d_{vn}	Reference diameter of virtual cylindrical gear in normal section	mm
d^*_v	Pitch diameter of virtual cylindrical gear acc. FVA411	mm
E	Modulus of elasticity (Young's modulus)	N/mm^2
E	Auxiliary factor for tooth form factor	–
e_{fn}	Tooth slot width at the bottom of the tooth space	mm
F_1, F_2	Auxiliary factors for mid-zone factor	–
F_{ax}	Axial force	N
F_{mt}	Tangential force at mean diameter	N
F_{mtv}	Tangential force of virtual cylindrical gears	N
F_n	Normal force	N
F_p	Cumulative pitch deviation, total	µm
F_{pT}	Cumulative pitch tolerance, total	µm
F_R	Friction force	N
F_r	Runout deviation	µm
F_{rad}	Radial force	N
F_{rT}	Runout tolerance	µm

Symbols	Definition	Units
F_t	Tangential force	N
F_x	Index deviation	µm
F'_i	Single-flank composite deviation, total	µrad/µm
F''_i	Double-flank composite deviation, total	µm
F''_r	Double-flank runout deviation	µm
f	Distance from mid-point M to any line of contact	mm
f_m	Distance from mid-point M to the mean line of contact	mm
f_{max}	Maximum distance to the middle line of contact	mm
f_{pt}	Single pitch deviation	µm
f_{ptT}	Single pitch tolerance	µm
$f_{\alpha lim}$	Influence factor of limit pressure angle	–
f_r	Distance from mid-point M to the root line of contact	mm
f_t	Distance from mid-point M to the tip line of contact	mm
f'_i	Tooth-to-tooth single-flank composite deviation	µrad/µm
f'_k	Short-wave part of single-flank deviation	µrad/µm
f'_l	Long-wave part of single-flank deviation	µrad/µm
f''_i	Tooth-to-tooth double-flank composite deviation	µm
f''_e	Eccentricity of double-flank composite deviation	µm
G	Movement of axis in case of sliding	mm
G	Auxiliary factor for tooth form factor	N
G	Auxiliary value for integration of sliding velocity over path of contact	–
g_{an}	Length of the addendum part of the path of contact in the normal section	mm
g_{fn}	Length of the dedendum part of the path of contact in the normal section	mm
g_n	Distance of a contact point to the pitch or helix point on the contact line in the normal plane	mm
g_t	Distance of a contact point to the pitch or helix point on the contact line in the pitch plane	mm
$g_{v\alpha}$	Length of the path of contact of virtual cylindrical gears	mm
$g_{v\alpha n}$	Length of the path of contact of virtual cylindrical gears in the normal section	mm
H	Auxiliary factor for tooth form factor	–
HB	Brinell hardness	–
HRC	Rockwell hardness	–
HV	Vickers hardness	–
H_V	Tooth mesh loss factor	–
h_{ae}	Outer addendum	mm
h_{am}	Mean addendum	mm
h_{amc}	Mean chordal addendum	mm
h_{a0}	Tool addendum	mm
h_{fe}	Outer dedendum	mm
h_{fi}	Inner dedendum	mm

Symbols	Definition	Units
h_{fm}	Mean dedendum	mm
h_{Fa}	Arm of the bending moment for tooth root stress (load application at tooth tip)	mm
h_m	Mean whole depth	mm
h_{mw}	Mean working depth	mm
h_{t1}	Pinion whole depth perpendicular to root cone	mm
h_1, h_2	Auxiliary variables to determine the sliding velocity	–
I	Integral of sliding velocity distribution	–
j_{en}	Outer normal backlash	mm
j_{et}	Outer transverse backlash	mm
j_{mn}	Mean normal backlash	mm
j_{mt}	Mean transverse backlash	mm
K_A	Application factor	–
$K_{B\alpha}$	Transverse load factor for scuffing	–
$K_{B\beta}$	Face load factor for scuffing	–
K_{F0}	Lengthwise curvature factor for bending stress	–
$K_{F\alpha}$	Transverse load factor for bending stress	–
$K_{F\beta}$	Face load factor for bending stress	–
K_{gm}	Sliding factor for calculation of the friction coefficient	–
$K_{H\alpha}$	Transverse load factor for contact stress	–
$K_{H\beta}$	Face load factor for contact stress	–
K_{mp}	Number of mates meshing with the actual gear	–
K_v	Dynamic factor	–
k_c	Clearance factor	–
k_d	Depth factor	–
k_{hap}	Basic crown gear addendum factor (related to m_{mn})	–
k_{hfp}	Basic crown gear dedendum factor (related to m_{mn})	–
k_t	Circular thickness factor	–
k_1, k_2	Auxiliary constants for the calculation of tooth mesh loss factor	–
L	Auxiliary factor for the calculation of the dimensions of the contact ellipse	–
L_a	Auxiliary factor for correction factor	–
l_b	Length of contact line	mm
l_{bm}	Theoretical length of middle line of contact	mm
l'_{bm}	Projected length of middle line of contact in consideration of inclined lines of contact	mm
m_{et}	Outer transverse module	mm
m_{mn}	Mean normal module	mm
m_{mt}	Mean transverse module	mm
m_{sn}	Normal module of the virtual crossed axes helical gear	mm
N_L, N_I	Number of load cycles	–
n	Rotational speed	1/min
n_I	Number of load cycles of class I	–
n_p	Number of mates meshing with the actual gear	–

Symbols	Definition	Units
P	Nominal power	W
p_e	Base pitch	mm
p_{en}	Normal base pitch	mm
p_{et}	Transverse base pitch	mm
p^*	Related peak load	–
q_s	Notch parameter	–
R	Movement of axis in case of rolling	mm
Ra	= CLA = AA arithmetic average roughness	μm
R_e	Outer cone distance	mm
R_i	Inner cone distance	mm
R_m	Mean cone distance	mm
Rz	Mean roughness	μm
R	Distance of a point on a contact line to the helix axis	mm
r_{c0}	Tool radius	mm
r_s	Half diameter of crossed axes helical gear	mm
SA	Distance of contact points S and A at the contact line of the virtual crossed axes helical gear	mm
S_B	Safety factor for scuffing (contact temperature method)	–
SE	Distance between contact points S and E at the contact line of the virtual crossed axes helical gear	mm
S_F	Safety factor for bending stress (against breakage)	–
$S_{F\ min}$	Minimum safety factor for bending stress	–
S_{FZG}	Load stage in FZG- A/8,3/90-test	–
S_H	Safety factor for contact stress (against pitting)	–
$S_{H\ min}$	Minimum safety factor for contact stress	–
$S_{int\ S}$	Safety factor for scuffing acc. integral temperature method	–
$S_{S\ min}$	Minimum required safety factor for scuffing	–
s_{Fn}	Tooth root chord in calculation section	mm
s_{mn}	Mean normal circular tooth thickness	mm
s_{mnc}	Mean normal chordal tooth thickness	mm
s_{pr}	Amount of protuberance	mm
T	Torque	Nm
T_{1T}	Torque of load stage of the scuffing test	Nm
t_B	Mounting distance	mm
t_{xi}	Front crown to crossing point	mm
t_{xo}	Crown to crossing point (hypoid)	mm
t_z	Pitch apex beyond crossing point	mm
t_{zF}	Face apex beyond crossing point	mm
t_{zi}	Crossing point to inside point along axis	mm
t_{zm}	Crossing point to mean point along axis	mm
t_{zR}	Root apex beyond crossing point	mm
U	Voltage	V
u	Gear ratio	–
u_a	Equivalent gear ratio	–

Symbols	Definition	Units
V_L	Lubricant factor for calculation of the friction coefficient	–
V_R	Roughness factor for calculation of the friction coefficient	–
V_S	Slip factor for calculation of the friction coefficient	–
V_Z	Viscosity factor	–
v_{Bel}	Angle between sum of velocities and the pitch cone	°
v_F	Tangential velocity at the tooth flank	m/s
v_g	Sliding velocity	m/s
$v_{g,par}$	Sliding velocity parallel to the contact line	m/s
$v_{g\alpha}$	Sliding velocity in profile direction	m/s
$v_{g\beta}$	Sliding velocity in lengthwise direction	m/s
$v_{g\gamma}$	Total sliding velocity	m/s
v_{mt}	Tangential speed on the reference cone at mid-face width	m/s
v_t	Tangential speed at any contact point	m/s
$v_{\Sigma,C}$	Sum of velocities at pitch point C	m/s
$v_{\Sigma,h}$	Sum of velocities in the profile direction	m/s
$v_{\Sigma,m}$	Mean sum of velocities	m/s
$v_{\Sigma,s}$	Sum of velocities in the lengthwise direction	m/s
$v_{\Sigma,senk}$	Sliding velocity vertical to the contact line	m/s
$v_{\Sigma\alpha}$	Sum of velocities in the profile direction	m/s
$v_{\Sigma\beta}$	Sum of velocities in the lengthwise direction	m/s
$v_{\Sigma\gamma}$	Total sum of velocities	m/s
W_{m2}	Mean normal slot width of wheel	mm
w	Component of tangential speed in the tangential plane	m/s
w_{Bel}	Angle between contact line and the pitch cone	°
w_{Bn}	Normal unit load	N/mm
w_{Bt}	Transverse unit load	N/mm
$w_{Bt\ eff}$	Effective transverse unit load	N/mm
$w_{Bt\ max}$	Maximum transverse unit load	N/mm
w_α	Tangential speed component in the tangential plane in the profile direction	m/s
w_β	Tangential speed component in the tangential plane in the lengthwise direction	m/s
X_{BE}	Geometry factor (integral temperature method)	–
X_{Ca}	Tip relief factor	–
X_E	Running-in factor	–
X_G	Geometry factor (contact temperature method)	–
X_J	Mesh approach factor (contact temperature method)	–
X_L	Lubricant factor	–
X_M	Thermal flash factor	$\frac{N}{mm^{1/2} m^{1/2} s^{1/2} K}$
X_{mp}	Multiple mating factor	–
X_Q	Mesh approach factor (integral temperature method)	–
X_R	Roughness factor	–
X_S	Lubricant system factor	–

Symbols	Definition	Units
X_W	Structural (welding) factor	–
X_{WrelT}	Relative welding factor	–
X_Γ	Load sharing factor (scuffing)	–
$X_{\alpha\beta}$	Influence factor for pressure and spiral angle	–
X_ε	Contact ratio factor (scuffing)	–
x_{hm}	Profile shift coefficient	–
x_{sm}	Thickness modification coefficient (backlash included)	–
x_{smn}	Thickness modification coefficient (theoretical)	–
Y_{Fa}	Tooth form factor for load application at tip	–
Y_K	Bevel gear factor	–
Y_{LS}	Load sharing factor (bending strength)	–
Y_{NT}	Life factor of the standard test gear	–
$Y_{R\ rel\ T}$	Relative surface factor	–
Y_{Sa}	Stress correction factor for load application at tooth tip	–
Y_{ST}	Stress correction factor for the dimensions of standard test gear	–
Y_X	Size factor for tooth root stress	–
Y_ε	Contact ratio factor (tooth root)	–
$Y_{\sigma\ rel\ T}$	Relative sensitivity factor	–
Z_E	Elasticity factor	–
Z_F	Material factor for the calculation of the semi-width of the Hertzian contact band	$(N/mm^2)^{-1/3}$
Z_H	Zone factor	–
Z_{Hyp}	Hypoid factor	–
Z_K	Bevel gear factor (pitting stress)	–
Z_L	Lubricant factor	–
Z_{LS}	Load sharing factor (pitting)	–
$Z_{M\text{-}B}$	Mid-zone factor	–
Z_{NT}	Life factor of the standard test gear	–
Z_R	Roughness factor for contact stress	–
Z_S	Slip factor	–
Z_v	Speed factor	–
Z_W	Work hardening factor	–
Z_X	Size factor (pitting)	–
Z_β	Helix angle factor for pitting stress	–
z_0	Number of blade groups	–
z	Number of teeth	–
z_p	Number of crown gear teeth	–
z_v	Number of teeth of the virtual cylindrical gear	–
z_{vn}	Number of teeth of the virtual cylindrical gear in the normal section	–
α_{an}	Normal pressure angle at tooth tip	°
α_{dC}	Nominal design pressure angle on coast side	°
α_{dD}	Nominal design pressure angle on drive side	°

Symbols	Definition	Units
α_e	Effective normal pressure angle acc. [ISO23509]	°
α_{eC}	Effective pressure angle on coast side	°
α_{eD}	Effective pressure angle on drive side	°
α_{et}	Effective pressure angle in the transverse section acc. [ISO23509]	°
α_{Fan}	Load application angle at tip circle of the virtual spur gear	°
$\alpha_{Fan\Delta}$	Auxiliary angle for determination of the bending arm at tip	°
α_{lim}	Limit pressure angle	°
α_n	Normal pressure angle	°
α_{nD}	Generated normal pressure angle on drive side	°
α_{nC}	Generated normal pressure angle on coast side	°
α_{sn}	Normal pressure angle of the virtual crossed axes helical gear	°
α_{st}	Transverse pressure angle of the virtual crossed axes helical gear	°
α_t	Pressure angle in the transverse section	°
α_{vt}	Transverse pressure angle of the virtual cylindrical gear	°
α_{wn}	Normal working pressure angle	°
α_{wt}	Transverse working pressure angle	°
β_b	Spiral angle at the base circle	°
β_B	Inclination angle of the contact line	°
β_e	Outer spiral angle	°
β_i	Inner spiral angle	°
β_m, β_v	Mean spiral angle	°
β_s	Helix angle of the virtual crossed axes helical gear	°
β_{vb}	Helix angle at base circle of the virtual cylindrical gear	°
β_w	Working helix angle	°
Γ	Parameter on the line of action	–
γ	Angle of the tangential velocity relative to the contact line	°
γ	Auxiliary angle for the length of the line of contact	°
γ_α	Auxiliary angle for tooth form and tooth correction factors	°
Δa	Centre distance variation	μrad
$\Delta a''$	Double-flank centre distance variation	μrad
Δb_{x1}	Pinion face width increment	mm
Δg_{xi}	Increment along pinion axis from calculation point to inside	mm
Δg_{xe}	Increment along pinion axis from calculation point to outside	mm
ΔH	Pinion mounting distance variation	mm
ΔJ	Wheel mounting distance variation	mm
ΔV	Hypoid offset variation	mm
$\Delta\varphi$	Rotation angle variation	μrad
δ_a	Face angle	°
δ_f	Root angle	°
δ	Pitch angle	°
ε_a	Recess contact ratio	–
ε_f	Approach contact ratio	–

Symbols	Definition	Units
ε_n	Contact ratio in the normal section of virtual crossed axes helical gears	–
ε_v	Tip contact ratio of virtual cylindrical gears	–
ε_{vmax}	Tip contact ratio of virtual cylindrical gears (maximum value from pinion or wheel)	–
$\varepsilon_{v\alpha}$	Profile contact ratio of virtual cylindrical gears	–
$\varepsilon_{v\alpha n}$	Profile contact ratio of virtual cylindrical gears in the normal section	–
$\varepsilon_{v\beta}$	Overlap ratio of virtual cylindrical gears	–
$\varepsilon_{v\gamma}$	Total contact ratio of virtual cylindrical gears	–
ε_{α}	Profile contact ratio	–
ε_{β}	Overlap ratio	–
$\varepsilon_{\beta,Hyp}$	Overlap ratio of virtual cylindrical gears according to FVA411	–
ε_{γ}	Total contact ratio	–
$\varepsilon_{\gamma w}$	Effective total contact ratio	–
ζ_o	Pinion offset angle in face plane	°
ζ_m	Pinion offset angle in axial plane	°
ζ_{mp}	Pinion offset angle in pitch plane	°
ζ_R	Pinion offset angle in root plane	°
η	Wheel offset angle in axial plane	°
η	Efficiency	–
η	Auxiliary value for the minor half-axis of the contact ellipse	–
η_{oil}	Dynamic viscosity at oil temperature	mPas
θ_{a1}, θ_{a2}	Addendum angle	°
θ_{f1}, θ_{f2}	Dedendum angle	°
ϑ	Auxiliary value for the determination of the form factor	–
ϑ	Hertzian auxiliary angle of the contact ellipse	°
ϑ_{Bmax}	Maximum contact temperature	°C
ϑ_{fl}	Flash temperature	°C
$\vartheta_{fl\ max}$	Maximum flash temperature	°C
$\vartheta_{fla\ int}$	Weighted mean flank temperature	°C
$\vartheta_{fla\ int\ T}$	Weighted mean flank temperature of test gears	°C
$\vartheta_{fla\ int,h}$	Weighted mean flank temperature of hypoid gears	°C
ϑ_{flaE}	Flash temperature at pinion tooth tip when load sharing is neglected	°C
ϑ_{flm}	Average flash temperature	°C
ϑ_{int}	Integral temperature	°C
$\vartheta_{int\ S}$	Allowable integral temperature	°C
ϑ_M	Bulk temperature	°C
ϑ_{MT}	Test bulk temperature	°C
ϑ_{oil}	Oil temperature	°C
ϑ_S	Allowable scuffing temperature	°C
λ	Load sharing factor	–
λ_M	Heat conductivity	N/s K
μ_m	Mean coefficient of friction	–

Symbols	Definition	Units
μ_{mC}	Mean coefficient of friction at pitch point	–
μ_{mZ}	Mean coefficient of friction according to Wech	–
υ	Poisson's ratio	–
ν	Lead angle of cutter	°
ξ	Contact ellipse semi-minor axis auxiliary value	–
ρ	Curvature radius	mm
ρ_{a0}	Tool edge radius	mm
ρ_b	Epicycloid base circle radius	mm
ρ_C	Equivalent curvature radius at pitch point	mm
ρ_{Cn}	Equivalent curvature radius at pitch point in the normal plane	mm
ρ_E	Curvature radius at pinion tip	mm
ρ_{ers}	Equivalent curvature radius	mm
ρ_F	Fillet radius at point of contact of 30° tangent	mm
ρ_{lim}	Limit curvature radius	mm
ρ_M	Density	kg/mm^3
ρ_n	Curvature radius in normal section	mm
ρ_{P0}	Crown gear to cutter centre distance	mm
ρ_{red}	Relative radius of curvature	mm
ρ_{Yn}	Curvature radius at contact point Y in normal plane	mm
σ_F	Tooth root stress	N/mm^2
$\sigma_{F\ lim}$	Nominal stress number (bending)	N/mm^2
σ_{F0}	Local tooth root stress	N/mm^2
σ_{FE}	Allowable stress number (bending)	N/mm^2
σ_{FP}	Permissible tooth root stress	N/mm^2
σ_H	Contact stress	N/mm^2
$\sigma_{H\ lim}$	Allowable stress number for contact stress	N/mm^2
σ_{H0}	Nominal value of contact stress	N/mm^2
σ_{HP}	Permissible contact stress	N/mm^2
Σ	Shaft angle	°
$\Sigma\theta_f$	Sum of dedendum angles	°
$\Sigma\theta_{fC}$	Sum of dedendum angles for constant slot width taper	°
$\Sigma\theta_{fS}$	Sum of dedendum angles for standard taper	°
$\Sigma\theta_{fM}$	Sum of dedendum angles for modified slot width taper	°
$\Sigma\theta_{fU}$	Sum of dedendum angles for uniform depth taper	°
φ	Angle between two contact lines	°
φ	Rotation angle	μrad
ω	Angular speed, angular frequency	1/s

Typical Indices

Index	Definition
A, B, D, E	Characteristic points on path of contact
a	Tooth tip
b	Base circle
C	Coast side
C, S, M	Pitch point, helix point, mean point
D	Drive side
e	Outer cone (heel) or "effective"
f	Tooth root
i	Inner cone (toe)
m	Mean cone
n	Normal section
P	Reference profile or basic crown gear
s	Virtual crossed axes helical gear
t	Transverse section
v	Virtual cylindrical gear
x	Any point
Y	Any point on the path of contact
y	Any point
0	Generating tool
1	Pinion
2	Wheel

Typical Abbreviations

Abbreviation	Definition
A	Starting point of path of contact
B	Inner point of single-flank mesh
C	Pitch point
D	Outer point of single-flank mesh
E	End of path of contact
EWP	Single-flank test
FH	Face hobbing (continuous indexing)
FM	Face milling (single indexing)
KS	Structure-borne noise
KSP	Structure-borne noise test
ZWP	Double-flank test
2F	Two-face grinding
3F	Three-face grinding

The original version of this book was revised. An erratum to this book can be found at DOI 10.1007/978-3-662-43893-0_9.

Contents

List of Contributors

Christian Brecher Professor Dr.-Ing., Lehrstuhl für Werkzeugmaschinen, RWTH Aachen

Markus Brumm Dr.-Ing., Klingelnberg GmbH, Hückeswagen

Uwe Epler Dipl.-Ing., Klingelnberg GmbH, Hückeswagen

Adam Gacka Dr.-Ing., Caterpillar Global Mining Europe GmbH, Lünen, Germany

Bernd-Robert Höhn Professor Dr.-Ing., Lehrstuhl für Maschinenelemente, TU München

Carsten Hünecke Dr.-Ing., Klingelnberg GmbH, Hückeswagen

Roger Kirsch Dipl.-Ing., Reishauer AG, Wallisellen

Markus Klein Dr.-Ing., Airbus Helicopters, Donauwörth

Alexander Landvogt Dr.-Ing., Klingelnberg GmbH, Ettlingen

Klaus Michaelis Dr.-Ing., Lehrstuhl für Maschinenelemente, TU München

Hartmuth Müller Dr.-Ing., Klingelnberg GmbH, Hückeswagen

Karl-Martin Ribbeck Dipl.-Ing., Klingelnberg GmbH, Hückeswagen

Berthold Schlecht Professor Dr.-Ing., Institut für Maschinenelemente und Maschinenkonstruktion, TU Dresden

Frank Seibicke Dipl.-Ing., Klingelnberg GmbH, Ettlingen

Michael Senf Dr.-Ing., Institut für Maschinenelemente und Maschinenkonstruktion, TU Dresden

Joachim Thomas Dr. Ing., ZG Hypoid GmbH, Eching-Dietersheim

Hans-Jürgen Trapp Dr.-Ing., Klingelnberg GmbH, Hückeswagen

Olaf Vogel Dr. rer. nat., Klingelnberg GmbH, Ettlingen

Christian Wirth Dr.-Ing., ZG GmbH, Eching-Dietersheim

Chapter 1
Fields of Application for Bevel Gears

1.1 Historical Aspects

In the seventeenth century, workshop mechanization created what has now become an important economic sector—power drive technology. At the beginning of the industrial era, it was possible to use belt drives to successfully transmit power, but with steam engines achieving greater power and higher rotational speeds, the need for more effective drives became a necessity. From the middle of the nineteenth century, a gear manufacturing industry began to grow. At that time, a special relief-turned milling cutter was used for each cylindrical gear to cut the slots between the teeth.

Although in 1765 the Swiss mathematician and physicist Leonard Euler had already discovered in the involute curve a suitable tooth form for the kinematically correct transmission of rotation, there was still a long way to go before the design and making of gear cutting machines which could generate an involute profile. In 1856, Christian Schiele obtained a patent for a screw-shaped cutter to manufacture cylindrical gears, the forerunner of the modern hob.

Heinrich Schicht took up the idea of hobbing cylindrical gears and transferred it to bevel gears, using a conical hob instead of a cylindrical hob to manufacture spiral bevel gears. Schicht and Preis applied for a patent with this idea in 1921.

A different course was followed by Oscar Beale who, around 1900, developed a generating method for the production of bevel gears using two disc-shaped cutters, which could machine both flanks at the same time. Paul Böttcher improved this concept and, in 1910, presented a face mill cutter system which produced spiral-shaped teeth on bevel gears [KRUM50].

J. Klingelnberg (ed.), *Bevel Gear*, DOI 10.1007/978-3-662-43893-0_1

1.2 Vehicle Transmissions

Bevel gears first acquired substantial significance with the developing car industry at the beginning of the twentieth century. At that time, rear axle transmissions, including differential gears, was the normal vehicle power train concept in motor vehicles.

Figure 1.1 shows a typical rear axle drive, with the gear train shown in diagrammatic form. The smaller bevel gear, driven by the Cardan shaft, is called the "pinion" while the larger bevel gear driven by the pinion is called the "wheel".

Fig. 1.1 Principle of a rear axle drive

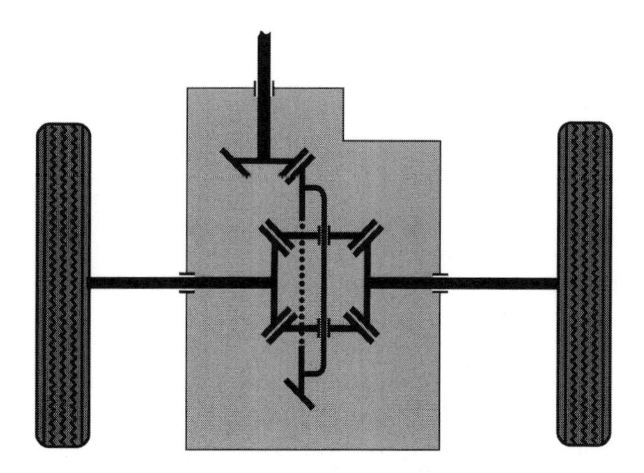

The wheel driving the axle is linked to a carrier with four differential gears. The two gears with horizontal axes each drive a tire. When one of the tires slows down, the other tire rotates faster and there is no slippage with the surface of the road.

The bevel gears in the differential are required to compensate the speed difference between the two tires, whereas the pinion and the wheel are responsible for the transmission of engine power to the tires.

Even today, this principle remains essentially unchanged in all heavy and medium-sized utility vehicles.

Passenger car transmission concepts rely on two different design principles: one in which the engine is mounted transverse to the driving direction, with front-wheel drive; the other in which the engine is mounted lengthwise to the driving direction, using front and/or rear-wheel drive. Vehicles with a transverse engine and front-wheel drive allow very efficient use of the available space, and require no further bevel gears apart from the differential gears.

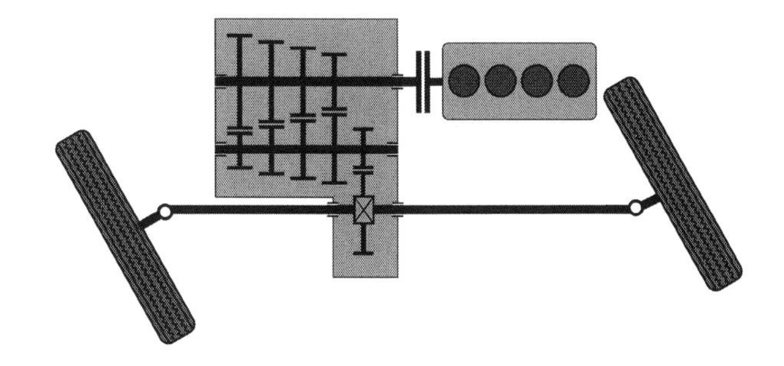

Fig. 1.2 Principle of a front-wheel drive with a transversely mounted engine

As shown in Fig. 1.2, the engine and gearbox are placed next to each other, at a right angle to the driving direction. For the sake of simplicity, the differential gear set is represented by a crossed box. The maximum length of the engine/gear unit is limited by the available mounting space. A further limitation on this concept is imposed by the traction power of the front wheel drive and its influence on the steering of the vehicle. The higher the drive torque, the more the associated forces pull on the steering and the more they affect comfort and driving safety.

Four-wheel drive vehicles have gained significant market ground in passenger vehicles providing improved traction power and road safety. Figure 1.3 shows a transverse engine in a four-wheel drive vehicle. Beside the cylindrical gears for the front axle drive, a power take off unit with bevel gears (PTO) is mounted on the output shaft of the gearbox and drives the rear wheels via a Cardan shaft and a second axle gear with a differential. The two axle drives are linked by means of a center differential, compensating differences in the rotation speed of the front and rear axles.

A common feature in all transverse engine concepts is that they place the power unit ahead of the front axle. This brings advantages in terms of space which is partly outweighed by the greater weight on the front axle. In upmarket cars, a longitudinal engine concept with rear-wheel or four-wheel drive prevails. The superior weight distribution has a positive effect on driving dynamics and safety.

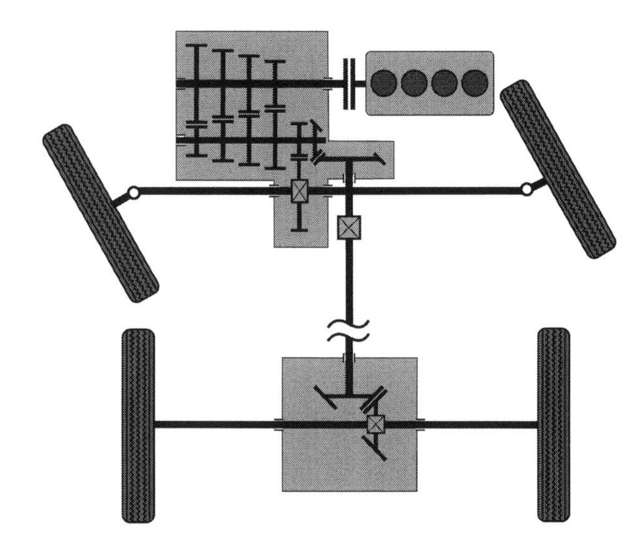

Fig. 1.3 Principle of a four-wheel drive with a transverse engine

Figure 1.4 shows the concept which consists of an engine mounted longitudinally to the driving direction, with front- and rear-wheel drive. The engine lies over the front axle and the front axle gear unit is beside or below the engine. This concept has advantages for bodywork design in terms of accident protection for passengers, as the heavy engine is mounted further back in the engine compartment leaving more space for an energy-absorbing design in the front of the vehicle.

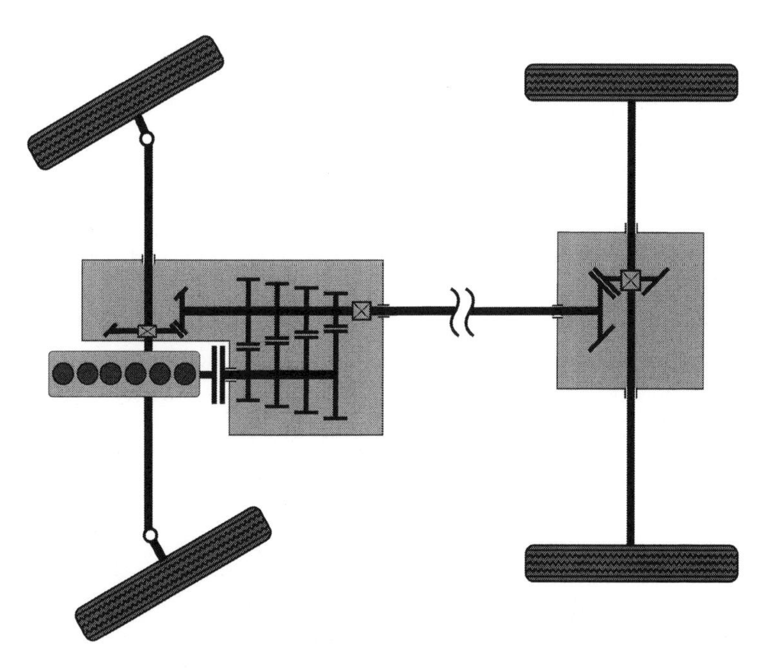

Fig. 1.4 Principle of a four-wheel drive with a longitudinal engine

1.3 Aircraft Engines

Although the volume of bevel gears used worldwide in the automotive sector is the largest, bevel gears also play an essential role in aircraft applications. Bevel gears are used wherever rotary movements have to be transmitted between two non-parallel axes. Typical applications include main rotor and tail rotor drives for helicopters, starter and hydraulic drives for aircraft turbines or flap actuators for aircraft wings.

1.3.1 Aircraft Turbines

Gas turbines have been used in aircraft engines for many decades. In turboprop engines, the shaft of the gas turbine acts through a gear train to power a propeller, while in a turbofan engine it drives a fan. Turbofan engines for large passenger aircraft have fan diameters of up to 3 m. Despite their very high power density, pure jet engines are not used in civilian aviation because of their poor efficiency and high noise emissions.

The actual engine, the gas turbine, is an internal combustion engine with a continuous gas through flow. At the front of the engine, air is compressed in one or several stages by axial or centrifugal compressors; it is then mixed with kerosene in the combustion chamber, ignited and burnt. Additional air is used for cooling. The resulting hot gas expands in the aft turbine section, converting thermal energy into mechanical energy which initially serves to drive the compressor ahead of the combustion chamber. In a pure jet engine, the remaining energy is used to accelerate the hot gas flow, and is thus converted into thrust. In a turboprop or turbofan engine, the remaining energy is converted into mechanical energy and used to drive either the propeller or the fan.

A small proportion of the energy in the turbine shaft is diverted to operate power consumers such as a generator or a hydraulic pump. The principle of the gear train is shown in Fig. 1.5. A first set of bevel gears, driving a shaft mounted at right angle to the turbine, appears on the turbine shaft, just ahead of the combustion chamber. At the lower end of this shaft is a second set of bevel gears, driving the consumer through a number of cylindrical gears to achieve the required rotational speed and direction of rotation. If this consumer is a generator, it can be reversed to act as an electric starter-motor to start the turbine. Because of the high rotation speeds and relatively low torques involved, these bevel gears are subject to quite different demands than those used for example in automotive axle gear drives.

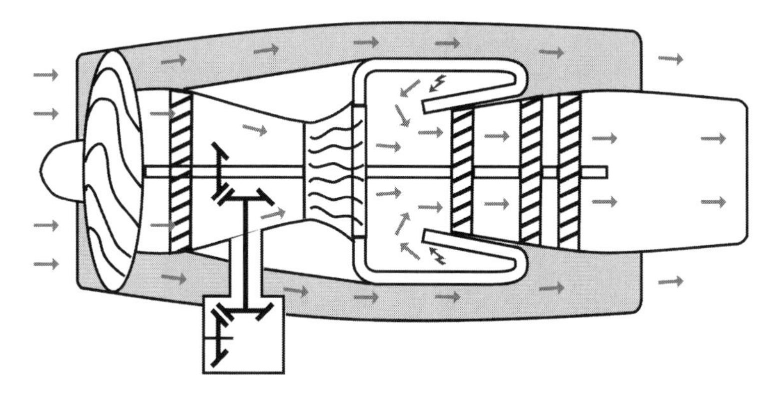

Fig. 1.5 Gas turbine with bevel gear trains

1.3.2 Helicopter Gears

Similarly to airplanes, helicopter engines are usually gas turbines and deliver the power used to drive the main and tail rotors while the residual exhaust jet supplements propulsion during flight. Because the shaft of gas turbines is invariably mounted horizontally, an angular gear train is needed to power the main rotor. The counter torque generated by the main rotor around the vertical axis of the helicopter is compensated by means of a tail rotor (Fig. 1.6).

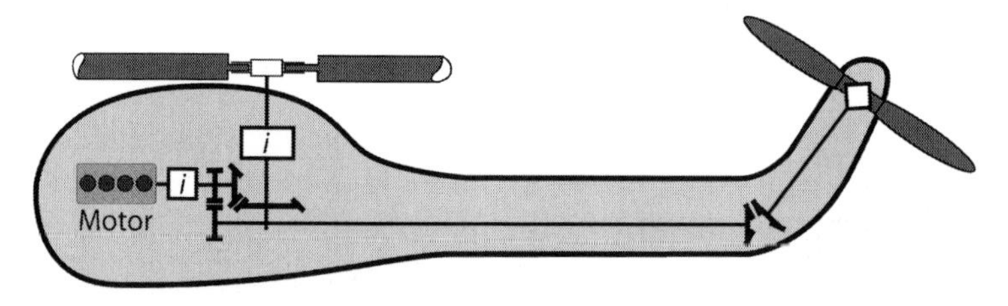

Fig. 1.6 Principle of a helicopter drive

The rotational speed of the main rotor is always chosen such that blade tip velocity at maximum forward flight speed is subsonic. Depending on the diameter of the rotor, this results in rotor speeds below 500 RPM. Typical power turbine shaft speeds being above 8,000 RPM, and thus significantly higher than that of the rotor, a reduction gearbox with a large transmission ratio is required. Planetary gears are ideal for such applications: the bevel gears are placed ahead of the input stage to the planetary gear train such that the gear designer is dealing with higher rotational speeds rather than higher torques.

Other bevel gear sets are necessary to drive the tail rotor. If a continuous shaft runs from the main gear unit to the tail rotor, one bevel gear set is required such that the longitudinal shaft can drive the tail rotor mounted at right angle. When it is not possible to install a continuous shaft (Fig. 1.6), a number of shaft sections are required. Further bevel gear sets are then needed to transmit rotation to each of the sections of the tail rotor drive.

1.3.3 Flap Drives on Aircraft Wings

Apart from rapidly rotating bevel gears in aircraft engines, another application is the actuation of wing flaps. In addition to ailerons controlling aircraft roll, the wings contain flaps to alter airfoil shape at take-off and landing. Flaps are at the trailing edge of the wing profile and as they extend aft out of the wing, their angle of attack increases. For safety reasons, all flap movements must be mechanically interlocked. This is achieved by means of a central shaft driving individual angular gear sets and a crank mechanism moving the flap backwards horizontally and at the same time increasing their angle of attack to the airflow.

For aerodynamic reasons, the wings of a modern aircraft are backswept and the airfoil varies along the wing. Because of flight safety requirements, all flap movements must originate from a central shaft. If the trailing edge of the wing is not straight, the central shaft actuating the flaps must be interrupted in several places. Figure 1.7 shows the rotary mechanisms, cranks and bevel gear drives used to actuate the flaps. At each change in direction along the central shaft, a bevel gear set is used to transmit the rotary motion which moves the flap. Unlike the high speed bevel gears in the turbine, flap drive bevel gears perform only slow speed servo movements and therefore require a completely different design.

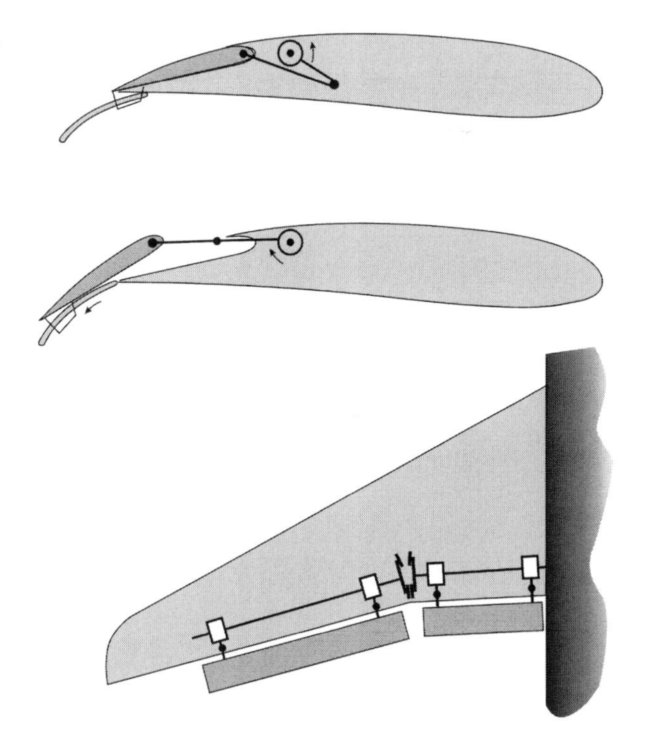

Fig. 1.7 Principle of a flap drive mechanism

1.4 Marine Drives

The classic power train concept for large vessels where the rudder is installed behind a propeller mounted on a shaft driven by the ship's engine is becoming less and less common. A desire for improved maneuverability first led to the development of the bow thruster where a tubular opening containing a propeller passes laterally through the entire width of the bow section of the ship, below the waterline. The propeller can move the bow to port or starboard either when the ship is stopped or moving ahead or astern at very slow speed. This is done by reversing the direction of rotation of the propeller or altering the pitch of the propeller blades. The propeller is powered by an electric or hydraulic motor installed in the ship.

The principle of the bow thruster is illustrated in Fig. 1.8. The motor installed inside the ship drives the propeller shaft through bevel gears. An alternative concept uses a direct drive electric motor for the bow thruster. Although the simplicity of this concept is attractive, the technical problems in sealing the direct drive sustainably under water are not easily solved. The classic bevel gear drive has advantages in terms of reliability and fail-safe properties since, even if the gearbox seal fails, a bow thruster with sufficient oil pressure in the gearbox will continue to operate for several weeks whereas a direct drive will sustain immediate seawater damage.

Fig. 1.8 Bow thruster unit

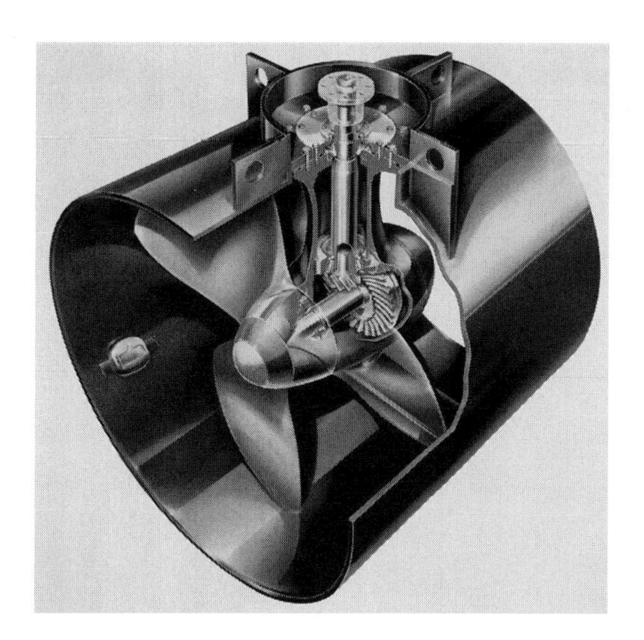

External thrust drives provide optimum maneuverability. Thrust drives are drive units mounted below the hull and capable of 360° rotation. Figure 1.9 shows such a thrust unit with two counter-rotating propellers. The counter-rotating propellers compensate turbulence in the water flow, thus providing increased efficiency. The bevel gear train contains two wheels driven by a common pinion.

Most thrust drives have only a single propeller. For a 10 MW thrust drive, the propeller diameter is approximately 5 m. The maximum power using this concept is limited by the bevel gears and is, at the moment, in the order of 15 MW.

External thrust drives are especially adapted to icebreakers since they can be used to push fragmented lumps of ice under the ice sheet; the edge of the sheet is then partially lifted out of the water when the bow of the ship next rides up over it and is thus easier to break since it is not resting on the surface of the water. Because of the expected propeller impacts on the ice, external thrust drive power for icebreakers is currently limited to around 7.5 MW [ROLLS].

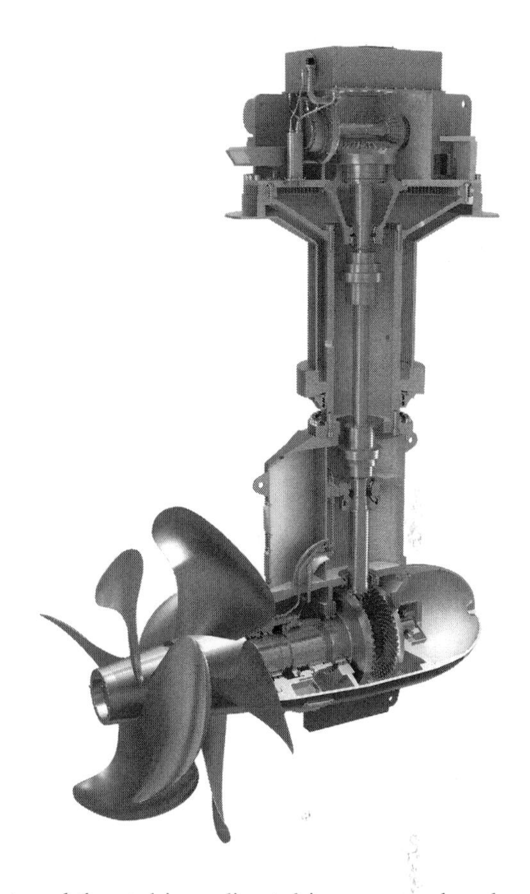

Fig. 1.9 External thrust drive

In order to increase power in external thrust drives, direct drives are employed. Instead of a bevel gear transmission from a motor above the waterline to the underwater propeller, an electric motor driving the propeller directly is fitted in the underwater gear housing. Drives of this kind can attain powers up to 20 MW [SCHOT]. Apart from the sealing problem mentioned above, these units differ from the geared variants through their much greater mass. Whereas, for example, a 10 MW gear-driven bow thruster has a mass of approximately 70 t, a direct drive concept of the same power will weigh 170 t [SCHOT]. The bending stresses on the ship's hull are many times higher, requiring a more complex hull structure and causing more complicated bearings for turning the thruster around its vertical axis (azimuth bearing).

1.5 Industrial Gears

Bevel gear use in general gear engineering is extremely varied. Whenever rotary motion is transmitted between two non-parallel axes, either a worm gear pair or a bevel gear set is used. The greater efficiency and comparatively simpler manufacture of bevel gears are often decisive factors when comparing to other skew axis gear transmissions.

At transmission ratios between 1:1 and 1:10, bevel gears are the preferred machine element. The higher the transmission ratio, the closer bevel gears come to their limits. Transmission ratios of 1:20 and more are realized in some bevel gear applications. However, this is possible only if a comparatively large pinion offset is envisaged. The pinion used in such cases is worm-shaped, with only two to three teeth which in practice are no longer supported by a sizeable gear body, thereby significantly restricting the torque that such gear sets can transmit.

References

[KRUM50] Krumme, W.: Klingelnberg Spiralkegelräder, 2nd edn. Springer, Heidelberg (1950)
[ROLLS] Rolls Royce Marine: Internet: www.rolls-royce.com/marine
[SCHOT] Schottel GmbH: Internet: www.schottel.de

Chapter 2
Fundamentals of Bevel Gears

2.1 Classification of Bevel Gears

Bevel gears can be classified according to various attributes.

These relate to:

- the progression of tooth depth (or height) along the face width,
- the type of tooth trace, i.e. straight or curved teeth,
- the form of the tooth trace curve,
- the pinion hypoid offset,
- the type of indexing operation, continuous or single indexing,
- the cutting method, generation or plunge cut,
- and the manufacturing method

Tooth depth along the face width can be constant or variable. With constant tooth depth, the face and root angles are of equal value, such that the depth of the tooth remains the same over the entire face width. On bevel gears with variable tooth depth, also known as tapered teeth, the face and root angles differ, causing a proportional change in tooth depth along the face width. At the small diameter of the bevel gear (toe), tooth depth is less than that at the large diameter (heel). Constant tooth depth may be regarded as a special case of tapered teeth (Fig. 2.1).

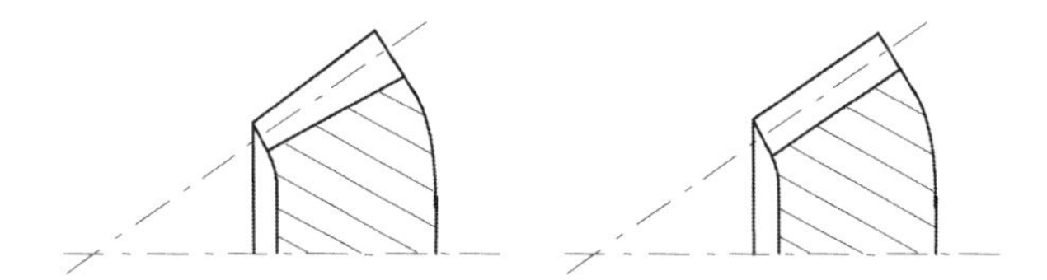

Fig. 2.1 Bevel gears with variable and constant tooth depth

The original version of this chapter was revised. The erratum to this chapter is available at DOI 10.1007/978-3-662-43893-0_9.

J. Klingelnberg (ed.), *Bevel Gear*, DOI 10.1007/978-3-662-43893-0_2

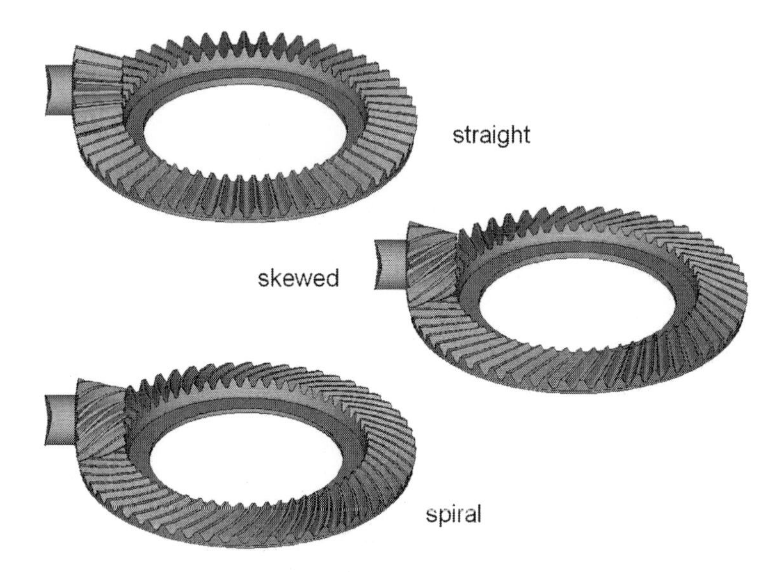

Fig. 2.2 Straight, skew and spiral bevel gears

Other bevel gear criteria are the type and form of the tooth trace on the basic crown gear (see Sect. 2.2.2). Depending on the type of tooth trace, bevel gears may be differentiated according to Fig. 2.2 into:

- straight bevel gears
- skew bevel gears
- spiral bevel gears

On spiral bevel gears, it is possible to draw a further distinction in terms of the form of the tooth trace, which may be:

- a circular arc,
- an elongated epicycloid,
- an involute or
- an elongated hypocycloid

Bevel gears may likewise be classified with respect to their hypoid offset. Bevel gears with no pinion offset have intersecting axes while bevel gears with pinion offset, known as hypoid gears, have crossed axes. In the latter case, a further distinction may be drawn between gears with positive or negative offset (see Fig. 2.3).

Positive offset:

- the pinion axis is displaced in the direction of the spiral angle of the wheel,
- the mean helix angle of the pinion is larger than that of the wheel,
- the diameter of the pinion increases when compared to that of an equivalent gear set with no offset.

Negative offset:

- the pinion axis is displaced in a direction opposite that of the spiral angle of the wheel,
- the mean helix angle of the pinion is smaller than that of the wheel,
- the diameter of the pinion decreases when compared to that of an equivalent gear set with no offset.

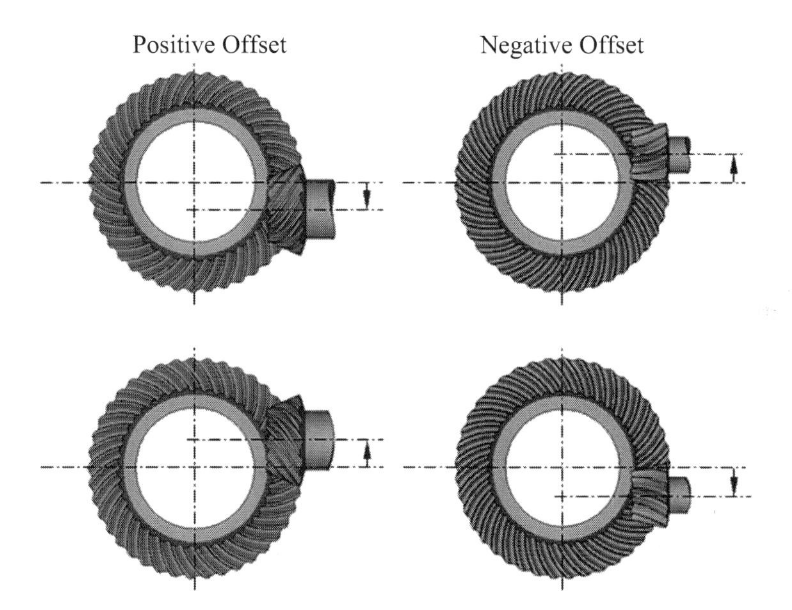

Fig. 2.3 Definition of hypoid offset

Spiral bevel gears, when manufactured in a metal-cutting process, can either be produced in single indexing or continuous indexing operations which govern the form of the tooth trace.

In the single indexing or face milling method, one tooth slot is cut, the tool is retracted and, after the work piece has been rotated by one pitch, the next tooth slot is cut until all the slots have been done. Since the cutting edges of the tool are arranged in a circle, e.g. on a face mill cutter, the tooth traces will show the form of a circular arc.

In the continuous indexing, or face hobbing method, the rotation of the cutter and that of the bevel gear being produced are coupled in such a way that at any time only one blade group passes through a particular tooth slot, the next blade group passing through the next slot etc. (see Fig. 2.4). Indexing is therefore continuous and all tooth slots are cut quasi-simultaneously. On the basic crown gear, these motions result in a tooth trace in the form of an elongated epicycloid. When the epicycloid is being machined, the ratio of the number of teeth to the number of starts on the cutter (number of blade groups) is equivalent to the ratio of the base circle radius to the roll circle radius. An elongated epicycloid occurs when the radius on which the cutting edges are positioned is greater than that of the rolling circle.

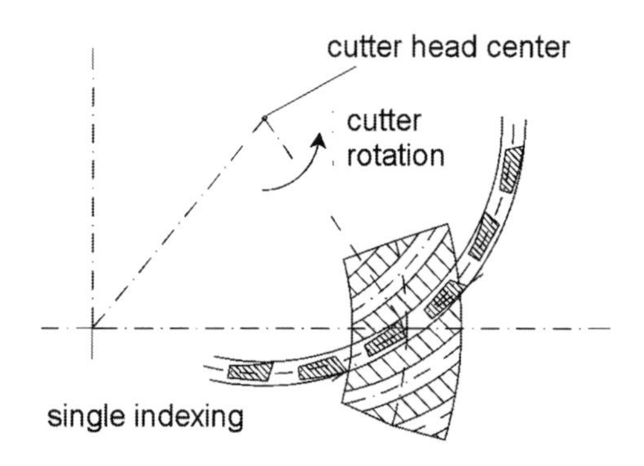

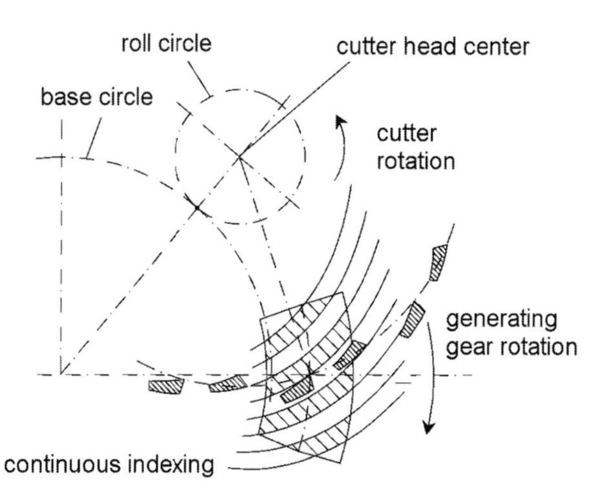

Fig. 2.4 Single indexing and continuous indexing methods

When manufacturing a generated spiral bevel gear, a curved tooth profile is created on both the pinion and wheel by a generating motion. An alternate method is to only generate the pinion and form cut the tooth spaces on the wheel by a simple plunge cut process, which is known as a non-generated method or FORMATE®. The process saves time in machining the wheel, and can be used for transmission ratios of about 2.5 and above. Since there is no generating motion when cutting the wheel, the tool profile is reproduced exactly in the tooth slot. The associated pinion is produced in a modified generating process (see Sect. 3.2.1) to ensure that it meshes properly with the wheel.

Spiral bevel gear tooth geometry depends on the manufacturing method employed, not only in terms of the classification criteria noted above, but also with respect to the final tooth flank and root topography. For example, it is impossible to pair a pinion produced with the Zyklo-Palloid® method to a wheel manufactured with the Spiroflex method, although both are made with a face hobbing process and match in terms of their macro geometry (e.g. normal module and elongated epicycloid).

Zyklo-Palloid® This is a continuous indexing method in which both bevel gears, the pinion and wheel, are always generated. Tooth depth is constant and the tooth trace is an elongated epicycloid. The special feature of this method is the use of a two-part cutter head (see Fig. 6.4) where one of the two interleaved parts of the cutter head holds inner blades cutting the convex tooth flanks, and the other part holds outer blades for the concave tooth flanks. Profile crowning is achieved by a spherical blade profile (cutter blade profile modification) while lengthwise crowning is obtained by a difference in radius between the inner and outer blades. The two-part cutter head allows the production of bevel gears in a so-called two-flank cut process, meaning that the two flanks of a bevel gear are manufactured in a single operation. A further advantage is that lengthwise crowning can be simply produced and can be adjusted continuously without tilting the cutter (see Sect. 3.2.1). The cutter blades are standardized and each size is employed for a specific range of modules. The Zyklo-Palloid method is used for both soft and hard cutting. Hard cutting processes are designated as HPG-S for small modules (≤ 8 mm) and HPG for larger modules. For spiral bevel gears with diameters above 1,200 mm, the Zyklo-Palloid method is currently the only process which allows hard finishing (see Sect. 6.4). Soft cutting is always performed using an oil based cooling lubricant, whereas hard finishing is a dry cutting operation.

Palloid® This process differs fundamentally from all other processes since the tool employed is a tapered hob similar to a "Christmas tree" (see Fig. 6.6) rather than a face cutter head. The method is used to generate spiral bevel gears of constant tooth depth in a conventional hobbing process which produces an involute tooth trace. In the normal section, the tooth thickness, space width and slot width are constant along the face width. This tooth form results in very light sensitivity of the contact pattern to the relative displacements of the mating gears (see Sect. 3.4.4), giving Palloid gears special merits over other spiral bevel gear types. Lengthwise crowning and some profile modifications are achieved by regrinding the tool. Additional flank modifications are made by means of the manufacturing kinematics. One disadvantage of this process lies in the special tool which needs to be adapted for different types of flank modification. For example, a change in tooth thickness requires a different tool. The largest available cutter module is currently limited to 8 mm. A further disadvantage is its lower productivity when compared to modern machining processes. The method is also limited because of the maximum possible hypoid offset and the choice of spiral angle. The only process used for hard finishing is lapping.

Zyklomet® This is a special form of the Zyklo-Palloid method, in which the tooth slots of the wheel are plunge cut with a face hobbing cutter head, and only the pinion is generated using a cutter which is inclined or tilted relative to the pitch plane (see Sect. 3.2.1). Lengthwise crowning is provided solely by the wheel while profile crowning is produced using spherical blade profiles. For the pinion, only one-part cutters are used while the wheel cutter is made of two parts and in some cases fitted with a larger number of blade groups. With one-part cutters, the lengthwise crowned wheel has to be manufactured in a single-flank process. For this purpose special bevel gear cutting machines with two cutters on their rolling cradles have been developed. This method is applied very rarely nowadays.

N-method This is a face hobbing method working with two-flank cuts. The tooth trace is an elongated epicycloid and tooth depth is constant. Both the pinion and the wheel are generated. The special feature of the N-method is that the epicycloid lengthwise curvature at the mean point corresponds to that of an involute, also termed 'involute case' or 'rectangular case' (see Sect. 3.4.4). This makes it possible to produce lengthwise crowning using a combination of face hobbing cutters with different resulting angles where inner and outer blades immediately follow each other on the cutter face. A drawback, however, is that the spiral angle of the gear then depends on the mean cone distance and the selected cutter diameter, and can no longer be chosen at will. Therefore, it is not possible to produce every spiral angle on a gear set with the available cutters, which have three blades per blade group (inner, middle and outer). Profile crowning is achieved using spherical blade profiles. This method is used only to a very limited extent now.

Spiroflex This is another face hobbing method in which the pinion and wheel are generated. Tooth depth is constant and the tooth trace is an elongated epicycloid. The pinion and wheel are manufactured in a two-flank cut. Profile crowning is inherent in the tool whose blades have a spherical profile. Lengthwise crowning is produced by tilting the cutter head (see Sect. 3.2.1). The pressure angles of the blades are adjusted to generate the specified flank pressure angles. The process is performed both wet and dry, whereby cutters with three blades (one roughing and two finishing blades) or with two finishing blades per group are employed. Cutters with only two blades per group (inner and outer blade only) are a recent development, allowing a larger number of starts (number of blade groups) with unchanged blade cross–section and tip radius. The method is used widely in the mass production of vehicle spiral bevel gears. Lapping is employed for hard finishing (see Sect. 6.5).

Spirac® Spirac is the name for the variant of the Spiroflex method in which the wheel is only plunge cut. It produces a bevel gear with constant tooth depth and an elongated epicycloid tooth trace. The tools used are the same as those for the Spiroflex method; lengthwise and profile crowning are also produced in the same way. Since forming, or plunge cut processes can only be used for gear ratios beyond 2.5, this variant is used widely in the automotive industry.

TRI-AC®/PENTAC®-FH These face hobbing methods are respectively equivalent to the Spiroflex and Spirac methods. They only differ in the design of the blades and cutters, which have one inner and one outer blade per blade group. Continuous indexing is used as well to produce spiral bevel gears with constant tooth depth and an elongated epicycloid tooth trace. Lengthwise crowning is obtained by tilting the cutter and profile crowning by using spherical blades. These methods are also applied in the automotive industry.

Kurvex Kurvex gears are face milled and have a circular arc tooth trace with a constant tooth depth. The cutter resembles that of the Zyklo-Palloid method, being divided in two parts, but only with inner and outer blades; there are no roughing or middle blades. The two-part cutter is necessary to produce an exact tapered space and slot width. Both the pinion and wheel are generated. Lengthwise crowning is created by a fixed specified difference in cutter radii. The method is characterized by a high degree of standardization, allowing a small number of tools to cover a large range of gear production. However, standardization also means that lengthwise crowning results from the choice of cutter size and cannot be modified freely. Machines are no longer produced for this method.

5-cut The term "5-cut" method derives from the steps in which the gears are manufactured. The wheel is machined in two cuts (roughing and finishing), i.e. both tooth flanks are cut simultaneously; the pinion is machined in three cuts, i.e. roughing of both tooth flanks, finishing of the convex tooth flank and finishing of the concave tooth flank. It is a single indexing or face milling method with a circular arc tooth trace and with tapered teeth. The generating motion is generally related to the root cone rather than to the pitch cone, as would be kinematically correct. As a feature of this method, the wheel is finished in a two-flank cut and the pinion in single flank cuts. The machine and tool settings for one flank of the pinion are independent of those for the other. One resulting advantage lies in the independent geometries of the concave and convex pinion tooth flanks, since any change in machine kinematics for subsequent flank modifications on one side has no effect on the other side. Lengthwise crowning is usually obtained by differences in cutter radii, while profile crowning is obtained by modifying the machine kinematics. The method is used for both generated and non-generated bevel gears. It is still used on a large scale in the aircraft industry, but has largely been replaced by the Completing method in other sectors.

Completing This face milling method is applied for mass production. The pinion and wheel are produced by two-flank cuts which gave this method its name. Tooth depth is variable and tooth trace is a circular arc. Lengthwise crowning is obtained by tilting the cutter, profile crowning by modifying machine kinematics and/or using spherical tools. As the resulting slot widths of the pinion and wheel are constant, the root and tip angles of the teeth depend on the chosen cutter diameter, and are not freely selectable. This method is characterized as a duplex bevel (see Fig. 2.13). Profile blades as well as stick blades are used in the face milling tools. The Completing method can be wet or dry cut. The hard finishing operation is usually grinding. This method is well established for ground gear sets in the automotive industry.

Arcoid This face milling method, comparable to the 5-cut and Completing processes, produces a bevel gear with tapered teeth and a circular arc as tooth trace. Differences affect the type of cutters, the milling technologies and the additional motions originally used to modify the tooth flanks. Only the so-called helical motion was employed (see Sect. 3.3.3), as this was the only additional motion the machines were equipped for.

Wiener 2-trace Bevel gears produced according to this 2-trace method by Wiener have constant tooth depth and circular arc tooth traces. This single indexing method is used mainly as a grinding operation in small series production. The term '2-trace' derives from the fact that the convex and concave tooth flanks of the pinion and wheel are each manufactured individually, using different tools and machine settings. Therefore, in order to perform the operation as productively as possible, grinding machines destined for this method are provided with a double spindle carrying two grinding wheels. The pinion and the wheel are produced by a generating process, lengthwise crowning results from appropriately chosen differences in tool radii and profile crowning from spherical tool profiles. The method is also used to grind gears which have been roughed with the Zyklo-Palloid method. The circular arc of the grinding wheel is then adjusted closely enough to the elongated epicycloid form for the different tooth traces to produce as little variance as possible in the grinding allowance.

Wiener 1-trace This is also a single indexing method for bevel gears with constant tooth depth. Unlike the Wiener 2-trace method, it produces the wheel in a two-flank operation, and the geometry of the individually produced pinion tooth flanks is adapted accordingly, e.g. the tooth slot is more tapered than in the Wiener 2-trace process. This method is used for both generating and forming processes in which the wheel is plunge cut.

Semi-completing This is a finishing method used for single flank grinding of bevel gears roughed with the Zyklo-Palloid method. Spiral bevel gears finished in this way have a constant tooth depth and a circular arc tooth trace. Both the pinion and the wheel are generated. A feature of this method is that the two flanks of the bevel gear are machined with different machine settings but using the same tool. This means that, unlike the Wiener 2-trace method, it allows the actual tool radii required for machining both the concave and convex flank to be accommodated on a single grinding wheel. This is achieved by modifying the machine kinematics, either by tilt, helical motion or modified roll (see Sect. 3.3.3), the latter being the more usual method. As only one tool is required, the first flank is finished by a generating motion in one direction on the machine, and the second flank by generating in the other direction. The machine settings are modified at the point of reversal. Thus, this method profits from the reverse motion which anyway is required to generate the opposite tooth flank, leading to advantages in productivity when compared to the Wiener 2-trace method. Lengthwise crowning is usually generated by means of different tool radii and profile crowning by means of a spherical tool profile (Table 2.1).

Table 2.1 Overview of the major cutting methods for spiral bevel gears

Manufacturing method	Indexing method	Tooth trace curve	Tooth depth	Slot width[a]		Profile crowning	Lengthwise crowning
				pinion	wheel		
Zyklo-Palloid®/ Zyklomet®	Continuous	Epicycloid	Constant	Variable	Variable	In the tool	Radius difference
Palloid®	Continuous	Involute	Constant	Constant	Constant	–[b]	In the tool
N-method	Continuous	Epicycloid	Constant	Variable	Variable	In the tool	Lead angle difference
Spiroflex/Spirac®	Continuous	Epicycloid	Constant	Variable	Variable	In the tool	Cutter tilt
TRI-AC®/PENTAC®-FH	Continuous	Epicycloid	Constant	Variable	Variable	In the tool	Cutter tilt
Kurvex	Single	Circular arc	Constant	Variable	Variable	–[c]	Radius difference
Arcoid	Single	Circular arc	Variable	Variable	Variable/ constant	Machine kinematics	Cutter tilt
5-cut	Single	Circular arc	Variable	Variable	Constant	Machine kinematics	Radius difference
Completing	Single	Circular arc	Variable	Constant	Constant	Machine kinematics	Cutter tilt
Wiener 2-Spur	Single	Circular arc	Constant	Variable	Variable	In the tool	Radius difference
Wiener 1-Spur	Single	Circular arc	Constant	Variable	Constant	In the tool	Radius difference
Semi-Completing	Single	Circular arc	Constant	Variable	Variable	In the tool	Radius difference

[a]At root cone in normal section
[b]Tool with tip relief
[c]Tool with protuberance

2.2 Gear Geometry

2.2.1 General

This chapter deals with the macro geometry of bevel gears, leaving aside modifications to the micro geometry which affect tooth contact (see Sect. 3.3). Because of their conical nature, the macro geometry of bevel gears alters continuously along the face width. Therefore, bevel gears cannot generally be described in such a simplified way as cylindrical gears. The pitch cones of a hypoid gear pair can be obtained from different definitions (see Fig. 2.3). In the past, numerous definitions were made with different perspectives in mind. The main geometry, for example, was described partly in the centre and partly at the heel of the tooth. Since 1997 an expert group in the International Organisation for Standardization has concentrated on the basic geometry of bevel gears, creating the ISO 23509 standard "Bevel and Hypoid Gear Geometry", which aims to unify all the commonly-used methods for the definition of bevel and hypoid gear geometry.

Amongst other things, it was agreed that the term "bevel gears" should be used as a generic name, embracing all sub-species like spiral bevel gears, non-offset bevel gears, Zerol® gears and hypoid gears. The specific hyponyms will be employed when the following text relates to one or more sub-type.

2.2.2 Basic Geometry

Every non-offset bevel gear pair has two cones which roll on each other without sliding, just like the pinion and the wheel. The apices of these 'basic bodies' or equivalent rolling cones meet at the point where the axes intersect and remain in contact along a common generatrix. The shaft angle Σ is delimited by the axes of the pitch cones $\delta_{1,2}$ of the bevel gear pair.

The generation of a cylindrical gear set may be illustrated by means of an imaginary (virtual) rack. In the case of a bevel gear, the rack is replaced by a usually planar virtual crown gear whose pitch angle is $\delta_P = 90°$. Figure 2.5 shows a bevel gear pair with a shaft angle $\Sigma = 90°$ and the associated virtual crown gear.

Figure 2.6 describes 3 bevel gear pairs with equal outside diameters but different shaft angles, along with the associated crown wheel in each case.

Fig. 2.5 Bevel gear pair with associated virtual crown gear

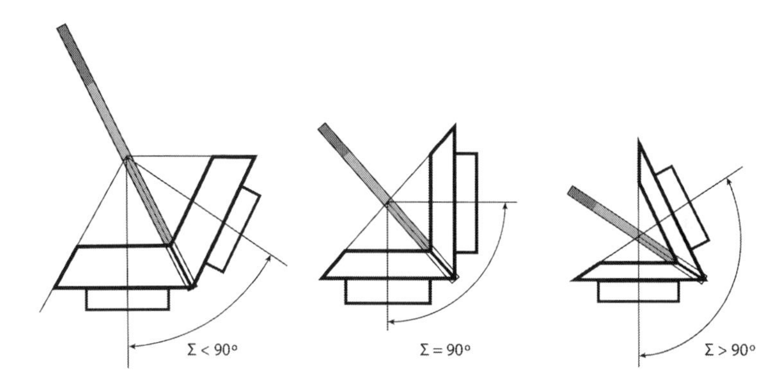

Fig. 2.6 Pairing options with different shaft angles

2.2.3 *Gear Dimensions*

The gear dimensions of non-offset bevel gears are shown in the axial section in Fig. 2.7, and their designations are listed in Table 2.2. Figure 2.8 shows Section A-A from Fig. 2.7. On bevel gears, Section A-A is defined as the transverse section and is always perpendicular to the pitch cone. Thus, Section A-A is not a plane section in itself, but rather corresponds to the complementary cone at the point under consideration. In Fig. 2.8, the complementary cone including the mean point is unrolled into an imaginary plane where the original mean pitch (cone) diameter d_m is transformed into the equivalent pitch diameter $d_v = d_m/\cos\,\delta$; all dimensions found in Fig. 2.8 are detailed in Table 2.3. The corresponding main dimensions for hypoid gears are given in Fig. 2.9 and described in Table 2.4.

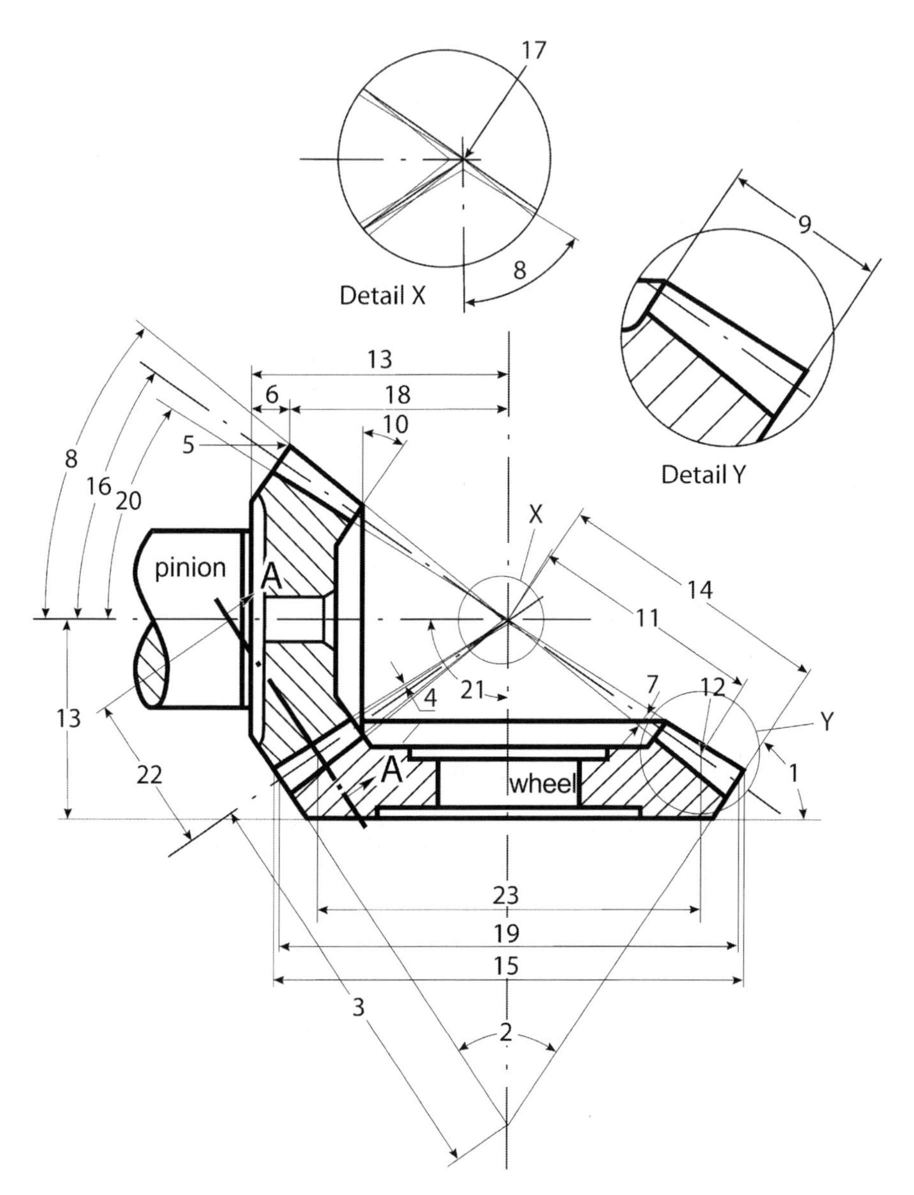

Fig. 2.7 Bevel gear geometry definition in the axial section [ISO23509]

Table 2.2 Key to Fig. 2.7

No.	Designation	No.	Designation
1	Back angle	13	Mounting distance, t_{B1}, t_{B2}
2	Back cone angle	14	Outer cone distance, R_e
3	Back cone distance	15	Outside diameter, d_{ae1}, d_{ae2}
4	Clearance, c	16	Pitch angle, δ_1, δ_2
5	Crown point	17	Pitch cone apex
6	Crown to back	18	Crown to crossing point, t_{xo1}, t_{xo2}
7	Dedendum angle, θ_{f1}, θ_{f2}	19	Outer pitch diameter, d_{e1}, d_{e2}
8	Face angle, δ_{a1}, δ_{a2}	20	Root angle, δ_{f1}, δ_{f2}
9	Face width, b	21	Shaft angle, Σ
10	Front angle	22	Equivalent pitch radius
11	Mean cone distance, R_m	23	Mean pitch diameter, d_{m1}, d_{m2}
12	Mean point		

Note See Fig. 2.8 for mean transverse section A-A

Table 2.3 Key to Fig. 2.8 (section A-A in Fig. 2.7)

No.	Designation	No.	Designation
1	(Mean) whole depth, h_m	7	Tooth thickness, s_{mc} (mean chordal)
2	Pitch point	8	Backlash
3	Clearance, c	9	(Mean) working depth, h_{mw}
4	Circular thickness, s_t	10	Addendum, h_{am}
5	Circular pitch	11	Dedendum, h_{fm}
6	Addendum h_{amc} (mean chordal)	12	Equivalent pitch radius

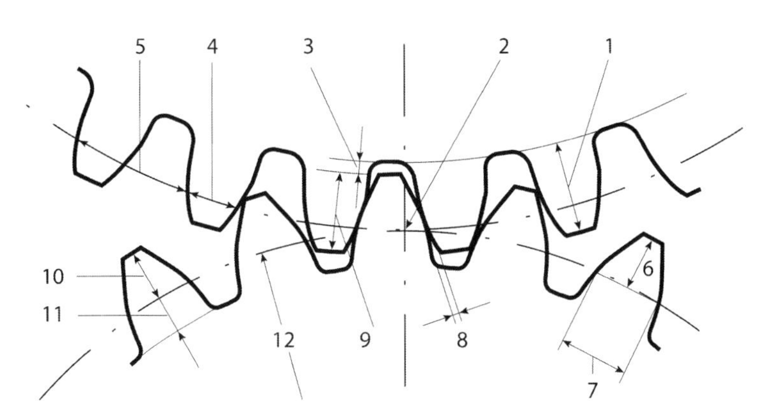

Fig. 2.8 Bevel gear geometry definition in the mean transverse section [ISO23509]

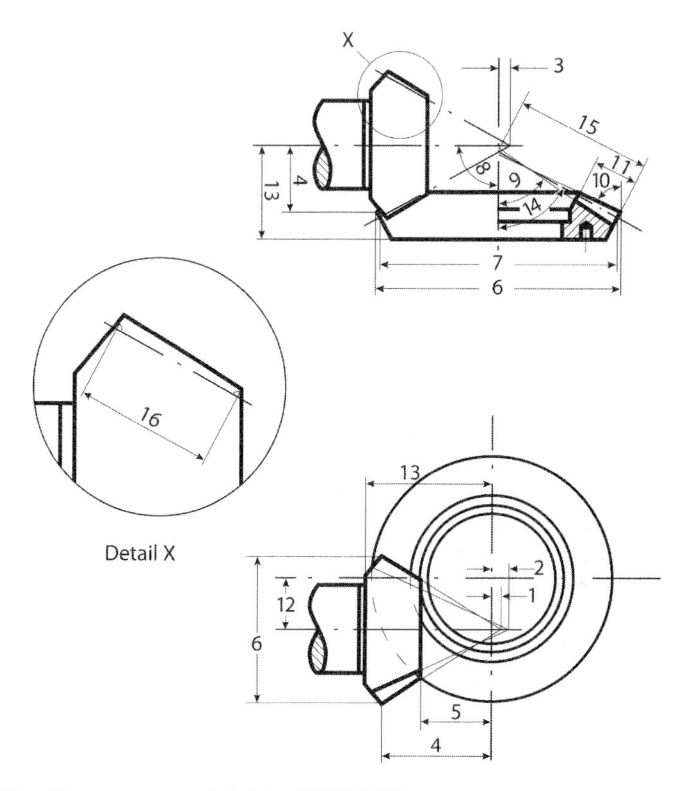

Fig. 2.9 Hypoid gear geometry definition [ISO23509]

Table 2.4 Key of Fig. 2.9

No.	Designation	No.	Designation
1	Face apex beyond crossing point, t_{zF1}	9	Root angle, δ_{f1}, δ_{f2}
2	Root apex beyond crossing point, t_{zR1}	10	Face angle, δ_{a1}, δ_{a2}
3	Pitch apex beyond crossing point, t_{z1}	11	Face width of the wheel, b_2
4	Crown to crossing point, t_{xo1}, t_{xo2}	12	Hypoid offset, a
5	Front crown to crossing point, t_{xi1}	13	Mounting distance, t_{B1}, t_{B2}
6	Outside diameter, d_{ae1}, d_{ae2}	14	Pitch angle of the wheel, δ_2
7	Outer pitch diameter, d_{e1}, d_{e2}	15	Outer cone distance, R_e
8	Shaft angle, Σ	16	Face width of the pinion, b_1

Note Distances beyond the crossing point of the mate have positive values, distances before the crossing point of the mate have negative values

2.2.4 *Tooth Form*

2.2.4.1 Tooth Profile

If a non-offset bevel gear is rolled on its fixed mating gear, one point of the rotating tooth flank will move over a spherical surface whose center is the crossing point of their axes. The tooth profile corresponding to this point is obtained from the intersection of the bevel gear tooth and the surface of the sphere [NIEM86.3] or, with sufficient accuracy, from the unrolled complementary cone (see Sect. 2.2.3).

For bevel gears, as for cylindrical gears, a trapezoidal profile is preferred as the reference profile, i.e. as the tooth profile of the basic rack (see Fig. 2.20). Therefore, in the normal section, the tooth flanks of the virtual crown gear are straight and, in the bevel gear generating process, move as cutting edges along the specific tooth trace. The resulting tooth flanks of the so-called octoid gear are identical to the enveloping surfaces, on the bevel gear, generated by the straight flanks of the virtual crown gear when the pitch cones of the crown gear and bevel gear roll on each other. The cutting method used to generate octoid gear teeth is therefore equivalent to the generation of involute teeth on cylindrical gears. However, one consequence of rolling on a cone, as opposed to rolling on a cylinder, is that the line of action in case of meshing octoid gear teeth deviates slightly from a straight line. On the corresponding spherical surface, the projected line of action appears as a figure-eight curve (see Fig. 2.10). Despite the deviation of the line of action (E) from a straight course, the octoid gears are kinematically exact.

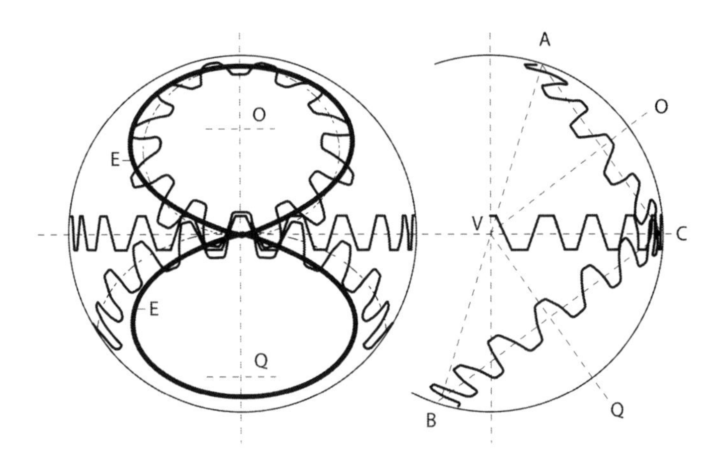

Fig. 2.10 Definition of an octoid gear [NIEM86.3]

What is termed as the spherical involute would also be suitable as a tooth profile for bevel gears. This tooth form has a virtual crown gear with a curved tooth flank profile whose direction of curvature changes on the pitch plane. The tooth flanks are created by unrolling a conical envelope from the base cone, but as they can only be manufactured point by point, this form of gear is of no practical interest.

2.2.4.2 Tooth Depth and Blank Geometry

The progression of tooth depth along the face width and the blank geometry of bevel gears depend essentially on the chosen manufacturing method (see Table 2.1). It is also impossible to determine the geometry of bevel gears without knowing what cutting method will be used. For simplification, Fig. 2.11 defines the most important tooth proportions on the basis of straight bevel gear teeth:

- The tooth depth may be constant—or uniform—along the face width or may increase continuously from toe to heel and is measured perpendicularly to the pitch cone. Tooth depth and root angle determine the size and shape of the blank.
- The tooth thickness is variable along the face width and is measured on the pitch cone in the transverse or normal section.
- The slot width is dependent on the cutting method and is usually tapered. Only in the cases of the Palloid® method—a continuous indexing process—and the Completing method—a single indexing process (see Sect. 2.1)—is the slot width constant in the normal section along the face width. In the Completing method, this is achieved by a tilted root line as indicated in Fig. 2.12, again showing straight bevel gear teeth for simplicity. In all other cases, the slot width is determined by the structure of the tool, the point width and the tool tip radius.
- Along the pitch cone of a bevel gear (i.e. in the pitch plane of the crown gear), the space width in the normal section is usually not constant. The exception is the Palloid process which has an involute as the tooth trace and is therefore self-equidistant in every normal section. Because of the tilted root line, the Completing method attains the desired constant slot width but tooth space width is tapered along the pitch cone (see Fig. 2.12).

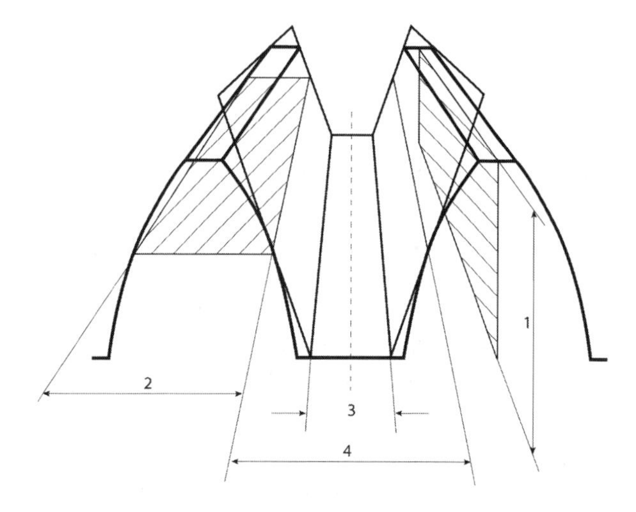

Fig. 2.11 Tooth form variables [ISO23509]
1 tooth depth
2 tooth thickness
3 slot width (crown gear)
4 space width in the pitch plane

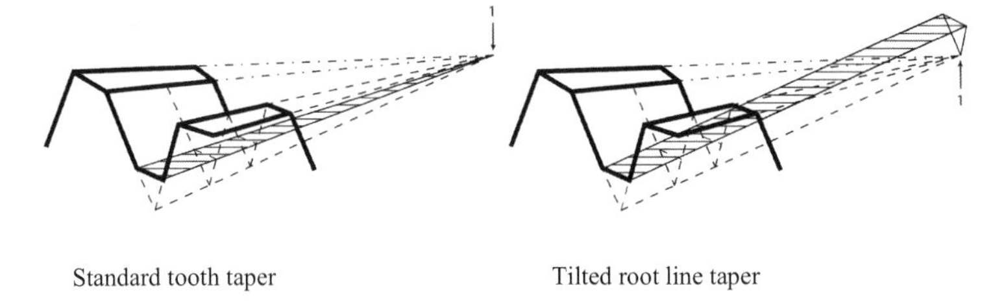

Fig. 2.12 Principle of the tilted root line [ISO23509]. 1 Apex of the pitch cone

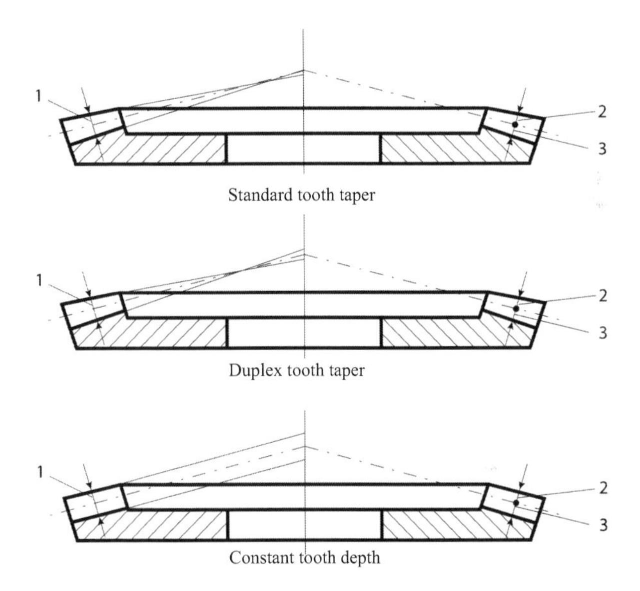

Fig. 2.13 Tooth depth wise taper. 1 mean whole depth, 2 mean addendum, 3 mean dedendum

Figure 2.13 shows the customary tooth depth variants, or depth wise taper, which are described briefly below. The formulae for the associated angles are given in Tables 2.12 and 2.13.

Standard depth tooth depth is directly proportional to the cone distance at any particular section along the tooth. The extended root line intersects the axis of the bevel gear at the pitch cone apex. The extended tip line intersects the axis at a different point, defined by the root line of the mating gear plus a constant clearance. The sum of the dedendum angles of the pinion and wheel is not dependent on the tool radius. Most straight bevel gears are of the standard depth type.

Constant slot width (Duplex) this tooth depth form occurs when the root line has to be tilted as required in the case of the Completing method, to obtain a constant slot width in the normal section of the pinion and wheel (see Fig. 2.12). The formulae in Tables 2.12 and 2.13 indicate that the tool radius r_{c0} has a significant effect on the tilt angle of the root line. Too large a tool radius being produces unreasonably small tooth depth at the toe and excessive tooth depth at the heel. As a result, tooth tips become too thin at the heel and there is a danger of undercutting at the root. It is therefore recommended that the tool radius r_{c0} should be no larger than the mean cone distance of the wheel R_{m2}. If the tool radius is too small, the opposite effect occurs, and therefore the selected value should not be lower than 1.1 R_{m2} sin β_{m2} (cf. Sect. 3.4.4).

Modified slot width in this case, the wheel member has a constant slot width as with the Duplex taper, but the pinion member does not. Therefore, the Completing method can be used only for the wheel (see Tables 2.12 and 2.13) and the amount of tilt is somewhat arbitrary.

Uniform depth if the height of the bevel gear tooth is constant along the face width, the face and root angles are of equal value and, except in the case of angular correction, both will be equal to the pitch angle. The tooth tip line then runs parallel to the tooth root line. Should the tooth tips become thinner than the limiting value at which full hardening or crack formation begins, a tip chamfering is performed (see Fig. 2.14).

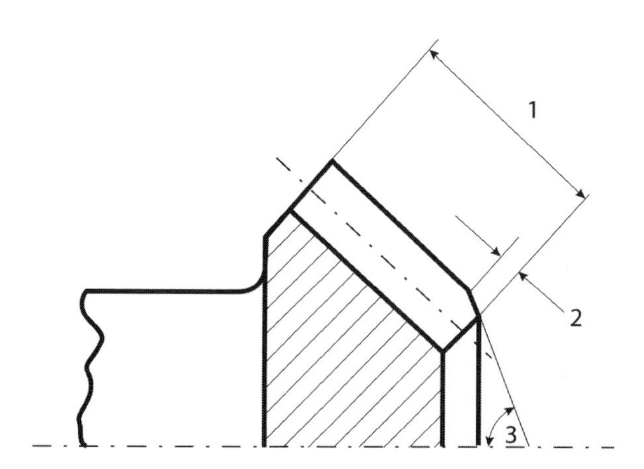

Fig. 2.14 Tooth tip chamfering. 1 face width b, 2 length of chamfer, 3 angle of chamfer

A dedendum angle modification may be provided in order to prevent the tool from colliding with a journal bearing or with the shaft of the work piece (see Fig. 2.15). This involves a rotation of the tooth tip and tooth root lines round the mean point of the gear (i.e. the design point), by an amount which generally should not exceed 5°. Unlike all other bevel gears, gears with angular corrections have constant tooth depth defined perpendicularly to the root cone rather than the pitch cone.

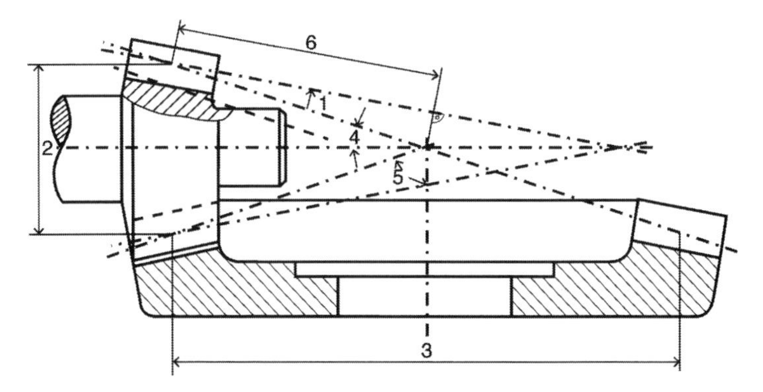

Fig. 2.15 Dedendum angle modification. 1 dedendum angle modification, 2 mean pitch diameter d_{m1} of the pinion, 3 mean pitch diameter d_{m2} of the wheel, 4 pitch angle δ_1 of the pinion, 5 pitch angle δ_2 of the wheel, 6 mean radius R_{mP} of the generating crown gear

2.2.4.3 Longitudinal Tooth Form

The various longitudinal tooth forms have been described in Sect. 2.1. The commonest forms today are the circular arc created with the single indexing method and the elongated epicycloid produced with the continuous indexing method. The first is referred to as face milling (FM), the second as face hobbing (FH).

2.2.4.4 Hand of Spiral

In order to define the hand of spiral of a bevel gear, one looks from the apex of the pitch cone toward a tooth which is at the 12 o'clock position. If the tooth, seen from front to back, curves to the right, the gear has a right hand of spiral, and vice versa. The concave flank is normally on the right side of the tooth on right handed bevel gears, and on the left side of the tooth on left handed bevel gears (sees Fig. 2.16). In the special case of an inverse spiral, this situation is reversed [SEIB03].

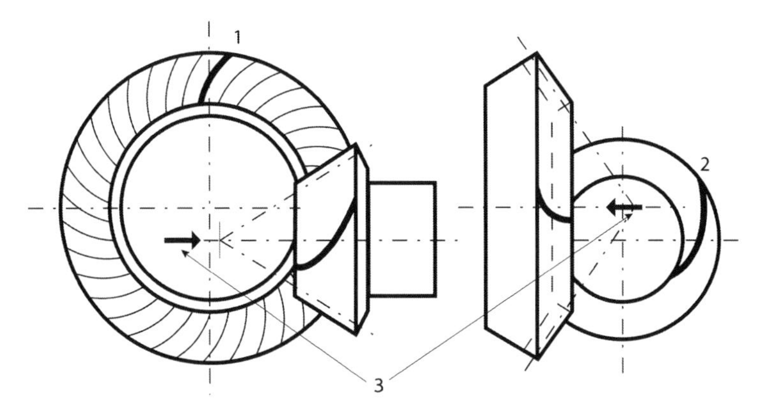

Fig. 2.16 Definition of the hand of spiral. 1 right hand, 2 left hand, 3 viewed from the pitch cone apex

2.2.4.5 Drive and Coast Tooth Flanks

On spiral bevel gears with a positive hypoid offset, load conditions are favorable if the concave tooth flank of the pinion drives the convex tooth flank of the wheel (see Sect. 3.4.2). Usage in car axle drives has led to the choice of this flank of the pinion tooth for the drive mode, i.e. when the engine is moving the vehicle forward. Inversely, the convex tooth flank of the pinion is loaded when in coast mode, i.e. when the engine brakes the moving vehicle. From the preceding considerations, the tooth flanks on all spiral bevel gear sets, with or without hypoid offset, receive the following designations:

pinion	concave flank	=	drive flank
	convex flank	=	coast flank
wheel	convex flank	=	drive flank
	concave flank	=	coast flank

2.2.4.6 Contact Ratio

Similarly to cylindrical gears, the contact ratio describes the average number of teeth which are simultaneously engaged. A distinction is drawn between the profile contact ratio ε_α, which results from the working height of the profile, and the overlap ratio ε_β, which is determined by the helix or spiral angle. The total contact ratio ε_γ is the sum of these two parameters. The relationships are illustrated in Fig. 2.17. The sum of the contact and overlap ratios constitutes the total contact ratio only if the tooth flanks are conjugate to each other. The calculation can be made by means of virtual cylindrical gears in the normal section (see Table 4.3). When bevel gear teeth are crowned because of their operating conditions (see Sect. 3.1), the total contact ratio is somewhat smaller than that of non-crowned bevel gear teeth. Using an elliptical contact pattern as a basis, it is possible to calculate the total contact ratio as the square root of the sum of the profile contact and overlap ratios squared [COLE52]. This approach is generally used when calculating the load capacity according to standards (see Table 4.2), and it represents a good approximation.

The total contact ratio can be determined more exactly using tooth contact analysis (see Sect. 3.3). The actual effective total contact ratio (see Fig. 5.4) depends on tooth crowning and, in addition, depends to a significant extent on the actual load which changes the dimensions of the contact pattern. It can therefore only be calculated using an analytical method which accounts for the flattening of the tooth profiles in contact under load, the deflection of the teeth due to their flexibility and, where appropriate, the deflections of the axes of rotation. Best suited for this task are the usual modern methods of calculation using FEM or BEM for tooth contact analysis under load (see Sect. 4.4).

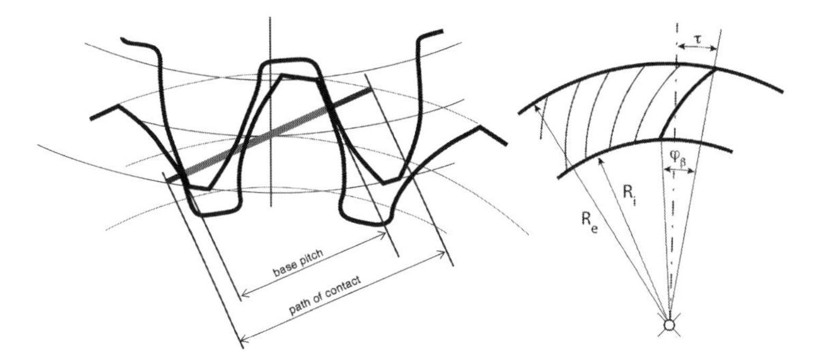

Fig. 2.17 Contact ratio

profile contact ratio $\varepsilon_\alpha = \dfrac{\text{path of contact}}{\text{base pitch}}$ (2.1)

overlap ratio $\varepsilon_\beta = \dfrac{\text{included angle of the face width}}{\text{included angle of one axial pitch}} \dfrac{\varphi_\beta}{\tau}$ (2.2)

total contact ratio $\varepsilon_\gamma = \varepsilon_\alpha + \varepsilon_\beta$ valid for conjugate tooth flanks (2.3)

total contact ratio $\varepsilon_\gamma = \sqrt{\varepsilon_\alpha^2 + \varepsilon_\beta^2}$ for an elliptical contact pattern (2.4)

2.2.5 *Hypoid Gears*

2.2.5.1 Hypoid Offset

If an offset is included in a bevel gear set, the starting point is usually the wheel and the pinion axis is displaced by the desired offset in such a way that, in a common tangential plane (the crown gear plane), the pitch cones of the two gears contact each other in the middle of their face widths at the mean point. The direction of the offset results from the hand of spiral of the wheel, depending on whether a positive offset (larger pinion diameter than without offset) or a negative offset (smaller pinion diameter than without offset) is desired (cf. Sect. 2.1 and Fig. 2.3).

Amongst other things, offset is a design element allowing the gear engineer to reconcile the demands of ground clearance of cars, gearbox space, load capacity and noise behavior. It is useful to define a relative offset a_{rel} for the assessment of hypoid offsets, irrespectively of the size of the wheel.

$a_{rel} = 2a/d_{m2}$
$a_{rel} = 0$ applies to bevel gears without hypoid offset
$a_{rel} = 1$ applies, for example, to crossed axes helical gears

The sign is determined by the sign rule for hypoid offsets (see Fig. 2.3). With a positive relative offset, there is a corresponding increase not only in the diameter, the spiral angle and the pitch angle of the pinion, but also in the overlap ratio and the axial force. With a negative relative offset, the parameters behave in exactly the opposite way until, in the extreme case ($a_{rel} = -1$), the pinion becomes cylindrical. The effects on load capacity, efficiency and noise behavior are described in Chaps. 4 and 5.

2.2.5.2 Hypoid Geometry

Hypoid gears represent the general case of bevel gears. Instead of rolling cones which are the 'basic bodies' of gears without offset, the 'basic bodies' of hypoid gears are two one-sheet hyperboloids (hence hypoid gears, see Fig. 2.18). They contact each other along a straight line, the helix axis, and roll together with simultaneous sliding in the lengthwise direction (helical motion).

For these gears to be economically produced, and since bevel gear teeth usually include profile and lengthwise crowning, the helical pitch surfaces are approximated by means of conical surfaces. As a result, only the mean point P on the pinion and wheel exactly satisfies the condition required by the helical motion. This provides a "scaffolding" for a hypoid gear set in which, starting from the mean point P, the pitch cones of the pinion and wheel span a common plane of contact and the gear axes do not intersect but cross at centre distance a (see Fig. 2.19).

The associated pitch angles depend not only on the gear ratio and shaft angle (as in the special case of non-offset bevel gears), but also on the offset, tool radius and other factors. The pitch angle can only be calculated iteratively (see Sect. 2.3.2). The mean point P is also termed the calculation or design point.

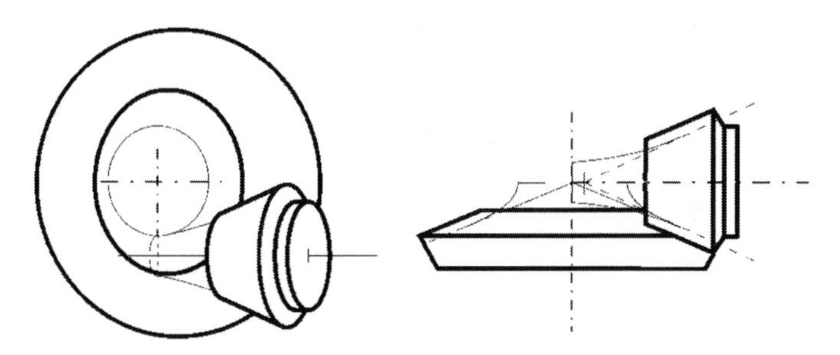

Fig. 2.18 Hypoid gear set

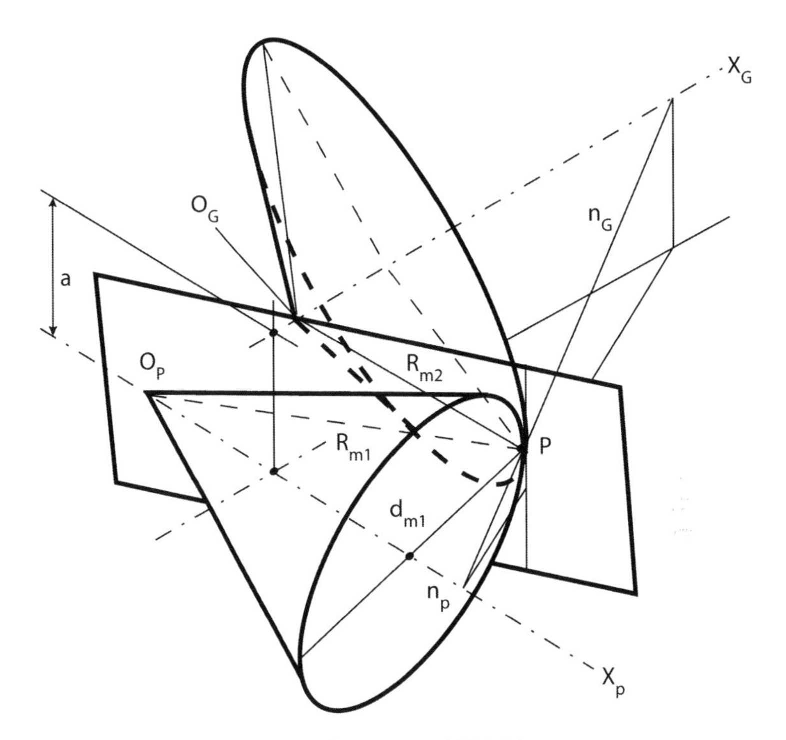

Fig. 2.19 Schematic elements of a hypoid gear set [ISO23509]

2.2.5.3 Pressure Angle

The pressure angles (see Sect. 3.1) chosen for a bevel gear set will be referred to below as design pressure angles α_d. They are not necessarily identical on both tooth flanks. In particular, hypoid gears are made with different pressure angles on the drive and coast sides (see Sect. 2.2.4.5) in order to balance the sliding conditions on the two flanks. In general, the smallest theoretically feasible pressure angle is not equal to 0°, as in the case of cylindrical gears and non-offset bevel gears, but is rather equal to the limit pressure angle α_{lim}. It is highly dependent on parameters such as the hypoid offset, spiral angle and tool radius (see Formula (2.35)). In this formula, expression "R_{m1} sin β_{m1}–R_{m2} sin β_{m2}" may become zero for certain hypoid gear designs, in which case the limit pressure angle will also be $\alpha_{lim} = 0°$. However, when α_{lim} is not equal to 0°, the sliding conditions and paths of contact on the drive and coast tooth flanks become different. This can be compensated by adding or subtracting the amount of α_{lim} to or from the design pressure angles on the drive and coast flanks respectively.

An influence factor of the limit pressure angle $f_{\alpha lim}$ is introduced so that the full amount of the limit pressure angle need not always be taken into account when calculating the blade angles of the tool. This is done partly because it is not always simple to change blade angles, for instance when standardized tools or form blades are used, and partly because of the use of stick blades or grinding wheels whose blade angles may differ but cannot be smaller than a limiting value if these tools are to be re-sharpened in their longitudinal direction. On form blades, a value of $f_{\alpha lim} = 0$ is employed meaning that there is neither compensation of the pressure angle and, hence, nor of the sliding conditions; however, if $f_{\alpha lim} = 1$, full compensation is applied.

For the Completing method, a value of $f_{\alpha lim} \approx 0.5$ is generally used because a large amount of tool tilt is required. This tilt must also be allowed for in both blade angles of the tool, potentially entailing very low values for one of the blade angles and, thus, making it impossible to re-sharpen the tool.

The design pressure angle α_d is modified by the limit pressure angle α_{lim}, multiplied by the influence factor of the limit pressure angle $f_{\alpha lim}$, to obtain the flank angle α_n of the generating crown gear (see Formulae (2.55) and (2.56)). The pressure angle, being determinant for the sliding conditions, is known as the effective pressure angle α_e and is always equivalent to the blade angle α_n minus the limit pressure angle (see Formulae (2.57) and (2.58).

These interrelationships are described in the formulae contained in Table 2.10.

2.3 Bevel Gear Geometry Calculation

2.3.1 Structure of the Calculation Method

Once the input data have been established, it is possible to calculate the pitch cone parameters of the pinion and wheel. The calculation is either for a non-offset bevel gear pair or for a hypoid gear pair. In principle, there are various methods to calculate the pitch cone parameters of hypoid gear pairs, for instance the one described in [AGMA2005], the Oerlikon method and the method defined in the Klingelnberg Standard [KN3029]. The results of these different approaches are quite similar.

The iterative calculation of hypoid geometry data proceeds either from the given geometry parameters or from those selected on the basis of load capacity considerations (see Sect. 3.1). Depending on the chosen method, geometry parameters include the shaft angle Σ, hypoid offset a, number of teeth of the pinion and wheel $z_{1/2}$, mean or the outer pitch diameter of the wheel d_{m2} or d_{e2}, spiral angle of the pinion or wheel $\beta_{m1/2}$, tool radius r_{c0} and, if necessary, number of blade groups z_0 in the tool. The parameters for the pitch cones of the pinion and wheel are calculated on the basis of these inputs, allowing the schematic of a hypoid gear pair to be constructed as shown in Fig. 2.19. The equations used to calculate bevel gear geometry without offset have a closed solution, i.e. a complete geometry calculation is also possible on the basis of other geometry data.

In both cases, a closed solution of the remaining gear dimensions is possible using the pitch cone parameters. The necessary additional geometry inputs can be made according to the factor system typically employed in Europe (data type A) or the AGMA system (data type B). The appropriate factors may be converted one to the other, allowing their use in describing the same gear dimensions.

Both methods for the determination of the pitch cone parameters and the different data types to calculate the remaining gear dimensions can be, in principle, combined independently of each other, irrespectively of the gear manufacturing process employed.

2.3.2 Calculation of Pitch Cone Parameters

Input data The following sections describe one method to calculate non-offset bevel gears and one method to calculate hypoid gears. Other methods may be consulted in ISO 23509. Table 2.5 compiles the necessary input data. These inputs must be selected in such a way that they result in a gear with the necessary load capacity (see Sect. 3.1).

Table 2.5 Input data to calculate the pitch cone parameters

Symbol	Description	Method 0 (non-offset gears)	Method 1 (hypoid gears)
Σ	Shaft angle	X	X
a	Hypoid offset	0,0	X
$z_{1,2}$	Number of teeth	X	X
d_{e2}	Outer pitch diameter of the wheel	X	X
b_2	Face width of the wheel	X	X
β_{m1}	Mean spiral angle of the pinion	–	X
β_{m2}	Mean spiral angle of the wheel	X	–
r_{c0}	Tool radius	X	X
z_0	Number of blade groups (only for face hobbing methods)	X	X

The formulae described below yield the pitch cone parameters R_{m1}, R_{m2}, δ_1, δ_2, β_{m1}, β_{m2} and c_{be2}. Parameter c_{be2}, the face width factor, describes the relationship $(R_{e2}–R_{m2})/b_2$. This factor is required because the calculation point P is not always at the center of the face width of the wheel. For the sake of simplicity, however, it is recommended that this factor should be set as $c_{be2} = 0.5$ such that the calculation point lies at exactly half the face width of the wheel and the calculated results cannot be influenced by other parameters.

Table 2.6 contains the formulae to calculate the pitch cone parameters for bevel gears without offset. The formulae can easily be adapted to allow the use of input data other than those given in Table 2.5. The face width factor is always set at $c_{be2} = 0.5$ for non-offset bevel gears.

Table 2.6 Calculation of pitch cone parameters for bevel gears without offset

Designation	Formula	No.
Gear ratio	$u = z_2/z_1$	(2.5)
Pitch angle, pinion	$\delta_1 = \arctan\left(\frac{\sin\Sigma}{\cos\Sigma + u}\right)$	(2.6)
Pitch angle, wheel	$\delta_2 = \Sigma - \delta_1$	(2.7)
Outer cone distance	$R_{e1,2} = \frac{d_{e2}}{2\sin\delta_2}$	(2.8)
Mean cone distance	$R_{m1,2} = R_{e2} - 0.5b$	(2.9)
Spiral angle, pinion	$\beta_{m1} = \beta_{m2}$	(2.10)
Face width factor	$c_{be2} = 0.5$	(2.11)

Table 2.7 provides the formulae to calculate hypoid gear pitch cone parameters. As the calculation can only be performed iteratively, a number of auxiliary values are determined in the course of calculation.

Table 2.7 Calculation of pitch cone parameters for hypoid gears

Designation	Formula	No.
Gear ratio	$u = \frac{z_2}{z_1}$	(2.12)
Desired spiral angle, pinion	$\beta_{\Delta 1} = \beta_{m1}$	(2.13)
Approximate pitch angle, wheel	$\delta_{\text{int2}} = \arctan\left(\frac{u \sin \Sigma}{1.2\,(1 + u \cos \Sigma)}\right)$	(2.14)
Mean pitch radius, wheel	$r_{mpt2} = \frac{d_{e2} - b_2 \sin \delta_{\text{int2}}}{2}$	(2.15)
Approximate pinion offset angle, in the pitch plane	$\varepsilon_i' = \arcsin\left(\frac{a \sin \delta_{\text{int2}}}{r_{mpt2}}\right)$	(2.16)
Approximate hypoid dimension factor	$K_1 = \tan \beta_{\Delta 1} \sin \varepsilon_i' + \cos \varepsilon_i'$	(2.17)
Approximate pitch radius, pinion	$r_{mn1} = \frac{r_{mpt2} K_1}{u}$	(2.18)
Wheel offset angle, in the axial plane	$\eta = \arctan\left[\frac{a}{r_{mpt2}(\tan \delta_{\text{int2}} \sin \Sigma + \cos \Sigma) + r_{mn1}}\right]$	(2.19)
Start of iteration		
Intermediate offset angle in the axial plane, pinion	$\varepsilon_2 = \arcsin\left(\frac{a - r_{mn1} \sin \eta}{r_{mpt2}}\right)$	(2.20)
Intermediate pitch angle, pinion	$\delta_{\text{int1}} = \arctan\left(\frac{\sin \eta}{\tan \varepsilon_2 \sin \Sigma} - \frac{\cos \eta}{\tan \Sigma}\right)$	(2.21)
Intermediate offset angle in the pitch plane, pinion	$\varepsilon_2' = \arcsin\left(\frac{\sin \varepsilon_2 \sin \Sigma}{\cos \delta_{\text{int1}}}\right)$	(2.22)
Intermediate mean spiral angle, pinion	$\beta_{\text{m int1}} = \arctan\left(\frac{K_1 - \cos \varepsilon_2'}{\sin \varepsilon_2'}\right)$	(2.23)
Increment in hypoid dimension factor	$\Delta K = \sin \varepsilon_2' (\tan \beta_{\Delta 1} - \tan \beta_{\text{m int1}})$	(2.24)
Mean radius increment, pinion	$\Delta r_{\text{mpt1}} = r_{\text{mpt2}} \frac{\Delta K}{u}$	(2.25)

Offset angle in the axial plane, pinion	$\varepsilon_1 = \arcsin\left(\sin \varepsilon_2 - \frac{\Delta r_{\mathrm{mpt1}}}{r_{\mathrm{mpt2}}} \sin \eta\right)$	(2.26)
Pitch angle, pinion	$\delta_1 = \arctan\left(\frac{\sin\eta}{\tan\varepsilon_1 \sin\Sigma} - \frac{\cos\eta}{\tan\Sigma}\right)$	(2.27)
Offset angle in the pitch plane, pinion	$\varepsilon_1' = \arcsin\left(\frac{\sin\varepsilon_1 \sin\Sigma}{\cos\delta_1}\right)$	(2.28)
Mean spiral angle, pinion	$\beta_{\mathrm{m1}} = \arctan\left(\frac{K_1 + \Delta K - \cos \varepsilon_1'}{\sin \varepsilon_1'}\right)$	(2.29)
Mean spiral angle, wheel	$\beta_{\mathrm{m2}} = \beta_{\mathrm{m1}} - \varepsilon_1'$	(2.30)
Pitch angle, wheel	$\delta_2 = \arctan\left(\frac{\sin\varepsilon_1}{\tan\eta \sin\Sigma} - \frac{\cos\varepsilon_1}{\tan\Sigma}\right)$	(2.31)
Mean cone distance, pinion	$R_{\mathrm{m1}} = \frac{r_{\mathrm{mn1}} + \Delta r_{\mathrm{mpt1}}}{\sin \delta_1}$	(2.32)
Mean cone distance, wheel	$R_{\mathrm{m2}} = \frac{r_{\mathrm{mpt2}}}{\sin \delta_2}$	(2.33)
Mean radius pinion	$r_{mpt1} = R_{m1} \sin\delta_1$	(2.34)
Limit pressure angle	$\alpha_{\mathrm{lim}} = \arctan\left[-\frac{\tan\delta_1 \tan \delta_2}{\cos \varepsilon_1'}\left(\frac{R_{\mathrm{m1}} \sin \beta_{\mathrm{m1}} - R_{\mathrm{m2}} \sin \beta_{\mathrm{m2}}}{R_{\mathrm{m1}} \tan \delta_1 + R_{\mathrm{m2}} \tan \delta_2}\right)\right]$	(2.35)
Limit radius of curvature	$\rho_{\mathrm{lim}} = \frac{\sec \alpha_{\mathrm{lim}}(\tan \beta_{\mathrm{m1}} - \tan \beta_{\mathrm{m2}})}{-\tan \alpha_{\mathrm{lim}}\left(\frac{\tan \beta_{\mathrm{m1}}}{R_{\mathrm{m1}} \tan \delta_1} + \frac{\tan \beta_{\mathrm{m2}}}{R_{\mathrm{m2}} \tan \delta_2}\right) + \frac{1}{R_{\mathrm{m1}} \cos \beta_{\mathrm{m1}}} - \frac{1}{R_{\mathrm{m2}} \cos \beta_{\mathrm{m2}}}}$	(2.36)
Formulae (2.37) to (2.42) apply only to face hobbed gear sets:		
Number of crown gear teeth	$z_P = \frac{z_2}{\sin\delta_2}$	(2.37)
Lead angle of cutter	$\nu = \arcsin\left(\frac{R_{m2} z_0}{r_{c0} z_P} \cos\beta_{m2}\right)$	(2.38)
First auxiliary angle	$\lambda = 90° - \beta_{m2} + \nu$	(2.39)
Crown gear to cutter centre distance	$\rho_{P0} = \sqrt{R_{m2}^2 + r_{c0}^2 - 2R_{m2} r_{c0} \cos\lambda}$	(2.40)

(continued)

Table 2.7 (continued)

Designation	Formula	No.
Second auxiliary angle	$\eta_1 = \arccos\left[\frac{R_{m2}\cos\beta_{m2}}{\rho_{P0}z_P}(z_P + z_0)\right]$	(2.41)
Mean radius of lengthwise tooth curvature	$\rho_{m\beta} = \mathrm{R}_{m2}\cos\beta_{m2}\left[\tan\beta_{m2} + \frac{\tan\eta_1}{1 + \tan\nu(\tan\beta_{m2} + \tan\eta_1)}\right]$	(2.42)
Formula (2.4.3) applies only to face milled gear sets:		
Mean radius of length- wise tooth curvature	$\rho_{m\beta} = r_{c0}$	(2.43)
The iterative calculation from Equation (2.20) to Equation (2.43) is repeated first with 1.1 η and then with interpolated values for η, until the condition $\left\lvert\frac{\rho_{m\beta}}{\rho_{\lim}} - 1\right\rvert \leq 0.01$ is satisfied.		
End of iteration		
Face width factor	$c_{be2} = \frac{\frac{d_{e2}}{2\sin\delta_2} - R_{m2}}{b_2}$	(2.44)

2.3.3 *Calculation of Gear Dimensions*

Additional input data Once the pitch cone parameters have been determined, a number of additional input data are needed in order to calculate gear dimensions (see Table 2.8). Data for bevel and hypoid gears may be given either as "data type A" or as "data type B".

Table 2.8 Additional input data to calculate gear dimensions

Data type A		Data type B	
Symbol	**Description**	**Symbol**	**Description**
α_{dD}	Nominal design pressure angle, drive side[a]	α_{dD}	Nominal design pressure angle, drive side[a]
α_{dC}	Nominal design pressure angle, coast side[a]	α_{dC}	Nominal design pressure angle, coast side[a]
$f_{\alpha lim}$	Influence factor of limit pressure angle[a]	$f_{\alpha lim}$	Influence factor of limit pressure angle[a]
x_{hm1}	Profile shift coefficient[b]	c_{ham}	Mean addendum factor[b] of wheel
k_{hap}	Basic crown gear addendum factor[b]	k_d	Depth factor [b]
k_{hfp}	Basic crown gear dedendum factor	k_c	Clearance factor [b]
x_{smn}	Thickness modification coefficient [b]	k_t W_{m2}	Thickness factor[b] or Mean slot width of the wheel
	For data types A and B:		
j_{mn}, j_{mt2}, j_{en}, j_{et2}	Backlash (one of four)		
θ_{a2}	Addendum angle, wheel		
θ_{f2}	Dedendum angle, wheel		

[a]The pressure angles on the drive and coast sides of hypoid gears are generally balanced. Certain applications, however, may be performed without this compensation (see Sect. 2.2.5.3)
[b]All dimensionless factors are related to the mean normal module m_{mn}

Permissible data for the values in Table 2.8 are predefined at the design stage (see Sect. 3.1). For this purpose, data type A values can be converted into data type B values and vice versa.

Table 2.9 summarises the relations between the two data types. Figure 2.20 presents the tooth profile of the basic rack for the wheel and the tooth profile with shift. All variables for both data types are plotted.

Table 2.9 Relations between data types A and B

Data type A	Data type B	No.
$x_{hm1} = k_d\left(\frac{1}{2} - c_{ham}\right)$	$c_{ham} = \frac{1}{2}\left(1 - \frac{x_{hm1}}{k_{hap}}\right)$	(2.45)
$k_{hap} = \frac{k_d}{2}$	$k_d = 2k_{hap}$	(2.46)
$k_{hfp} = k_d\left(k_c + \frac{1}{2}\right)$	$k_c = \frac{1}{2}\left(\frac{k_{hfp}}{k_{hap}} - 1\right)$	(2.47)
$x_{smn} = \frac{k_t}{2} = \frac{1}{2}\left(\frac{W_{m2}}{m_{mn}} + k_d\left(k_c + \frac{1}{2}\right)(\tan\alpha_{nD} + \tan\alpha_{nC}) - \frac{\pi}{2}\right)$	$k_t = 2x_{smn}$	(2.48)

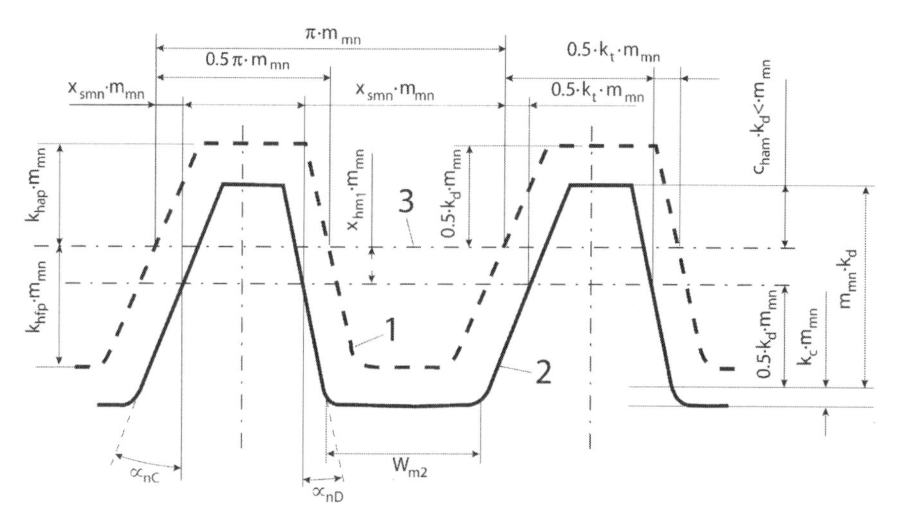

Fig. 2.20 Tooth profile of the basic rack for the wheel [ISO23509].
1 tooth profile of the basic rack, 2 tooth profile with profile shift and thickness modification, 3 reference line

Determination of basic data and tooth depth at the calculation point Table 2.10 contains the basic data for the gear dimensions. The formulae to calculate tooth depth at the calculation point are given in Table 2.11.

Table 2.10 Calculation of basic data

Designation	Formula	No.
Mean pitch diameter, pinion	$d_{m1} = 2R_{m1} \sin \delta_1$	(2.49)
Mean pitch diameter, wheel	$d_{m2} = 2R_{m2} \sin \delta_2$	(2.50)
Offset angle in the axial plane, pinion	$\zeta_m = \arcsin \left(\dfrac{2a}{d_{m2} + d_{m1} \dfrac{\cos \delta_2}{\cos \delta_1}} \right)$	(2.51)
Offset angle in the pitch plane	$\zeta_{mp} = \arcsin \left(\dfrac{\sin \zeta_m \sin \Sigma}{\cos \delta_1} \right)$	(2.52)
Offset in the pitch plane	$a_p = R_{m2} \sin \zeta_{mp}$	(2.53)
Mean normal module	$m_{mn} = \dfrac{2R_{m2} \sin \delta_2 \cos \beta_{m2}}{z_2}$	(2.54)
Generated normal pressure angle on the drive side	$\alpha_{nD} = \alpha_{dD} + f_{\alpha\lim} \alpha_{\lim}$	(2.55)
Generated normal pressure angle on the coast side	$\alpha_{nC} = \alpha_{dC} - f_{\alpha\lim} \alpha_{\lim}$	(2.56)
Effective pressure angle on the drive side	$\alpha_{eD} = \alpha_{nD} - \alpha_{\lim}$	(2.57)
Effective pressure angle on the coast side	$\alpha_{eC} = \alpha_{nC} + \alpha_{\lim}$	(2.58)
Outer pitch cone distance, wheel	$R_{e2} = R_{m2} + c_{be2} b_2$	(2.59)
Inner pitch cone distance, wheel	$R_{i2} = R_{e2} - b_2$	(2.60)
Outer pitch diameter, wheel	$d_{e2} = 2R_{e2} \sin \delta_2$	(2.61)
Inner pitch diameter, wheel	$d_{i2} = 2R_{i2} \sin \delta_2$	(2.62)
Outer transverse module, wheel	$m_{et2} = d_{e2}/z_2$	(2.63)
Face width from calculation point to outside (heel), wheel	$b_{e2} = R_{e2} - R_{m2}$	(2.64)
Face width from calculation point to inside (toe), wheel	$b_{i2} = R_{m2} - R_{i2}$	(2.65)
Crossing point to calculation point along wheel axis	$t_{zm2} = \dfrac{d_{m1} \sin \delta_2}{2 \cos \delta_1} + \dfrac{a}{\tan \zeta_m \tan \Sigma}$	(2.66)
Crossing point to calculation point along pinion axis	$t_{zm1} = \dfrac{d_{m2}}{2} \cos \zeta_m \sin \Sigma + t_{zm2} \cos \Sigma$	(2.67)
Pitch apex beyond crossing point along axis	$t_{z1,2} = R_{m1,2} \cos \delta_{1,2} - t_{zm1,2}$	(2.68)

Table 2.11 Tooth depth at the calculation point

Designation	Formula	No.
Mean working depth	$h_{mw} = 2m_{mn} k_{hap}$	(2.69)
Mean addendum, wheel	$h_{am2} = m_{mn} \left(k_{hap} - x_{hm1} \right)$	(2.70)
Mean dedendum, wheel	$h_{fm2} = m_{mn} \left(k_{hfp} + x_{hm1} \right)$	(2.71)
Mean addendum, pinion	$h_{am1} = m_{mn} \left(k_{hap} + x_{hm1} \right)$	(2.72)
Mean dedendum, pinion	$h_{fm1} = m_{mn} \left(k_{hfp} - x_{hm1} \right)$	(2.73)
Clearance	$c = m_{mn} \left(k_{hfp} - k_{hap} \right)$	(2.74)
Mean tooth depth or mean whole depth	$h_m = h_{am1,2} + h_{fm1,2}$	(2.75)
	$h_m = m_{mn} \left(k_{hap} + k_{hfp} \right)$	(2.76)

Determination of the addendum and dedendum angles For a given depth wise tooth taper, the sum of dedendum angles can be calculated according to Table 2.12, and the addendum and dedendum angles according to Table 2.13.

Table 2.12 Sum of dedendum angles, $\Sigma\theta_f$

Depth wise taper	Sum of dedendum angles (°)		No.
Standard depth	$\Sigma\theta_{fS} = arctan\left(\frac{h_{fm1}}{R_{m2}}\right) + arctan\left(\frac{h_{fm2}}{R_{m2}}\right)$		(2.77)
Uniform depth	$\Sigma\theta_{fU} = 0$		(2.78)
Constant slot width (Duplex)	$\Sigma\theta_{fC} = \left(\frac{90 m_{et}}{R_{e2}\tan\ \alpha_n \cos\beta_m}\right)\left(1 - \frac{R_{m2}\sin\beta_{m2}}{r_{c0}}\right)$		(2.79)
Modified slot width	$\Sigma\theta_{fM} = \Sigma\theta_{fC}$ or $\Sigma\theta_{fM} = 1,3\Sigma\theta_{fS}$	whichever is smallest	

Table 2.13 Addendum angle, θ_{a2} and dedendum angle, θ_{f2}, for the wheel

Depth wise taper	Angle (°)	No.
Standard depth	$\theta_{a2} = \arctan\left(\frac{h_{fm1}}{R_{m2}}\right)$	(2.80)
	$\theta_{f2} = \Sigma\theta_{fS} - \theta_{a2}$	(2.81)
Uniform depth	$\theta_{a2} = \theta_{f2} = 0$	(2.82)
Constant slot width (Duplex)	$\theta_{a2} = \Sigma\theta_{fC}\frac{h_{am2}}{h_{mw}}$	(2.83)
	$\theta_{f2} = \Sigma\theta_{fC} - \theta_{a2}$	(2.84)
Modified slot width	$\theta_{a2} = \Sigma\theta_{fM}\frac{h_{am2}}{h_{mw}}$	(2.85)
	$\theta_{f2} = \Sigma\theta_{fM} - \theta_{a2}$	(2.86)

Pitch cone angles and pitch cone apex distances (see Table 2.14) In order to calculate the pitch cone angles and the distances from the pitch cone apex to the crossing point, it is also necessary to know offset angles ζ_R and ζ_0.

Table 2.14 Pitch cone angles and pitch cone apex distances

Designation	Formula	No.
Face angle, wheel,	$\delta_{a2} = \delta_2 + \theta_{a2}$	(2.87)
Root angle, wheel	$\delta_{f2} = \delta_2 - \theta_{f2}$	(2.88)
Auxiliary angle to calculate pinion offset in the root plane	$\varphi_R = \arctan\left(\frac{-a\cot\Sigma\cos\delta_{f2}}{R_{m2}\cos\theta_{f2} - t_{z2}\cos\delta_{f2}}\right)$	(2.89)
Auxiliary angle to calculate pinion offset in the face plane	$\varphi_o = \arctan\left(\frac{-a\cot\Sigma\cos\delta_{a2}}{R_{m2}\cos\theta_{a2} - t_{z2}\cos\delta_{a2}}\right)$	(2.90)
Pinion offset angle in the root plane	$\zeta_R = \arcsin\left(\frac{a\ \cos\varphi_R\ \sin\delta_{f2}}{R_{m2}\ \cos\theta_{f2} - t_{z2}\ \cos\delta_{f2}}\right) - \varphi_R$	(2.91)

(continued)

Table 2.14 (continued)

Designation	Formula	No.
Pinion offset angle in the face plane	$\zeta_o = \arcsin\left(\dfrac{a\cos\varphi_o \sin\delta_{a2}}{R_{m2}\cos\theta_{a2} - t_{z2}\cos\delta_{a2}}\right) - \varphi_o$	(2.92)
Face angle, pinion	$\delta_{a1} = \arcsin\left(\sin\Sigma\cos\delta_{f2}\cos\zeta_R - \cos\Sigma\sin\delta_{f2}\right)$	(2.93)
Root angle, pinion	$\delta_{f1} = \arcsin\left(\sin\Sigma\cos\delta_{a2}\cos\zeta_o - \cos\Sigma\sin\delta_{a2}\right)$	(2.94)
Addendum angle, pinion	$\theta_{a1} = \delta_{a1} - \delta_1$	(2.95)
Dedendum angle, pinion	$\theta_{f1} = \delta_1 - \delta_{f1}$	(2.96)
Wheel face apex beyond the crossing point along wheel axis	$t_{zF2} = t_{z2} - \dfrac{R_{m2}\sin\theta_{a2} - h_{am2}\cos\theta_{a2}}{\sin\delta_{a2}}$	(2.97)
Wheel root apex beyond the crossing point along wheel axis	$t_{zR2} = t_{z2} + \dfrac{R_{m2}\sin\theta_{f2} - h_{fm2}\cos\theta_{f2}}{\sin\delta_{f2}}$	(2.98)
Pinion face apex beyond the crossing point along pinion axis	$t_{zF1} = \dfrac{a\sin\zeta_R\cos\delta_{f2} - t_{zR2}\sin\delta_{f2} - c}{\sin\delta_{a1}}$	(2.99)
Pinion root apex beyond the crossing point along pinion axis	$t_{zR1} = \dfrac{a\sin\zeta_o\cos\delta_{a2} - t_{zF2}\sin\delta_{a2} - c}{\sin\delta_{f1}}$	(2.100)

Determination of the pinion tooth face width (see Table 2.15 and Fig. 2.21) Whereas calculation of the pinion face width is trivial for non-offset bevel gears, its value being generally equal to that of the wheel, for hypoid gears it is a function of the offset and can be calculated using three different methods (A to C) as presented below.

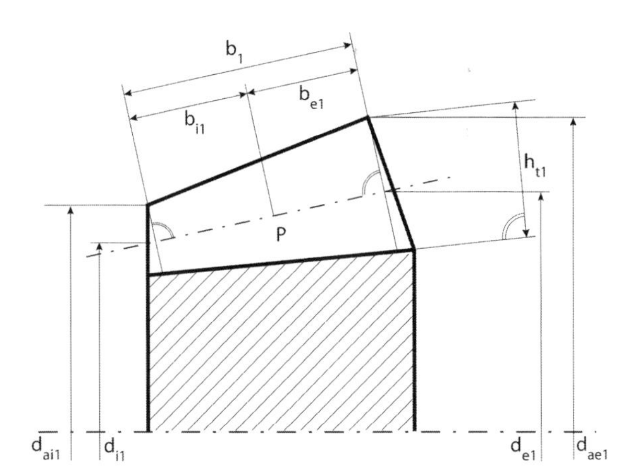

Fig. 2.21 Pinion face width, inner and outer diameter [ISO23509]

Table 2.15 Calculation of the pinion face width

Designation	Formula	No.
Face width in the pitch plane, pinion	$b_{p1} = \sqrt{R_{e2}^2 - a_p^2} - \sqrt{R_{i2}^2 - a_p^2}$	(2.101)
Face width from calculation point to front crown, pinion	$b_{1A} = \sqrt{R_{m2}^2 - a_p^2} - \sqrt{R_{i2}^2 - a_p^2}$	(2.102)
Formulae (2.103) to (2.105) apply to non-offset bevel gears:		
Face width, pinion	$b_1 = b_2$	(2.103)
Face width from calculation point to outside (heel), pinion	$b_{e1} = c_{be2} b_1$	(2.104)
Face width from calculation point to inside (toe), pinion	$b_{i1} = b_1 - b_{e1}$	(2.105)
The following formulae apply to hypoid gears. There are three possible methods:		
Method A		
Auxiliary angle	$\lambda' = \arctan\left(\frac{\sin\zeta_{mp}\cos\delta_2}{u\cos\delta_1 + \cos\delta_2\cos\zeta_{mp}}\right)$	(2.106)
Starting value for pinion face width	$b_{reri1} = \frac{b_2\cos\lambda'}{\cos\left(\zeta_{mp} - \lambda'\right)}$	(2.107)
Pinion face width increment along pinion axis	$\Delta b_{x1} = h_{mw}\sin\zeta_R\left(1 - \frac{1}{u}\right)$	(2.108)
Increment along pinion axis from calculation point to outside	$\Delta g_{xe} = \frac{c_{be2} b_{reri1}}{\cos\theta_{a1}}\cos\delta_{a1} + \Delta b_{x1} - \left(h_{fm2} - c\right)\sin\delta_1$	(2.109)
Increment along pinion axis from calculation point to inside	$\Delta g_{xi} = \frac{(1 - c_{be2}) b_{reri1}}{\cos\theta_{a1}}\cos\delta_{a1} + \Delta b_{x1} + \left(h_{fm2} - c\right)\sin\delta_1$	(2.110)
Pinion face width from calculation point to outside	$b_{e1} = \frac{\Delta g_{xe} + h_{am1}\sin\delta_1}{\cos\delta_{a1}}\cos\theta_{a1}$	(2.111)
Pinion face width from calculation point to inside	$b_{i1} = \frac{\Delta g_{xi} - h_{am1}\sin\delta_1}{\cos\delta_1 - \tan\theta_{a1}\sin\delta_1}$	(2.112)
Pinion face width along the pitch cone	$b_1 = b_{i1} + b_{e1}$	(2.113)
Method B		
Pinion face width along the pitch cone	$b_1 = b_2\left(1 + \tan^2\zeta_{mp}\right)$	(2.114)
Pinion face width from calculation point to outside	$b_{e1} = c_{be2} b_1$	(2.115)
Pinion face width from calculation point to inside	$b_{i1} = b_1 - b_{e1}$	(2.116)

(continued)

Table 2.15 (continued)

Designation	Formula	No.
Method C		
Pinion face width along the pitch cone	$b_1 = \text{int}\left(b_{p1} + 3m_{mn}\tan\left\lvert\zeta_{mp}\right\rvert + 1\right)$	(2.117)
Additional pinion face width	$b_x = \dfrac{b_1 - b_{p1}}{2}$	(2.118)
Pinion face width from calculation point to inside	$b_{i1} = b_{1A} + b_x$	(2.119)
Pinion face width from calculation point to outside	$b_{e1} = b_1 - b_{i1}$	(2.120)

Determination of inner and outer spiral angles (see Table 2.16) The spiral angle changes continuously along the face width on spiral bevel gears. Accordingly, the inner and outer spiral angle values are calculated from those at the mean point. This calculation is not necessary in the case of straight bevel gears.

Table 2.16 Calculation of the inner and outer spiral angles (for spiral bevel gears only)

Designation	Formula	No.
Formulae (2.121) to (2.135) apply to all pinions:		
Cone distance of crown gear to outer boundary of pinion (possibly > R_{e2})	$R_{e21} = \sqrt{R_{m2}^2 + b_{e1}^2 + 2R_{m2}b_{e1}\cos\zeta_{mp}}$	(2.121)
Cone distance of crown gear to inner boundary of pinion (possibly < R_{i2})	$R_{i21} = \sqrt{R_{m2}^2 + b_{i1}^2 - 2R_{m2}b_{i1}\cos\zeta_{mp}}$	(2.122)
Lead angle of cutter	$\nu = \arcsin\left(\dfrac{z_0 m_{mn}}{2r_{c0}}\right)$	(2.123)
Basic crown gear to cutter centre distance	$\rho_{P0} = \sqrt{R_{m2}^2 + r_{c0}^2 - 2R_{m2}r_{c0}\sin\left(\beta_{m2} - \nu\right)}$	(2.124)
Epicycloid base circle radius	$\rho_b = \dfrac{\rho_{P0}}{1 + \dfrac{z_0}{z_2}\sin\delta_2}$	(2.125)
Auxiliary angle	$\varphi_{e21} = \arccos\left(\dfrac{R_{e21}^2 + \rho_{P0}^2 - r_{c0}^2}{2R_{e21}\rho_{P0}}\right)$	(2.126)
Auxiliary angle	$\varphi_{i21} = \arccos\left(\dfrac{R_{i21}^2 + \rho_{P0}^2 - r_{c0}^2}{2R_{i21}\rho_{P0}}\right)$	(2.127)
Spiral angle of the basic crown gear at the outer boundary point	$\beta_{e21} = \arctan\left(\dfrac{R_{e21} - \rho_b\cos\varphi_{e21}}{\rho_b\sin\varphi_{e21}}\right)$	(2.128)
Spiral angle of the basic crown gear at the inner boundary point	$\beta_{i21} = \arctan\left(\dfrac{R_{i21} - \rho_b\cos\varphi_{i21}}{\rho_b\sin\varphi_{i21}}\right)$	(2.129)
Formulae (2.130) and (2.131) are obtained for the face milled pinion by inserting $z_0 = 0$ in Formulae (2.123) to (2.129):		
Spiral angle of the basic crown gear at the outer boundary point	$\beta_{e21} = \arcsin\left(\dfrac{2R_{m2}r_{c0}\sin\beta_{m2} - R_{m2}^2 + R_{e21}^2}{2R_{e21}r_{c0}}\right)$	(2.130)
Spiral angle of the basic crown gear at the inner boundary point	$\beta_{i21} = \arcsin\left(\dfrac{2R_{m2}r_{c0}\sin\beta_{m2} - R_{m2}^2 + R_{i21}^2}{2R_{i21}r_{c0}}\right)$	(2.131)

(continued)

Table 2.16 (continued)

Designation	Formula	No.
The following formulae apply to both face hobbed and face milled pinions:		
Pinion offset angle in the pitch plane at the outer boundary point	$\zeta_{ep21} = \arcsin\left(a_p / R_{e21}\right)$	(2.132)
Pinion offset angle in the pitch plane at the inner boundary point	$\zeta_{ip21} = \arcsin\left(a_p / R_{i21}\right)$	(2.133)
Outer spiral angle, pinion	$\beta_{e1} = \beta_{e21} + \zeta_{ep21}$	(2.134)
Inner spiral angle, pinion	$\beta_{i1} = \beta_{i21} + \zeta_{ip21}$	(2.135)
Formulae (2.136) to (2.139) apply to the face hobbed wheel and with $z_0=0$ to the face milled wheel:		
Auxiliary angle	$\varphi_{e2} = \arccos\left(\dfrac{R_{e2}^2 + \rho_{P0}^2 - r_{c0}^2}{2R_{e2}\rho_{P0}}\right)$	(2.136)
Auxiliary angle	$\varphi_{i2} = \arccos\left(\dfrac{R_{i2}^2 + \rho_{P0}^2 - r_{c0}^2}{2R_{i2}\rho_{P0}}\right)$	(2.137)
Outer spiral angle, wheel	$\beta_{e2} = \arctan\left(\dfrac{R_{e2} - \rho_b \cos\varphi_{e2}}{\rho_b \sin\varphi_{e2}}\right)$	(2.138)
Inner spiral angle, wheel	$\beta_{i2} = \arctan\left(\dfrac{R_{i2} - \rho_b \cos\varphi_{i2}}{\rho_b \sin\varphi_{i2}}\right)$	(2.139)

Determination of tooth depth and thickness Tooth depth is calculated according to the formulae in Table 2.17 (see Fig. 2.22), tooth thickness according to Table 2.18.

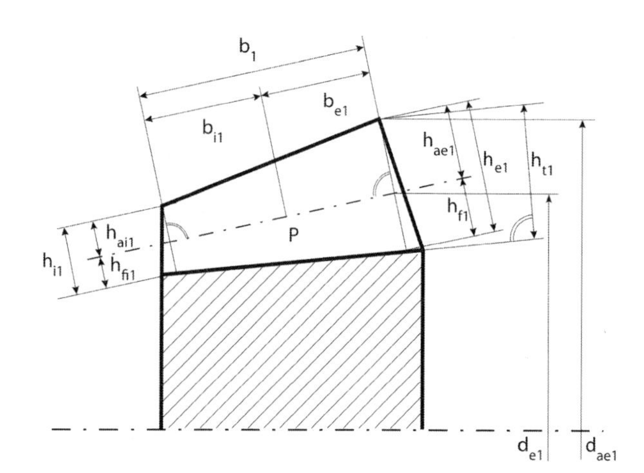

Fig. 2.22 Tooth depth on the pinion

Table 2.17 Calculation of tooth depth

Designation	Formula	No.
Outer addendum	$h_{ae1,2} = h_{am1,2} + b_{e1,2} \tan\theta_{a1,2}$	(2.140)
Outer dedendum	$h_{fe1,2} = h_{fm1,2} + b_{e1,2} \tan\theta_{f1,2}$	(2.141)
Outer tooth depth	$h_{e1,2} = h_{ae1,2} + h_{fe1,2}$	(2.142)
Inner addendum	$h_{ai1,2} = h_{am1,2} - b_{i1,2} \tan\theta_{a1,2}$	(2.143)
Inner dedendum	$h_{fi1,2} = h_{fm1,2} - b_{i1,2} \tan\theta_{f1,2}$	(2.144)
Inner tooth depth	$h_{i1,2} = h_{ai1,2} + h_{fi1,2}$	(2.145)

Table 2.18 Calculation of tooth thickness

Designation	Formula	No.
Average normal pressure angle	$\alpha_n = \frac{(\alpha_{nD} + \alpha_{nC})}{2}$	(2.146)
Pinion thickness modification coefficient, calculated using:		
...outer normal backlash	$x_{sm1} = x_{smn} - j_{en} \frac{1}{4m_{mn} \cos\alpha_n} \frac{R_{m2} \cos\beta_{m2}}{R_{e2} \cos\beta_{e2}}$	(2.147)
...outer transverse backlash	$x_{sm1} = x_{smn} - j_{et2} \frac{R_{m2} \cos\beta_{m2}}{4m_{mn} R_{e2}}$	(2.148)
...mean normal backlash	$x_{sm1} = x_{smn} - j_{mn} \frac{1}{4m_{mn} \cos\alpha_n}$	(2.149)
...mean transverse backlash	$x_{sm1} = x_{smn} - j_{mt2} \frac{\cos\beta_{m2}}{4m_{mn}}$	(2.150)
Mean normal circular tooth thickness, pinion	$s_{mn1} = 0.5m_{mn}\pi + 2m_{mn}(x_{sm1} + x_{hm1} \tan\alpha_n)$	(2.151)
Wheel thickness modification coefficient, calculated using:		
...outer normal backlash	$x_{sm2} = -x_{smn} - j_{en} \frac{1}{4m_{mn} \cos\alpha_n} \frac{R_{m2} \cos\beta_{m2}}{R_{e2} \cos\beta_{e2}}$	(2.152)
...outer transverse backlash	$x_{sm2} = -x_{smn} - j_{et2} \frac{R_{m2} \cos\beta_{m2}}{4m_{mn} R_{e2}}$	(2.153)
...mean normal backlash	$x_{sm2} = -x_{smn} - j_{mn} \frac{1}{4m_{mn} \cos\alpha_n}$	(2.154)
...mean transverse backlash	$x_{sm2} = -x_{smn} - j_{mt2} \frac{\cos\beta_{m2}}{4m_{mn}}$	(2.155)
Mean normal circular tooth thickness, wheel	$s_{mn2} = 0.5m_{mn}\pi + 2m_{mn}(x_{sm2} - x_{hm1} \tan\alpha_n)$	(2.156)
Mean transverse circular tooth thickness	$s_{mt1.2} = s_{mn1.2} / \cos\beta_{m1.2}$	(2.157)
Mean normal diameter	$d_{mn1.2} = \frac{d_{m1.2}}{(1 - \sin^2\beta_{m1.2} \cos^2\alpha_n) \cos\delta_{1.2}}$	(2.158)
Mean normal chordal tooth thickness	$s_{\text{mnc}1.2} = d_{\text{mn}1.2} \sin(s_{\text{mn}1.2} / d_{\text{mn}1.2})$	(2.159)
Mean chordal addendum	$h_{\text{amc}1.2} = h_{am1.2} + 0.5d_{\text{mn}1.2} \cos\delta_{1.2} \left[1 - \cos\left(\frac{s_{mn1.2}}{d_{mn1.2}}\right)\right]$	(2.160)

Determination of the remaining dimensions Calculation of further inner and outer cone dimensions, such as cone distances, tip cone and root cone diameters, is given in Table 2.19 (cf. Figs. 2.21 and 2.22).

Table 2.19 Calculation of further dimensions

Designation	Formula	No.
Outer pitch cone distance, pinion	$R_{e1} = R_{m1} + b_{e1}$	(2.161)
Inner pitch cone distance pinion	$R_{i1} = R_{m1} - b_{i1}$	(2.162)
Outer pitch diameter, pinion	$d_{e1} = 2R_{e1} \sin \delta_1$	(2.163)
Inner pitch diameter, pinion	$d_{i1} = 2R_{i1} \sin \delta_1$	(2.164)
Outside diameter (outer face cone diameter)	$d_{ae1,2} = d_{e1,2} + 2h_{ae1,2} \cos \delta_{1,2}$	(2.165)
Outer root cone diameter	$d_{fe1,2} = d_{e1,2} - 2h_{fe1,2} \cos \delta_{1,2}$	(2.166)
Inner face cone diameter	$d_{ai1,2} = d_{i1,2} + 2h_{ai1,2} \cos \delta_{1,2}$	(2.167)
Inner root cone diameter	$d_{fi1,2} = d_{i1,2} - 2h_{fi1,2} \cos \delta_{1,2}$	(2.168)
Crossing point to crown along axis	$t_{xo1,2} = t_{zm1,2} + b_{e1,2} \cos \delta_{1,2} - h_{ae1,2} \sin \delta_{1,2}$	(2.169)
Crossing point to front crown along axis	$t_{xi1,2} = t_{zm1,2} - b_{i1,2} \cos \delta_{1,2} - h_{ai1,2} \sin \delta_{1,2}$	(2.170)
Pinion tooth depth, perpendicular to root cone	$h_{t1} = \frac{t_{zF1} + t_{xo1}}{\cos \delta_{a1}} \sin\left(\theta_{a1} + \theta_{f1}\right) - \left(t_{zR1} - t_{zF1}\right) \sin \delta_{f1}$	(2.171)

2.3.4 Undercut Check

Depending on the chosen profile parameters, undercut may occur on bevel gears as it occurs on cylindrical gears. The following section provides formulae to check whether undercut will occur on the teeth and, if so, what areas along the face width are endangered. In principle, the pinion is most at risk. Absence of undercut may here become the decisive criterion in the choice of an adequate profile shift (see Sect. 3.1), especially on bevel gears with uniform tooth depth.

Since profile shifts on bevel pinions and wheels usually mean that their sum is zero,[1] $x_{hm2} = -x_{hm1}$, the check for the absence of undercutting on the pinion (see Table 2.20) must always be followed by an undercut check on a generated wheel (see Table 2.21). The following tables do not apply to gear sets with a plunge cut wheel (forming process, see Sect. 2.1). Table 2.20 can, however, be used as a first approximation for the pinion. Plunge cut wheels do not require an undercut check.

[1] Other sums in profile shift occur in practice only if the same tool is used for the pinion and wheel. In terms of the effect, however, the actual gears will again have a shift sum of zero.

Table 2.20 Undercut check on the pinion

Designation	Formula	No.
The point on the face width at which undercut check should take place is marked by an "x". All following formulae relate to this checkpoint		
Pinion cone distance to checkpoint	$R_{i1} \leq R_{x1} \leq R_{e1}$	(2.172)
Wheel cone distance to corresponding point (R_{x2} may be $< R_{i2}$ or $> R_{e2}$)	$R_{x2} = \sqrt{R_{m2}^2 + (R_{m1} - R_{x1})^2 - 2R_{m2}(R_{m1} - R_{x1})\cos\zeta_{mp}}$	(2.173)
Formulae (2.174) and (2.175) apply to face hobbing, and using $z_0 = 0$, also apply to face milling:		
Auxiliary angle	$\varphi_{x2} = \arccos\left(\frac{R_{x2}^2 + \rho_{P0}^2 - r_{c0}^2}{2R_{x2}\rho_{P0}}\right)$	(2.174)
Spiral angle, wheel	$\beta_{x2} = \arctan\left(\frac{R_{x2} - \rho_b \cos\varphi_{x2}}{\rho_b \sin\varphi_{x2}}\right)$	(2.175)
Offset angle in the pitch plane, pinion	$\zeta_{xp2} = \arcsin\left(\frac{a_p}{R_{x2}}\right)$	(2.176)
Spiral angle, pinion	$\beta_{x1} = \beta_{x2} + \zeta_{xp2}$	(2.177)
Pitch diameter, pinion	$d_{x1} = 2R_{x1}\sin\delta_1$	(2.178)
Pitch diameter, wheel	$d_{x2} = 2R_{x2}\sin\delta_2$	(2.179)
Normal module at the checkpoint	$m_{xn} = \frac{d_{x2}}{z_2}\cos\beta_{x2}$	(2.180)
Effective diameter, pinion	$d_{Ex1} = d_{x2}\frac{z_1\cos\beta_{x2}}{z_2\cos\beta_{x1}}$	(2.181)
Corresponding pinion cone distance	$R_{Ex1} = \frac{d_{Ex1}}{2\sin\delta_1}$ a	(2.182)
Intermediate value	$z_{nx1} = \frac{z_1}{(1 - \sin^2\beta_{x1}\cos^2\alpha_n)\cos\beta_{x1}\cos\delta_1}$	(2.183)
Limit pressure angle at the checkpoint	$\alpha_{\lim x} = -\arctan\left[\frac{\tan\delta_1\tan\delta_2}{\cos\zeta_{mp}}\left(\frac{R_{Ex1}\sin\beta_{x1} - R_{x2}\sin\beta_{x2}}{R_{Ex1}\tan\delta_1 + R_{x2}\tan\delta_2}\right)\right]$	(2.184)
Effective pressure angle (drive tooth flank)	$\alpha_{eDx} = \alpha_{nD} - \alpha_{\lim x}$	(2.185)
Effective pressure angle (coast tooth flank)	$\alpha_{eCx} = \alpha_{nC} + \alpha_{\lim x}$	(2.186)
The smaller of the two effective pressure angles is used in the following calculations:		
	if $\alpha_{eCx} < \alpha_{eDx}$: $\alpha_{e\min x} = \alpha_{eCx}$	(2.187)
	if $\alpha_{eCx} \geq \alpha_{eDx}$: $\alpha_{e\min x} = \alpha_{eDx}$	(2.188)
Calculation of the minimum profile shift coefficient on the pinion		
Working tool addendum at the checkpoint	$k_{hapx} = k_{hap} + \frac{(R_{x2} - R_{m2})\tan\theta_{a2}}{m_{mn}}$	(2.189)
Minimum profile shift coefficient, pinion	$x_{hx1} = 1{,}1k_{hapx} - \frac{z_{nx1}m_{xn}\sin^2\alpha_{e\min x}}{2m_{mn}}$	(2.190)
Minimum profile shift coefficient at the calculation point, pinion	$x_{hm\min x1} = x_{hx1} + \frac{(d_{Ex1} - d_{x1})\cos\delta_1}{2m_{mn}}$	(2.191)
Pinion undercut at checkpoint is avoided if $x_{hm1} > x_{hm\min x1}$		

Table 2.21 Undercut check on a generated wheel

Designation	Formula	No.
The point on the face width at which the undercut check should take place is marked by an "x". All following formulae relate to this check point		
Wheel cone distance to checkpoint	$R_{i2} \leq R_{x2} \leq R_{e2}$	(2.192)
Formulae (2.193) and (2.194) apply to face hobbing, and using $z_0 = 0$, also apply to face milling:		
Auxiliary angle	$\varphi_{x2} = \arccos\left(\frac{R_{x2}^2 + \rho_{P0}^2 - r_{c0}^2}{2R_{x2}\rho_{P0}}\right)$	(2.193)
Spiral angle, wheel	$\beta_{x2} = \arctan\left(\frac{R_{x2} - \rho_b \cos\varphi_{x2}}{\rho_b \sin\varphi_{x2}}\right)$	(2.194)
Pitch diameter, wheel	$d_{x2} = 2R_{x2}\sin\delta_2$	(2.195)
Normal module	$m_{xn} = \frac{d_{x2}}{z_2}\cos\beta_{x2}$	(2.196)
Intermediate value	$z_{nx2} = \frac{z_2}{(1 - \sin^2\beta_{x2}\cos^2\alpha_n)\cos\beta_{x2}\cos\delta_2}$	(2.197)
The smaller of the two generated normal pressure angles on the wheel is used in the following calculations:		
	if $\alpha_{nC} < \alpha_{nD}: \quad \alpha_{eminx} = \alpha_{nC}$	(2.198)
	if $\alpha_{nC} \geq \alpha_{nD}: \quad \alpha_{eminx} = \alpha_{nD}$	(2.199)
Calculation of the maximum profile shift coefficient at the pinion to avoid undercut at the wheel, since: $x_{hm2} = -x_{hm1}$		
Working tool addendum at the checkpoint, wheel	$k_{hapx} = k_{hap} + \frac{(R_{x2} - R_{m2})\tan\theta_{f2}}{m_{mn}}$	(2.200)
Maximum profile shift coefficient at the calculation point, pinion	$x_{hmmaxx1} = -\left(1,1k_{hapx} - \frac{z_{nx2}m_{xn}\sin^2\alpha_{eminx}}{2m_{mn}}\right)$	(2.201)
Wheel undercut at the checkpoint is avoided if: $x_{hm1} < x_{hm\,max\,x1}$		

2.4 Sliding Velocities and Sum Velocities

2.4.1 General

In this chapter, the velocity conditions at a randomly chosen point on the pinion and wheel tooth flanks are derived from the gear kinematics. These velocities decisively affect the lubrication and friction parameters on the mating flanks and, hence, influence the load capacity and efficiency of the bevel gear set.

The coordinate system for the following derivations is defined according to Fig. 2.23.

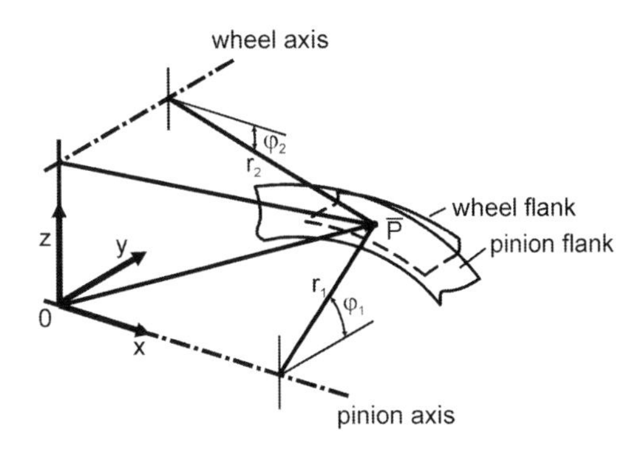

Fig. 2.23 Coordinate system to calculate surface velocities

2.4.2 *Absolute Velocities*

The absolute velocities of the tooth surfaces at a point of contact between the pinion and wheel correspond to the circumferential velocities around the relevant axis. These can be calculated by means of the distances between the contact points and the axis of rotation and the relevant angular velocities, as shown in Table 2.22.

Table 2.22 Calculation of circumferential velocities

Designation	**Formula**	**No.**
Magnitude of the circumferential velocity	$v_{t1,2} = \lvert\overline{v}_{t1,2}\rvert = r_{1,2}\omega_{1,2}$	(2.202)
Circumferential velocity, pinion	$\overline{v}_{t1} = v_{t1}\begin{pmatrix} 0 \\ \sin\varphi_1 \\ -\cos\varphi_1 \end{pmatrix}$	(2.203)
Circumferential velocity, wheel	$\overline{v}_{t2} = v_{t2}\begin{pmatrix} -\sin\varphi_2 \\ 0 \\ -\cos\varphi_2 \end{pmatrix}$	(2.204)

2.4.3 *Sliding Velocities*

The sliding velocity is equivalent to the difference between the circumferential velocities of the pinion and gear tooth flanks at a specified contact point. Figure 2.24 depicts the relevant vector relationships.

The sliding velocity according to Table 2.23 is always perpendicular to the normal vectors of the mating tooth flanks at the specified contact point. On bevel gears without hypoid offset, pure rolling without sliding between the tooth flanks takes place on the pitch cones and, as on cylindrical gears, sliding in the profile direction occurs in the remaining tooth flank regions. In the case of hypoid gears, additional tooth lengthwise sliding, produced by the offset, is superimposed in all tooth flank regions.

Table 2.23 Calculation of the sliding velocities

Designation	Formula	No.
Sliding velocity	$\overline{v}_{g1} = \overline{v}_{t1} - \overline{v}_{t2}; \quad \overline{v}_{g2} = \overline{v}_{t2} - \overline{v}_{t1}$	(2.205)

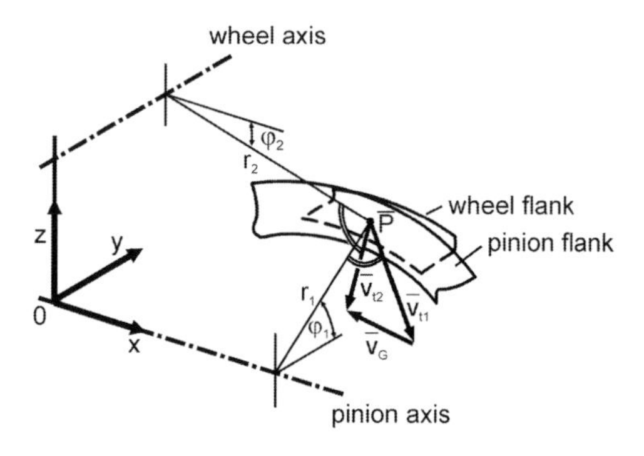

Fig. 2.24 Velocities at a contact point $\overline{\mathrm{P}}$

2.4.4 *Sum Velocity*

The term "sum velocity" may be understood as the sum of the surface velocities of the pinion and wheel tooth flanks at any contact point. The sum velocity may be regarded as the transport velocity for the lubricant film on the mating tooth flanks, and is therefore directly related to the formation of the lubricant film. The sum velocity is consequently included when calculating the coefficient of friction (see Sects. 4.2.6 and 4.3.3) and the lubricant film thickness.

Figure 2.25a shows the velocities on the pinion tooth flank and Fig. 2.25b on the corresponding wheel flank. At contact point $\overline{\mathrm{P}}$, vector $\overline{n}_{p_t}$ is perpendicular to tangential plane T. In this meshing position, the contact line has a tangent $t - t$ at point P where the major half-axis of the ellipse of contact is regarded as the contact line. The tangential velocity $\overline{w}_t$ on the tooth flank is the component of circumferential velocity $\overline{v}_t$ in the tangential plane T. The respective tooth flank tangential velocities of the pinion and wheel are therefore $\overline{w}_{t1}$ and $\overline{w}_{t2}$.

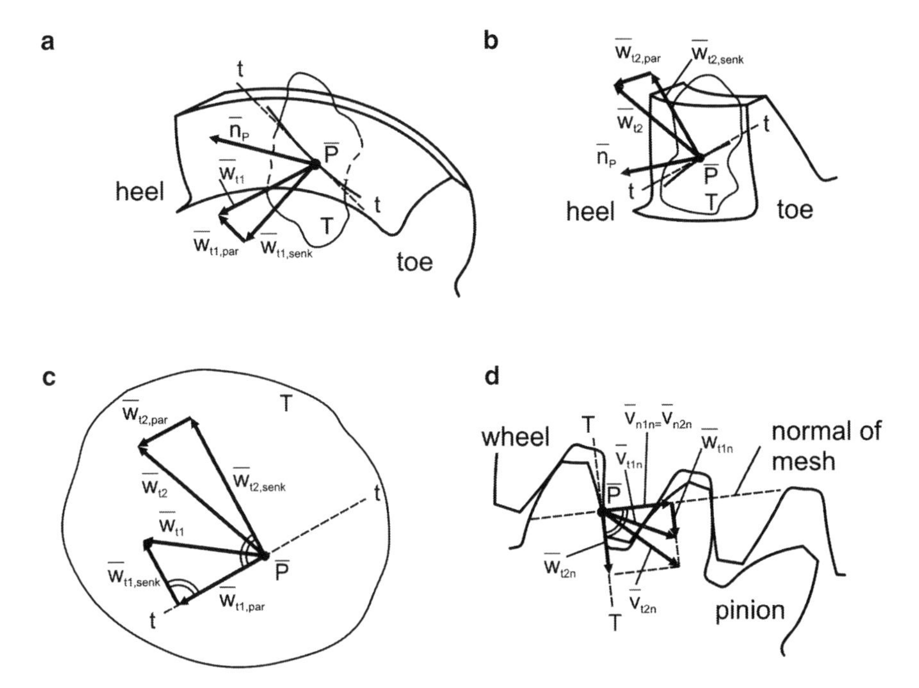

Fig. 2.25 Velocities for the determination of the sum velocity

For the derivation of specific sliding in Sect. 2.4.5 (below), it is necessary to know the components of the tooth flank tangential velocity perpendicular to the contact line. In order to make calculations possible such as the flank load capacity according to Sect. 4.2.5, it is also necessary to know the components of the tangential velocity parallel to, and the components of the sum velocity perpendicular to the contact line. Figure 2.25 depicts the components of the velocity vectors for the pinion and wheel in the tangential plane. The corresponding relationships are listed in Table 2.24. The direction vector of contact line $\overline{t}_B$ at any point may be approximated by the difference in coordinates between two consecutive contact points along the contact line.

As shown in Fig. 2.25, the velocities may also be projected in the normal section. The circumferential velocities $\overline{v}_{t1,2n}$ can be analyzed into a component $\overline{w}_{t1,2n}$ perpendicular to, and a component $\overline{v}_{n1,2n}$ parallel to, the contact normal. Since tooth flanks do not interpenetrate or separate when meshing, $\overline{v}_{n1n} = \overline{v}_{n2n}$ must necessarily apply.

Table 2.24 Calculation of the tangential and sum velocities

Designation	Formula	No.
Auxiliary vector	$\overline{h}_{1,2} = \left(\overline{n}^0 \circ \overline{v}_{,2}\right)$	(2.206)
Tooth flank tangential velocity	$\overline{w}_{t1,2} = \overline{v}_{t1,2} - \overline{h}_{1,2}$	(2.207)
Components of tooth flank tangential velocity parallel to contact line	$\overline{w}_{t,par1,2} = \frac{\overline{w}_{t1,2} \circ \overline{t}_B}{\overline{t}_B^2} \overline{t}_B$	(2.208)
Components of tooth flank tangential velocity perpendicular to contact line	$\overline{w}_{t,senk1,2} = \overline{w}_{t1,2} - \overline{w}_{tpar1,2}$	(2.209)
Sum velocity	$\overline{v}_\Sigma = \overline{w}_{t1} + \overline{w}_{t2}$	(2.210)
Component of sum velocity perpendicular to contact line	$\overline{v}_{\Sigma,senk} = \overline{w}_{t1,senk} + \overline{w}_{t2,senk}$	(2.211)

2.4.5 *Specific Sliding*

A clear-cut definition of specific sliding, of the kind which applies to cylindrical gears and bevel gears without offset, is not possible in the case of hypoid gears, where neither tooth flank tangential velocities nor sliding velocities are collinear. In order to calculate the effects of specific sliding on tooth flank load capacity according to Sect. 4.2, the velocity relationships are therefore considered in a plane perpendicular to the contact line (see Table 2.25). On bevel gears without offset, this definition is equivalent to the usual definition on cylindrical gears. The specific sliding parallel to the contact line which takes place on hypoid gears, and the associated influence on load capacity are taken into account in Sect. 4.2.5.2.

Table 2.25 Calculation of specific sliding

Designation	Formula	No.
Pinion specific sliding perpendicular to contact line	$\zeta_{1,senk} = 1 - \frac{\overline{w}_{t2,senk}}{\overline{w}_{t1,senk}}$	(2.212)
Wheel specific sliding perpendicular to contact line	$\zeta_{2,senk} = 1 - \frac{\overline{w}_{t1,senk}}{\overline{w}_{t2,senk}}$	(2.213)

2.5 Tooth Forces

2.5.1 *Tooth Force Analysis*

Tooth forces resulting from gear geometry and torque may be divided into tangential, axial and radial components, which are required in order to determine the forces and moments acting on the shafts and bearings. The axial and radial forces are dependent on the tooth geometry of the loaded tooth flank.

2.5.2 Calculation of Tooth Forces

Tooth forces are calculated with the aid of the formulae listed in Table 2.26. The directions of the forces strongly depend on the magnitude of the spiral angle and the direction of rotation, as illustrated in Fig. 2.26.

Table 2.26 Calculating tooth forces

Designation	Formula	No.
Tangential force		
on the wheel	$F_{mt2} = \frac{2 \cdot T_2}{d_{m2}} \cdot 1000$	(2.214)
on the pinion	$F_{mt1} = \frac{F_{mt2} \cos \beta_{m1}}{\cos \beta_{m2}} = \frac{2 \cdot T_1}{d_{m1}} \cdot 1000$ Factor 1000 above results from the conversion of Nm into Nmm. The units are defined in the Symbols and Units section.	(2.215)
Axial force	**Loaded flank: drive flank**	
on the pinion	$F_{ax1,D} = \left(\tan \alpha_{nD} \frac{\sin \delta_1}{\cos \beta_{m1}} + \tan \beta_{m1} \cos \delta_1 \right) F_{mt1}$	(2.216)
on the wheel	$F_{ax2,D} = \left(\tan \alpha_{nD} \frac{\sin \delta_2}{\cos \beta_{m2}} - \tan \beta_{m2} \cos \delta_2 \right) F_{mt2}$	(2.217)
Axial force	**Loaded flank: coast flank**	
on the pinion	$F_{ax1,C} = \left(\tan \alpha_{nC} \frac{\sin \delta_1}{\cos \beta_{m1}} - \tan \beta_{m1} \cos \delta_1 \right) F_{mt1}$	(2.218)
on the wheel	$F_{ax2,C} = \left(\tan \alpha_{nC} \frac{\sin \delta_2}{\cos \beta_{m2}} + \tan \beta_{m2} \cos \delta_2 \right) F_{mt2}$	(2.219)
Positive axial forces (+) are those acting on the bevel gear pushing it away from the crossing point of the axes and hence out of engagement. Negative axial forces (−) are those acting on the bevel gear pulling it towards the crossing point and hence into engagement (see Fig. 2.26)		

Designation	Formula	No.
Radial force	**Loaded flank: drive flank**	
on the pinion	$F_{rad1,D} = \left(\tan \alpha_{nD} \frac{\cos \delta_1}{\cos \beta_{m1}} - \tan \beta_{m1} \sin \delta_1 \right) F_{mt1}$	(2.220)
on the wheel	$F_{rad2,D} = \left(\tan \alpha_{nD} \frac{\cos \delta_2}{\cos \beta_{m2}} + \tan \beta_{m2} \sin \delta_2 \right) F_{mt2}$	(2.221)
Radial force	**Loaded flank: coast flank**	
on the pinion	$F_{rad1,C} = \left(\tan \alpha_{nC} \frac{\cos \delta_1}{\cos \beta_{m1}} + \tan \beta_{m1} \sin \delta_1 \right) F_{mt1}$	(2.222)
on the wheel	$F_{rad2,C} = \left(\tan \alpha_{nC} \frac{\cos \delta_2}{\cos \beta_{m2}} - \tan \beta_{m2} \sin \delta_2 \right) F_{mt2}$	(2.223)
Positive radial forces (+) are those acting on the bevel gear pushing it away from the crossing point of the axes and hence out of engagement. Negative radial forces (−) are those acting on the bevel gear pulling it towards the crossing point and hence into engagement (see Fig. 2.26)		

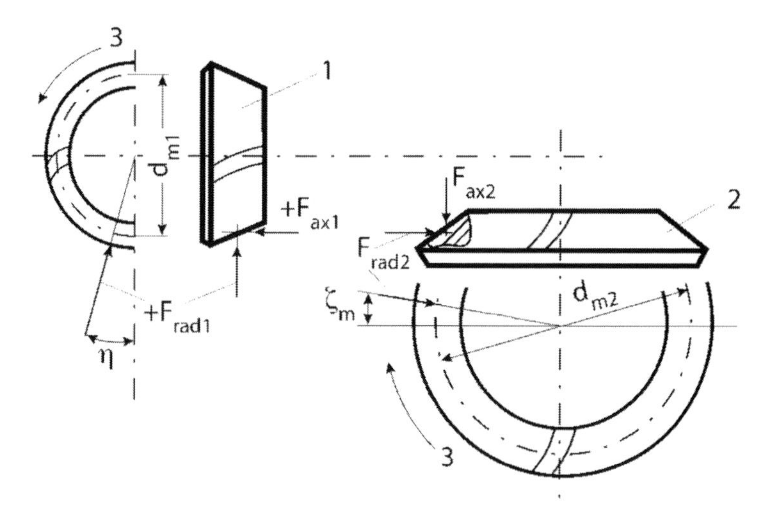

Fig. 2.26 Tooth forces. 1 pinion, 2 wheel, 3 direction of rotation

2.5.3 *Bearing Forces*

Forces acting on bearing may be calculated from tooth forces and additional external forces. The radial force on the bearings thus contains components of the tangential, axial and radial forces of the gear mesh and additional external force components. The axial force on the bearing is the axial tooth force plus external force components.

References

[AGMA2005] ANSI/AGMA 2005-D03: Design Manual for Bevel Gears (2003)

[COLE52] Coleman, W.: Improved method for estimating fatigue life of bevel and hypoid gears. SAE Q. Trans. **6**(2S), 314–331 (1952)

[ISO23509] ISO 23509: Bevel and Hypoid Gear Geometry (2006)

[KN3029] Auslegung von Hypoid-Getrieben mit Klingelnberg Zyklo-Palloid-Verzahnung: Klingelnberg GmbH (1995)

[NIEM86.3] Niemann, G., Winter, H.: Maschinenelemente Band III. Springer, New York (1986)

[SEIB03] Seibicke, F.: Dimensionierung, Auslegung und Herstellung von Kegelradsätzen mit inverser Spirale; Dresdner Maschinenelemente Kolloquium 2003, Tagungsband; Verlagsgruppe Mainz GmbH (2003)

Chapter 3
Design

3.1 Starting Values for the Geometry

Apart from high functional reliability, modern bevel gear sets are subjected to considerable demands in terms of transmitted torque, small mass, low production costs and low noise.

The most important starting values influencing the basic geometrical design of bevel gears are:

- the transmission ratio u
- the shaft angle Σ
- the hypoid offset a
- the transmitted torque T
- the installation space and hence the outside diameter d_{e2} of the wheel

Torque and required installation space naturally influence each other as, depending on the design (geometry and material), a particular gear size will only be able to transmit a specific maximum torque.

The variables describing the basic geometry are interrelated in a manner apparent from the following equation.

$$d_{e2} - b_2 \cdot \sin\delta_2 = u \cdot z_1 \cdot \frac{m_{mn}}{\cos\beta_{m2}} \tag{3.1}$$

Basic data known from previous gear sets or other experience should form the basis of the draft design.

Outer pitch diameter d_{e2} The outer pitch diameter determines the size of the gear unit. The available installation space is usually also determined in the

The original version of this chapter was revised. The erratum to this chapter is available at DOI 10.1007/978-3-662-43893-0_9.

J. Klingelnberg (ed.), *Bevel Gear*, DOI 10.1007/978-3-662-43893-0_3

specifications. The necessary pitch diameter depends on the transmitted torque, the transmission ratio and the material employed.

In [NIEM86.3], it is suggested that on the basis of known designs, factor K_K* should be determined by means of the respective virtual cylindrical gear data. This factor should be used to calculate the outer pitch diameter of the pinion, employing the following draft formula:

$$d_{e1} = \sqrt[3]{\frac{18500 \cdot T_1}{u \cdot K_K^*}} \tag{3.2}$$

$$K_K^* = \left(\frac{F_{mt}}{b_1 \cdot d_{v1}}\right) \cdot \left(\frac{u_v + 1}{u_v}\right) \tag{3.3}$$

If no empirical values exist, Eq. 3.4 according to [KN3028] supplies good working values for a design on the safe side.

$$d_{e2} = \sqrt[2,8]{1000 \cdot T_1 \cdot \left(\frac{u^3}{u^2 + 1}\right) \cdot \sqrt[5]{n_1}} \tag{3.4}$$

Face width b When choosing the face width, it has proved useful to keep to certain size relationships. Two criteria may be employed:

The ratio of the outer cone distance of the wheel to its face width should meet the condition below:

$(3.0 \leq R_{e2}/b_2 \leq 5.0)$

and the ratio of the face width of the wheel to the mean normal module should meet the following condition:

$(7.0 \leq b_2/m_{mn} \leq 14.0)$.

The first criterion is useful only for bevel gears with a shaft angle of roughly 90°; in other cases, the criterion for b_2/m_{mn} should be satisfied since, otherwise, the teeth will be too wide.

Number of teeth z Several criteria need to be considered when choosing the number of teeth for the pinion and wheel. The number of teeth will influence not only the profile curvature and the whole depth but also usability in terms of undercutting and pointed teeth. As is known from cylindrical gears, the profile shift needed to avoid undercutting will decrease with higher tooth numbers if the transmission ratio is kept constant.

Various approaches are possible to select the number of pinion teeth. The first approach is to maintain a specific minimum number of crown gear teeth $z_{Pmin} \geq 25$ ... 35.

$$z_1 = \frac{z_{P\min} \cdot \sin \Sigma}{\sqrt{1 + u^2 - 2 \cdot \cos \Sigma}} \tag{3.5}$$

The following rule-of-thumb formula to determine the number of pinion teeth can be deduced for non-offset bevel gears with a 90° shaft angle and a given ratio R_{e2}/b_2.

$$z_1 = \frac{d_{e2} \cdot \left(2 \cdot \frac{R_{e2}}{b_2} - 1\right) \cdot \cos\beta_{m2}}{2 \cdot u \cdot m_{mn} \cdot \left(\frac{R_{e2}}{b_2}\right)} \tag{3.6}$$

A rough calculation of the mean normal module can be made from ratios R_{e2}/b_2 and b_2/m_{mn}.

$$m_{mn} = \frac{R_{e2}}{\frac{R_{e2}}{b_2} \cdot \frac{b_2}{m_{mn}}} \tag{3.7}$$

For the selection of the number of teeth, it is also necessary to observe the limitations of the machine tools manufacturing the gears in terms of feasible minimum and maximum tooth numbers and maximum transmission ratios. In continuous indexing face hobbing processes, the number of teeth and the number of blade groups should not be dividable by a common number. If this were the case, the same blade group would always machine the same slot such that inaccuracies in the blade group would be reflected directly in the pitch accuracy of the bevel gear. A periodicity corresponding to the number of blade groups would then be evident in the measured pitch deviation.

Mean spiral angle β_{m2} In most manufacturing methods, it is possible to select the mean spiral angle (see Sect. 2.1). Apart from the contact ratio, the spiral angle influences tooth forces and hence bearing loads. Sometimes the spiral angle also has an effect on the face and root angles (Duplex taper for example, see Sect. 2.2.4.2). For non-offset spiral bevel gears, the mean spiral angle should lie in a range of 30° to 45° if there is no conflicting experience or requirement. In most cases, a spiral angle of 35° is usually chosen. For hypoid gears, the mean spiral angle on the wheel should be selected such as to ensure a pinion spiral angle less than 50° in order to avoid problems at the toe or heel.

Tool radius r_{c0} Tools for the different manufacturing methods being supplied in various sizes and categories, the choice of the tool radius or diameter also depends on the chosen manufacturing method. Therefore, different tool radii will be optimal for different manufacturing methods. The criterion is the ratio of the tool radius to the mean cone distance of the wheel. Table 3.1 provides guide values for advantageous ratios of r_{c0} to R_{m2}. However, depending on requirements and tool availability, it may be necessary to deviate from these values.

Table 3.1 Reference values for ratio r_{c0}/R_{m2}

	Single-indexing, tapered tooth depth (e.g. Completing)	**Single-indexing, constant tooth depth (e.g. Kurvex)**	**Continuous indexing stick blade (e.g. Spirac®)**	**Continuous indexing profile blade (Zyklo-Palloid®)**
r_{c0}/R_{m2}	1,0	0,8	0,8	0,6

The tool radius determines the radius of curvature of the tooth trace, and, therefore, greatly influences the response of the contact pattern to relative displacements of the mating gears (see Sect. 3.4.5). We may thus refer to a design as a "small cutter" or a "large cutter" design. Consequently, there will be effects on the inner and outer spiral angles of the gears and on the space width taper in the normal section along the tooth face width.

Profile shift coefficient x_{hm} Bevel gear sets are always designed as "zero shift gears", i.e. gears for which the sum of the profile shifts is zero or, in other words, for which the amount of pinion profile shift is equal and opposed in sign to that of the wheel. This limitation in the choice of profile shift, when compared to cylindrical gears, is partly compensated by the fact that the necessary tooth root thickness to satisfy the desired root stress condition can be adapted by tooth thickness modification. This can be realized without special tooling or additional effort in all but the Palloid® method.

Criteria for the selection of profile shift are:

- undercutting prevention
- influence on tooth flank load capacity
- avoidance of pointed teeth

At this stage, designing to prevent undercutting can only be approximate (see Sect. 2.3.4) because the geometry and, hence undercutting, depend on the manufacturing method. Since bevel gears are designed as zero shift gear sets, undercut on a generated wheel or pinion may be unavoidable in the worst case. If no modification to the geometry is possible, such as an increase in tooth number, change in spiral angle or selection of a different tool radius, an even undercutting distribution should be ensured. On plunge cut wheels there is no undercut, as it occurs only on generated gears. This means however that larger undercutting occurs on the associated pinions. Therefore, a larger profile shift coefficient is necessary to avoid undercutting than would be the case with a pinion whose wheel is generated. Only a tooth flank generating software (see Sect. 3.3.1) will allow to determine exactly whether undercutting will occur, and how much will be present.

The choice of the profile shift coefficient also influences the surface load capacity. The radii of profile curvature of the contacting tooth flanks are of crucial importance for the Hertzian pressure. Apart from the magnitude of the contact stress, profile shift influences tooth flank sliding which affects load capacity and the occurrence of surface damage such as pitting or micro-pitting (see Sects. 4.1.4 and 4.2.5). A further criterion for the choice of the profile shift coefficient is therefore the ability to compensate the maximum specific sliding values (see Sect. 2.4.5) on the pinion and wheel.

Too large a profile shift can also produce pointed teeth in the heel region (see Fig. 3.1).

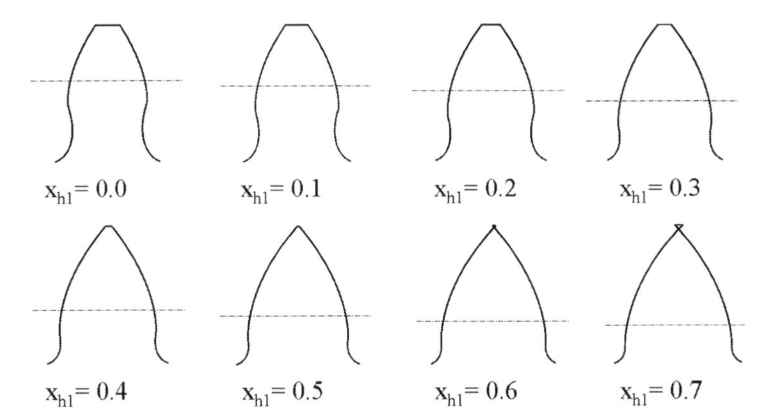

Fig. 3.1 Pinion profiles for different profile shift coefficients ($z_1 = 9$, $z_2 = 41$)

Thickness modification coefficient x_{smn} As noted above for the profile shift coefficient x_{hm}, the thickness modification coefficient for spiral bevel gears may be varied freely within certain limits. Thickness modification is used to balance tooth root strength between the pinion and the wheel. The change in tooth thickness alters the slot width of the gears which in turn affects the maximum edge radius which can be chosen for the tool. In unfavorable cases, the increase in tooth thickness and, hence, the decrease in nominal stresses, may be compensated by a greater stress concentration at the tooth root due to a necessarily smaller tool edge radius such that the local root stress remains roughly the same.

Design pressure angle α_d The design pressure angle for ground gears and those manufactured using a stick blade cutter head (see Sect. 6.2.3.4) can be chosen as desired, without entailing the use of special tools. For other methods, e.g. Zyklo-Palloid®, pressure angles are specified as 17.5°; 20° and 22.5°, depending on the available standard blades. As in the case of cylindrical gears, a value of 20° has emerged as a useful compromise between profile contact ratio and tooth root strength. Depending on load capacity, noise behavior etc., pressure angles will typically lie in a range from 16° to 24°. Pressure angles differing significantly from these values frequently lead to geometrical constraints which cannot be satisfied either by the gear set or by the tool.

Addendum factor k_{hap} The addendum factor is defined on the basic rack (see Fig. 2.20). Any value may be selected, 1.0 being the usual figure. It should also be remembered that manufacturing operations with standard blades (e.g. the Zyklo-Palloid® method) are attuned to this 1.0 value in terms of profile height and spherical radius of the cutter edge. The effects of the addendum factor on bevel gears are similar to those on cylindrical gears, determining the mean whole depth. A rise in the addendum factor entails an increase in the profile contact ratio, an increased risk of pointed teeth and an increase in the length of the bending moment arm.

Clearance c (theoretical) Projected into the axial section, the clearance is the minimum distance between the tooth tip and the bottom line of the mating gear. The distance from the tooth tip to the tooth flank of the mating gear is referred to below as the lateral clearance or interference. The clearance c on spiral bevel gears is usually between 0,2 and 0,3 times the mean normal module. On gears cut with tool tilt, a larger clearance is used to prevent the tip from running into the root of the mating gear whose root trace is curved owing to the tilt (see Fig. 3.2), such that the fillet may project beyond the clearance. The actual clearance depends on manufacturing and assembly tolerances (Table 3.2).

Table 3.2 Standard clearance factor values

Process	Zyklo-Palloid®	Palloid®	Continuous	Kurvex	5-cut	Completing	Wiener
k_{hap}	1.0	1.0	1.0	0.9	1.0	1.0	1.0
c/m_{mn}	0.25	0.3	0.25	0.2	0.3	0.35	0.25

With Palloid—reference is made to cutter module m_0 instead of m_{mn}

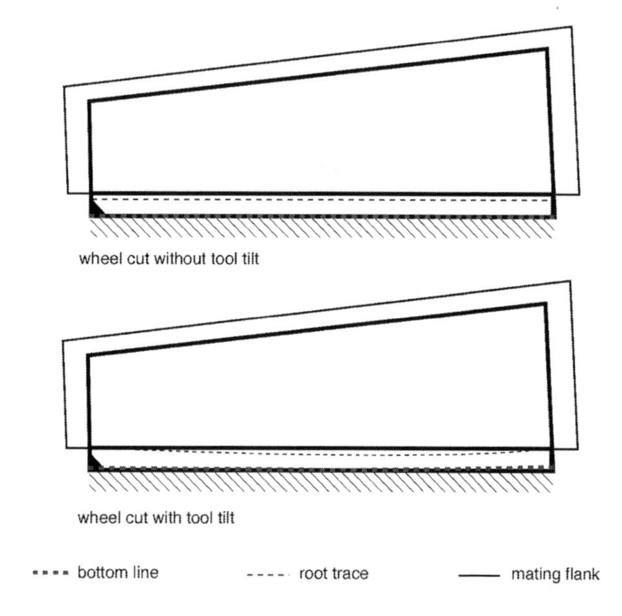

Fig. 3.2 Actual fillet on gears with and without tilt

Backlash j Backlash is necessary such that a gear set, which in practice will always have manufacturing deviations, remains operational. Without backlash, deviations in pitch and thickness or in mounting and bearing tolerances could cause meshing interference or jamming of the gears. In addition, relative displacements of the pinion and wheel not only cause displacements of the contact pattern but also a change in backlash. If the backlash is too small, there is risk of jamming, while too large a backlash will reduce the tooth thickness unnecessarily and increase the free travel during load reversals.

Backlash must therefore be chosen as a function of gear size, the manufacturing method, i.e. the attainable gear quality, and the application. Failing empirical know-how or defaults, the backlash measured on the outer cone should therefore lie in the following range:

$$j_{et,\min} = 0.02 \cdot m_{mn} + 0.03 \text{ mm} \tag{3.8}$$

$$j_{et,\max} = 0.024 \cdot m_{mn} + 0.06 \text{ mm} \tag{3.9}$$

Tool edge radius ρ_{a0} The tool edge radius directly affects the fillet generated on the bevel gear and, hence, stress concentration in the tooth root. It also influences tool life. On non-generated bevel gears the edge radius is reproduced directly as the fillet radius, whereas on generated gears the machine kinematics enlarge the fillet.

The tool edge radius is pre-defined for methods employing standardized tools. In other cases it can be chosen as desired, but its maximum value must meet two limits:

- the maximum tool edge radius, limited by the clearance c

$$\rho_{a0,\lim 1,2} = \frac{c_{1,2}}{1 - \sin \alpha_{nD,C1,2}} \tag{3.10}$$

- the maximum tool edge radius, limited by the minimum slot width $e_{fn,min}$

$$\rho_{a0,\lim 1,2} = \frac{0,5 \cdot e_{fn\min} \cdot \cos \alpha_{nD,C1,2}}{1 - \sin \alpha_{nD,C1,2}} \tag{3.11}$$

The calculation is made on the virtual crown gear for identical edge radii on the concave and convex tooth flanks. To prevent meshing interference or cut-off in the tooth root, the chosen tool edge radius should be smaller than the lesser of the two above limits.

Crowning This is differentiated into profile and lengthwise crowning (see Fig. 3.3). High crowning values lead to less sensitivity of the contact pattern to displacements of the gears, but also to a smaller contact pattern, a concentration of loads entailing high flank pressures as well as higher local tooth root stresses. An optimal modification of bevel gear tooth flanks can be established only if numerical simulation is available to calculate the exact tooth geometry, load-free and loaded tooth contact analysis. The latter must also include the relative displacements of the pinion and wheel induced by deflections. Apart from the absolute values of the modifications, their form and location also have a considerable influence on the sensitivity of the contact pattern to displacements. This can be evaluated by means of an ease-off analysis (see also Sect. 3.3.1).

If there is no empirical know-how and no means are available to evaluate the contact pattern displacement under load, the following face-width-dependent guide values for lengthwise crowning have proved valuable:

- for normal displacement: $b_2/250$ to $b_2/600$
- for slight displacement: $b_2/350$ to $b_2/800$

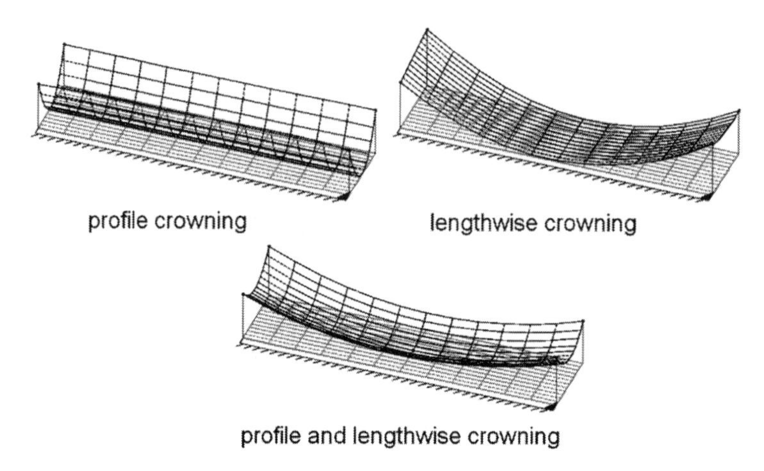

Fig. 3.3 Profile and lengthwise crowning as parts of the ease-off

Machinability The limits in machinability set by pointed teeth and undercutting can be calculated with sufficient accuracy using a virtual cylindrical gear (see Sect. 4.2.2). With spiral bevel gears, and especially with flat wheels, the cutter head may hit the gear rim at another point than the intended tooth slot, thus damaging the teeth. This effect, named as back cutting or interference at the rear, is more likely to occur with a diminishing tool radius. However, it depends on many parameters and must therefore be calculated differently for each manufacturing method.

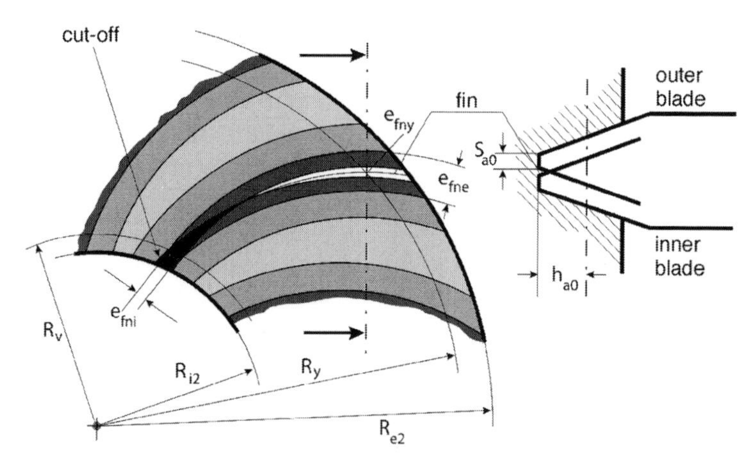

Fig. 3.4 Cut-off or fin (schematic on a virtual crown gear)

Apart from leaving a fin on the bottom of the tooth slot, cut-off may also occur on the tooth flanks (see Fig. 3.4). This is caused by the variable width of the tooth slot in the normal section, creating a minimum at one location and a maximum at another which are not necessarily at the inner and outer ends of the slot. To avoid cut-off, the point width s_{a0} of the tool must be smaller than or equal to the slot width at its narrowest point. On the other hand, the largest slot width $e_{fn,max}$ must be covered completely by the point widths of all blades in a group. A feature to be noted is that a number of manufacturing methods use standardized tools whose point widths cannot simply be changed as can be done when using stick blades or grinding wheels. In general, it may be stated that:

Avoidance of cut-off: $e_{fn,\min} \geq s_{a0}$
Avoidance of fin: $e_{fn,\max} \leq n \cdot s_{a0}$ with n = number of blades per group

A simplified calculation of the slot width may be made for the virtual crown gear at any desired radius R_{Py}. The following formula refers to the general case and sometimes requires adaption, e.g. for the Wiener 1-trace method.

$$e_{fny1,2} = \frac{\pi}{2} m_{yn} - 2 \cdot x_{sm1,2} \cdot m_{mn} - \left(k_{hfp} \cdot m_{mn} + \left(R_{Py} - R_{Pm}\right) \cdot \tan\theta_{f1,2}\right) \cdot \left(\tan\alpha_{nD} + \tan\alpha_{nC}\right) \quad (3.12)$$

3.2 Manufacturing Kinematics

3.2.1 *Basic Rack and Virtual Crown Gear (Generating Gear)*

On cylindrical gears, only a few variables (number of teeth, face width, pitch diameter, whole depth, profile shift, helix and pressure angles) are sufficient to determine exactly the relative motions between the tool and work piece. In generating methods, the relative motions are based on a basic rack with straight tooth profiles meshing with the work piece. If the rack is replaced by a hob, it is possible to determine the manufacturing motion which will replicate the motion of the rack.

On bevel gears, the same principle applies. Instead of a basic rack, a virtual basic crown gear is used which, like the rack for cylindrical gears, has straight tooth profiles. Figure 3.5 shows the basic crown gear also termed generating gear in a bevel gear manufacturing machine. The generating motion takes place when the cutter, rotating about its own axis, is simultaneously rotating about the axis of the generating gear while the work piece rotates about its own axis at a given gear ratio with the generating gear. The generating gear can be flat, as the one displayed in Fig. 3.5, or take a conical shape. For the different gear manufacturing methods described in Sect. 2.1, there are various ways in which a cutter can be substituted to the relevant virtual generating gear. Figure 3.6, for example, represents the Spirac® method in which a conical generating gear, used for a plunge cut wheel, is replaced by a tilted face cutter.

Determining the respective relative motions between the cutter and the bevel gear is therefore by no means a trivial concern, particularly since additional motions are superimposed to modify and improve tooth flank topography.

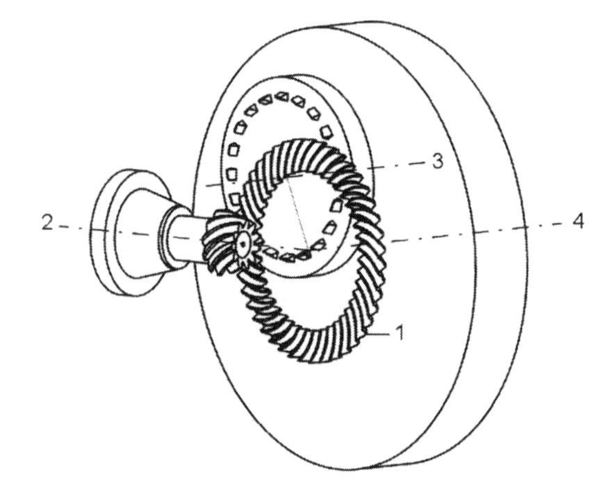

Fig. 3.5 Virtual crown gear in a bevel gear generator. 1 basic crown gear, 2 bevel gear axis, 3 cutter head axis, 4 cradle axis = basic crown gear axis

The following points will be noted with respect to the gear cutting machine: If, for example, a tooth of the virtual crown gear is represented by the cutting edges of a rotating cutter, the pitch plane of the cutter will usually be parallel to the pitch plane of the basic crown gear. However, in some manufacturing methods (see Table 2.1) the tool is tilted in order to modify the lengthwise crowning. If the aim is to increase lengthwise crowning, the cutter is tilted such that the tooth ends are cut deeper than in the middle of the face width. Thus, the effective radius of curvature of the outer blade is larger and that of the inner blade is smaller. If, on the other hand, the intention is to reduce lengthwise crowning, the cutter must be tilted in such a way that the depth of cut is greater in the middle of the face width. This corresponds to the difference in radii in a manufacturing method without tool tilt. In the gear cutting machine, the tool is tilted out of the pitch plane of the crown gear, around an axis whose direction coincides with the direction of the mean tooth slot on the crown gear. On a conventional mechanical cutting machine, the amount of tilt and direction of the tilt axis are set by means of the tilt and swivel angles (see Fig. 3.7).

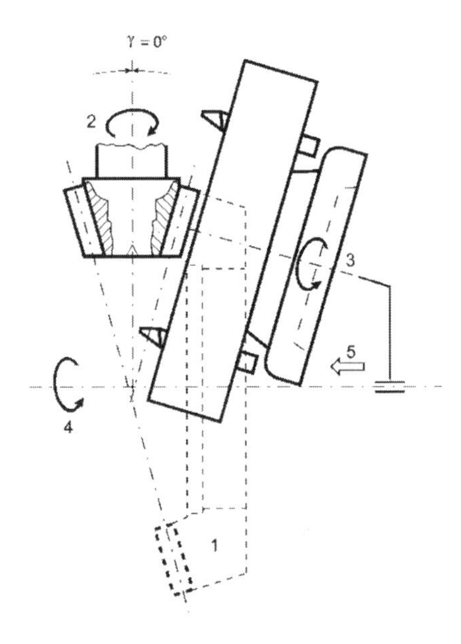

Fig. 3.6 Conical generating gear for the non-generating method. 1 conical generating gear, 2 bevel gear axis, 3 cutter axis, 4 cradle roll axis, 5 cutter depth feed γ machine root angle

The next three subsections provide a universal description of manufacturing kinematics for a virtual gear-cutting machine which can perform all important production processes without any geometrical restriction.

3.2.2 Model of a Virtual Bevel Gear Machine

A bevel gear machine is designed such that it imposes to the tool and bevel gear blank the relative motions of the generating gear meshing with the bevel gear. For this purpose one tooth of the generating gear is replaced by the straight flank profile of the tool. Figure 3.5 shows a tool used in the single indexing method. In order to roll the tool with the blank, i.e. to machine one tooth slot, and return to the starting position, the generating gear shown in the figure must rotate by a relatively small angle. On many mechanical machine tools, the cutter moves to and fro like a cradle, from the start of roll angle to the stop roll angle and back again.

Thus, conceptually, a virtual gear-cutting machine closely resembles the former mechanical machine, but with the significant advantage that the virtual machine is not subject to any limitation in terms of penetration, stability, damping, assembly, accessibility etc. Modern 6-axis CNC machines may be fundamentally different from earlier pure mechanical ones or NC models, but their relative generating motion is identical. The tool is guided using three translations and three rotations, thus a total of six axes in relation to the work piece.

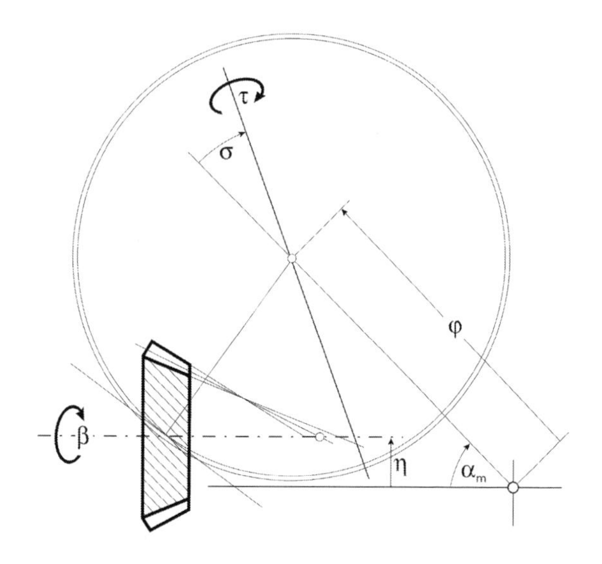

Fig. 3.7 Schematic view of the virtual machine along the cradle axis α.
φ radial distance,
σ swivel angle,
τ tilt angle,
η offset,
ß work piece axis

Figure 3.7 shows the virtual machine as seen along the axis of rotation of the cradle. For the historical reasons noted above, the term "cradle axis" is used in preference to virtual crown gear axis. The figure shows only one tooth of the generating gear, where a cup-type tool has been substituted.

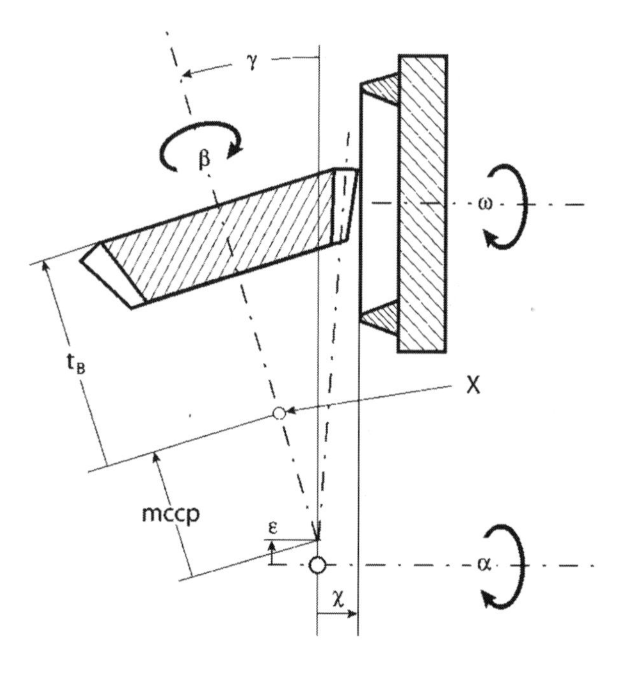

Fig. 3.8 Schematic view of the virtual machine when viewed in a direction perpendicular to the bevel gear and cradle axes.
α cradle roll axis,
β work piece axis,
γ machine root angle,
ε horizontal distance,
χ sliding base,
t_B mounting distance,
X crossing point of axes in the gearbox

3.2.3 *Calculation Model*

The motions between the tool and the work piece may be defined on the basis of Figs. 3.7 and 3.8. The cradle angle α and the cutter angle ω, which may be independent of each other, are used as command variables. The relative motion between one crown gear tooth and the work piece may be represented as follows:

radial distance	$\varphi = \text{constant}$
swivel angle	$\sigma = \text{constant}$
tilt angle	$\tau = \text{constant}$
hypoid offset	$\eta = \text{constant}$
machine root angle	$\gamma = \text{constant}$
horizontal distance	$\varepsilon = \text{constant}$
sliding base	$\chi = \text{constant}$
work rotation angle	$\beta = \beta(\alpha, \omega)$
machine center to X-point	$mccp = \text{constant}$

More generally, these motions may be described as functions of α and ω:

radial distance	$\varphi = \varphi(\alpha)$
swivel angle	$\sigma = \sigma(\alpha)$
tilt angle	$\tau = \tau(\alpha)$
hypoid offset	$\eta = \eta(\alpha)$
machine root angle	$\gamma = \gamma\ (\alpha)$
horizontal distance	$\varepsilon = \varepsilon(\alpha)$
sliding base	$\chi = \chi\ (\alpha)$
work rotation angle	$\beta = \beta(\alpha,\omega)$
machine center to X-point	$mccp = \text{constant}$

It will be evident that this description contains a number of redundancies. A radial motion, for example, may equally well be expressed as a combination of offset and horizontal motions. This redundant form is used since the classic relative motions between the virtual crown gear and work piece are easily apparent.

Nothing has yet been said about the types of functions to be used. Basically, all functions which display no singularities in the range from 0 to 2π are suitable. In practice, the Taylor series, centered about the mean cradle angle α_m, is used up to the sixth order as listed below in Table 3.3.

Table 3.3 Motion equations of the virtual machine expressed as Taylor series

Designation	Formula	No.
Radial motion	$\varphi(\alpha) = a_\varphi + b_\varphi(\alpha - \alpha_m) + c_\varphi(\alpha - \alpha_m)^2 + \ldots + g_\varphi(\alpha - \alpha_m)^6$	(3.13)
Vertical motion	$\eta(\alpha) = a_\eta + b_\eta(\alpha - \alpha_m) + c_\eta(\alpha - \alpha_m)^2 + \ldots + g_\eta(\alpha - \alpha_m)^6$	(3.14)
Angular motion	$\gamma(\alpha) = a_\gamma + b_\gamma(\alpha - \alpha_m) + c_\gamma(\alpha - \alpha_m)^2 + \ldots + g_\gamma(\alpha - \alpha_m)^6$	(3.15)
Horizontal motion	$\varepsilon(\alpha) = a_\varepsilon + b_\varepsilon(\alpha - \alpha_m) + c_\varepsilon(\alpha - \alpha_m)^2 + \ldots + g_\varepsilon(\alpha - \alpha_m)^6$	(3.16)
Helical motion	$\chi(\alpha) = a_\chi + b_\chi(\alpha - \alpha_m) + c_\chi(\alpha - \alpha_m)^2 + \ldots + g_\chi(\alpha - \alpha_m)^6$	(3.17)
Modified roll	$\beta(\alpha, \omega) = c\omega + a_\beta + b_\beta(\alpha - \alpha_m) + c_\beta(\alpha - \alpha_m)^2 + \ldots + g_\beta(\alpha - \alpha_m)^6$	(3.18)

The rotation of the work piece, controlled by modified roll, is the only function with two variables. For all tool types symmetrical about the axis of rotation, the cutter phase angle ω has no influence on the geometry of the tooth. For tools that are not symmetrical about their axis of rotation, Formula (3.18) is used with c as the constant transmission ratio z_0/z. This simple model may be employed to describe the manufacturing kinematics by means of coefficients.

The following section describes the calculation of the coefficients for orders zero and one in the Taylor series of Table 3.3. The influence of higher order coefficients is described in Sect. 3.3.

3.2.4 Sample Calculation of Machine Kinematics

The example presented below refers to the manufacturing kinematics for a hypoid pinion according to the single indexing method with constant depth, i.e. equal root, pitch and tip angles. The generating gear is a virtual crown gear and is identical for the pinion and wheel. The tool is either a circular face cutter or a cup-shaped grinding wheel. The tool radius r_{c0} for this case is measured in the pitch plane of the generating gear.

In the following, the machine kinematics for one tooth flank of the bevel gear are derived from the equations of motion defined in Table 3.3 The variables listed in Table 3.4 determine the gear blank as described in Sect. 2.3.

Table 3.4 Macro geometry of a bevel gear

Variable	Symbol
Number of teeth	z
Mean cone distance	R_m
Pitch angle	δ
Addendum	h_a
Dedendum	h_f
Profile shift	$x_{hm}m_{mn}$
Hypoid offset in the pitch plane	a_p
Mean spiral angle	β_m
Tool radius in the pitch plane	r_{c0}

Figure 3.9 shows the relative position of the bevel gear and the tool. The settings of the virtual machine can be derived from the formulae given in Table 3.5.

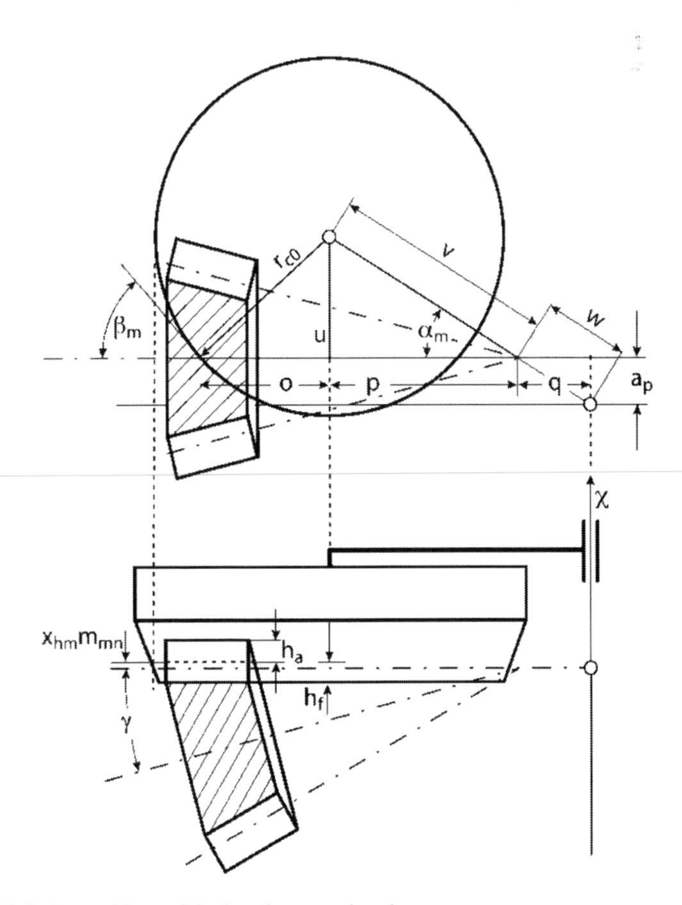

Fig. 3.9 Relative positions of the bevel gear and tool

Table 3.5 Equation of motion for a bevel gear with constant tooth depth, single indexing

Designation	Formula	No.
Auxiliary variables	$u = r_{c0} \cos \beta_m$	(3.19)
	$o = r_{c0} \sin \beta_m$	(3.20)
	$p = v / \tan \alpha_m$	(3.21)
	$q = a_p / \tan \alpha_m$	(3.22)
Mean cone distance of the crown gear	$R_m = o + p + q$	(3.23)
Mean cradle angle	$\tan \alpha_m = \frac{u + a_p}{R_m - o} = \frac{r_{c0} \cos \beta_m + a_p}{R_m - r_{c0} \sin \beta_m}$	(3.24)
Auxiliary variables	$v = \frac{r_{c0} \cos \beta_m}{\sin \alpha_m}$	(3.25)
	$w = \frac{a_p}{\sin \alpha_m}$	(3.26)
Radial motion	$\varphi(\alpha) = v + w = \frac{r_{c0} \cos \beta_m + a_p}{\sin \alpha_m}$	(3.27)
Vertical motion	$\eta(\alpha) = a_p$	(3.28)
Angular motion	$\gamma(\alpha) = \delta$	(3.29)
Horizontal motion	$\varepsilon(\alpha) = 0$	(3.30)
Helical motion	$\chi(\alpha) = x_{mh} m_{mn} - h_f$	(3.31)
Modified roll	$\beta(\alpha) = \frac{1}{\sin \delta}(\alpha - \alpha_m)$	(3.32)
Tilt	$\tau = 0$	(3.34)
Swivel	$\sigma = 0$	(3.35)
Machine center to crossing point	$mccp = 0$	(3.36)

3.3 Tooth Contact Analysis

3.3.1 Tooth Geometry Calculation

Tooth contact analysis is an important tool for the design, evaluation and optimization of bevel gears. It also forms the basis for more exact stress calculation methods. To conduct load-free tooth contact analysis, the tooth flanks are meshed by calculation in a load-free process. During meshing, it is possible to calculate the ease-off (see Sect. 3.3.3), the transmission error and the contact pattern.

In order to simulate tooth contact and calculate local loads and stresses, it is necessary to know the tooth flank and tooth root geometry including the transition between tooth flank and tooth root. It is also necessary to know the spatial position of the pinion and wheel. Tooth contact analysis can be done from the calculated geometry or from coordinates obtained with a three-dimensional gear measurement machine. Since tooth flank and tooth root geometry depends on the manufacturing method, as noted in Sect. 2.1, to calculate the tooth coordinates, it is necessary to simulate the gear manufacturing using a virtual gear-cutting machine. Software packages used for this purpose are often referred to as flank generators. Various approaches can be used for the calculation:

- calculation of the "enveloping surface" (generating method) or of the "motional surface" (non-generating method)
- calculation of the penetration
- calculation based on the law of gearing

The results of this simulation are exactly calculated tooth flank coordinates suitable as data points to calculate a fitting surface. It is also possible to perform an undercut calculation to determine the transition curve between the effective tooth flank and root zones.

Fitting surfaces calculation There are many important reasons to describe the calculated tooth flank and fillet points by means of fitting surfaces:

- minimized computational effort for subsequent calculation tasks
- independence from the gear making process
- various sources (support points) are possible as starting data
- tooth contact analysis and stress analysis are possible on any suitable grid
- any required number of discrete segments (profile sections) is possible on the tooth
- any point on each profile line can be calculated
- the description of the effective tooth flanks by means of fitting surfaces is sufficiently exact

Considering the tooth flank and root separately has proved helpful when determining the fitting surfaces [DUTS94].

3.3.2 Crowning

Mathematically, surfaces are called conjugate when they have line contact. Gears are termed conjugate when their tooth flanks have line contact in every meshing position. Contact then extends over the entire tooth flank. If the tooth profile on cylindrical gears is an involute, a change in center distance does not alter tooth contact as the involute is a self-equidistant curve (see Fig. 3.10). A small change in the parallel alignment of the gear axis will, however, cause the flanks to touch only at one edge of the tooth.

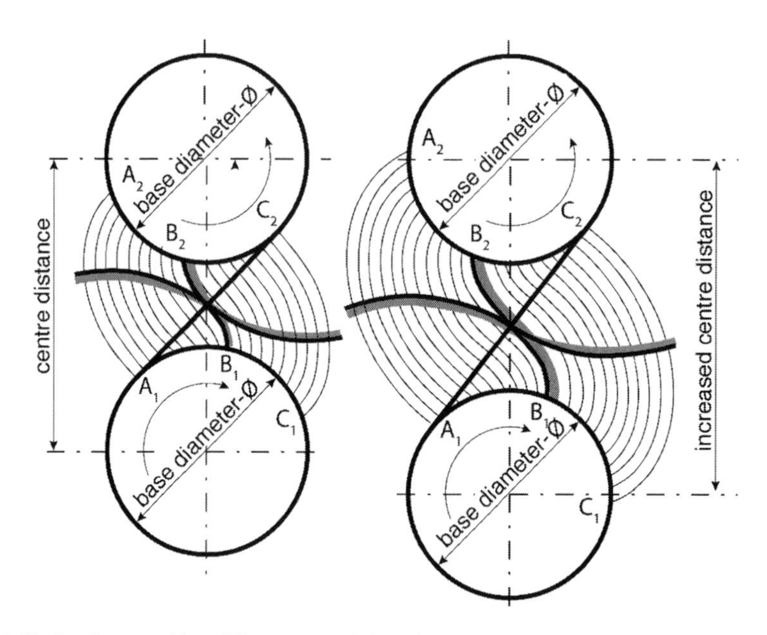

Fig. 3.10 Involute: a self-equidistant curve [MAAG90]

Unlike cylindrical gears, bevel gears possess tooth profiles that change continuously along the face width. The tooth profile is not an involute, and a displacement in the direction of the tooth depth leads to different meshing conditions. The transmitted torque causes deflections in the housing, the gear bodies and the teeth, thereby yielding different relative positions of the pinion and wheel for each load case. In order to ensure acceptable tooth contact under all operating conditions, bevel gears are never made with perfectly conjugate tooth flanks. The required insensitivity to relative displacements of the tooth flanks is achieved by superimposing crowning by which the tooth flanks deviate from their exactly conjugate form.

Section 3.4 is devoted to the effects of relative displacements between the meshing pinion and wheel. The necessary fundamentals of tooth flank design and the mechanisms that influence the tooth shape are presented in the following sections.

In designing bevel gears, a distinction is drawn between the macro- and micro geometries. The macro geometry includes all the typical gear variables, such as number of teeth, face width, pitch diameter, hypoid offset, tooth depth, profile shift, spiral and pressure angles and tool radius. Only the coefficients of orders zero and one are needed to calculate the basic manufacturing kinematics (see Sect. 3.2.3). The higher order coefficients affect mainly the micro geometry, causing only insignificant changes to the macro geometry.

3.3.3 Ease-off, Contact Pattern and Transmission Error

Ease-off calculation Contrary to cylindrical gears, the description of tooth flank modifications on bevel gears cannot be considered as deviations from a basic rack. Tooth contact is rather described in terms of the meshing parameters of the pinion and wheel. Crowning of the tooth flanks is no longer assigned to the wheel or the pinion, but rather relates to tooth contact between the wheel and the mating pinion of a bevel gear pair.

What is simulated is the load-free meshing of the tooth flanks (described, for example, by fitting surfaces) of a tooth pair at a theoretical transmission ratio (not including the subsequently calculated transmission error). Mating tooth flanks are separated by distances which vary for each meshing position *i* at a given meshing interval expressed by means of the pinion rotation angle φ_{1i} or the wheel rotation angle φ_{2i}. In meshing position *i*, arc distance ζ (r_2, υ_2, φ_1, k) exists between wheel flank 2 on tooth k, given by variables r_2, $\vartheta_{2,}$, and pinion flank 1 (Fig. 3.11). The function of variables r_2, ϑ_2 per meshing position *i* is termed the (instantaneous) ease-off function of φ_1 and k. The enveloping surface over all instantaneous ease-offs represents the minimum of all gape distances during one complete meshing cycle of a tooth pair. This distance is termed the contact distance or ease-off.

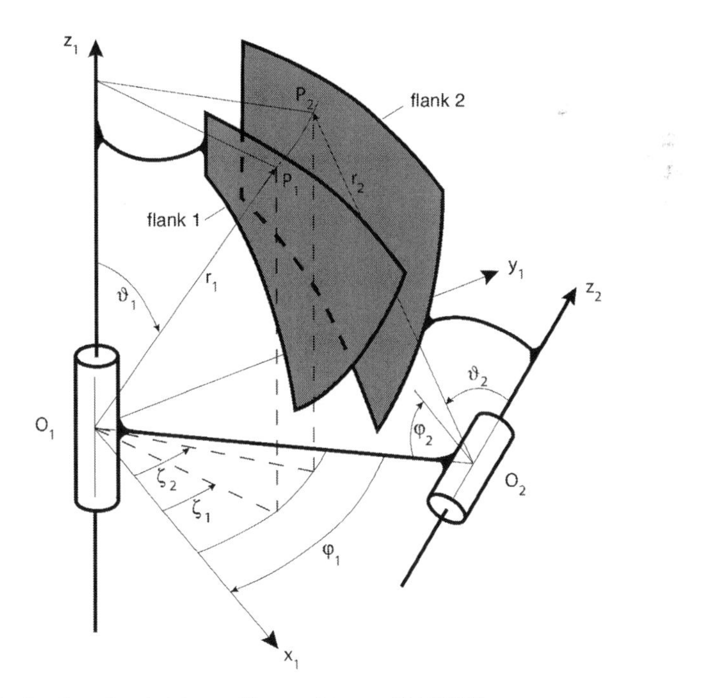

Fig. 3.11 Flanks of tooth pair *k* in meshing position φ_1 [BAER91]

This approach is by no means restricted to bevel gears but is also suitable to describe the contact parameters of any three-dimensional surfaces in rolling contact. Figure 3.12 shows the ease-off of a pair of meshing bevel gear tooth flanks. The display provides an immediate assessment of lengthwise and profile crowning, which is really no more than local modifications of the tooth flanks when compared to conjugate tooth flanks. It is useful to show contact distances as deviations from a plane rather than in a three-dimensional representation.

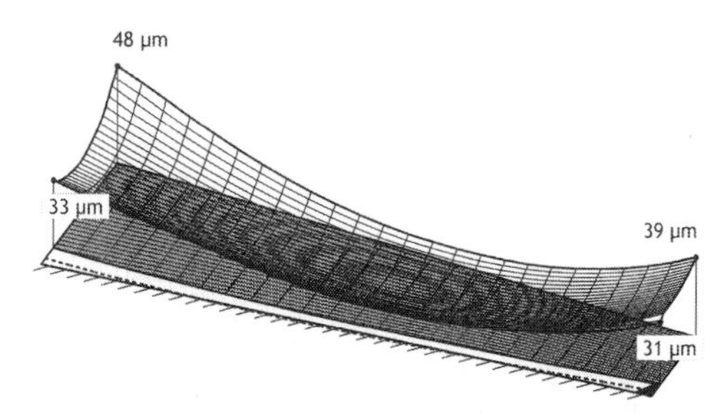

Fig. 3.12 Ease-off of a pair of bevel gear tooth flanks

Calculating the transmission error (load-free) Unloaded conjugate tooth flanks roll with kinematic exactness (see Sect. 3.3.2). As a result of modifications and deviations on the tooth flanks (including pitch deviations) and deviations in position caused by the environment, conjugate conditions do not exist anymore. A difference therefore occurs between the theoretical and the real instantaneous transmission ratios. This difference may be calculated if the associated pinion rotation angle, for a given wheel rotation angle, is corrected by an angular difference φ_{korr} until the pair of tooth flanks meet exactly at one point.

$$\varphi_{korr,i} = \varphi_{2,i} \cdot \frac{z_1}{z_2} - \varphi_{1,i} \tag{3.33}$$

This calculation of the difference angle of rotation is done iteratively and is performed for each mesh position i. This yields the variation in the angle of rotation or transmission error.

If pitch deviations are neglected, the transmission error of each tooth pair of a given gear set has the same shape. The transmission error curves of neighboring tooth pairs are then translated one pitch apart, as shown in Fig. 3.13.

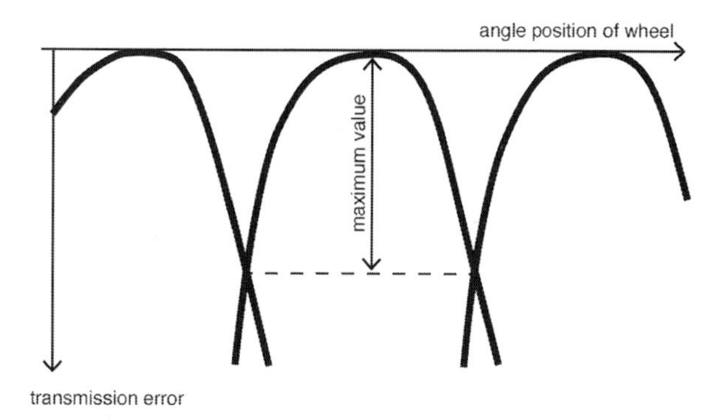

Fig. 3.13 Transmission error curves of three consecutive tooth pairs

Contact pattern calculation The contact pattern is a representation of all the contact lines during a complete mesh of a tooth pair. Load-free motion produces a different contact pattern than motion under load. The contact pattern which occurs under load is one result of the load distribution calculation (see Sect. 4.4.3.4).

The following points should be considered when calculating the load-free contact pattern:

- the tooth contour limiting the contact pattern,
- the simultaneous engagement of several teeth,
- specified or calculated deviations in relative position

Other parameters which may be considered are:

- pre-mesh or edge contact,
- specified pitch deviations

To determine the contact pattern, the bevel gear set rotates in such a way that at each moment one tooth flank pair is in contact. All meshing positions are therefore considered in predefined steps, taking the transmission error into account, with the currently active flank pair contacting at one point. As load increases, the tooth flanks deform creating a narrow contact ellipse in place of the load-free contact point. A profile line is formed by the points with the smallest contact distance which can be determined for each contact position (see Fig. 4.33). Points on a potential contact line which fall below a certain contact distance (usually 3 to 6 μm, depending on the thickness of the marking compound) at one meshing position at least, constitute the effective contact line and are part of the contact pattern.

Those points with the smallest contact distance on each contact line form what is termed the path of contact. Figure 3.14 shows a contact pattern in which the contact lines—the major axes of the contact pattern—are represented. In Fig. 3.14, the short lines perpendicular to the path of contact indicate the transfer points, or the points at which contact is transferred between neighboring tooth pairs.

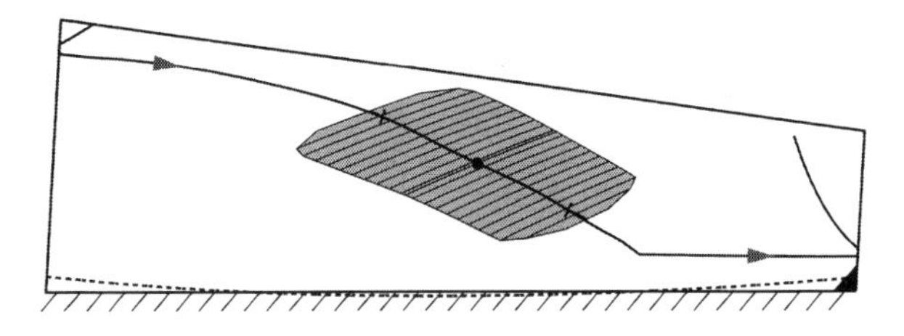

Fig. 3.14 Contact pattern with contact lines, path of contact and transfer points

Parameters of the ease-off topography The ease-off allows us to describe the contact geometry very clearly, and also to determine gear set relevant variables such as the position and size of the contact pattern and the transmission error. For this reason, the ease-off topography offers specialists complete information on meshing conditions. It is possible to define five parameters for the quantitative description of the ease-off topography, as indicated in Fig. 3.15.

There are practical reasons to analyze the ease-off topography using five parameters. The present case describes crowning in the directions of tooth depth and face width, the mean angular deviation in the profile and lengthwise direction, and twisting between toe and heel. These five parameters are especially suitable because they correlate directly with the form and position of the contact pattern. For example, to displace the contact pattern from the smaller to the larger diameter (toe to heel), the spiral angle difference is reduced; to lengthen the contact pattern in the face width direction, the lengthwise crowning is reduced.

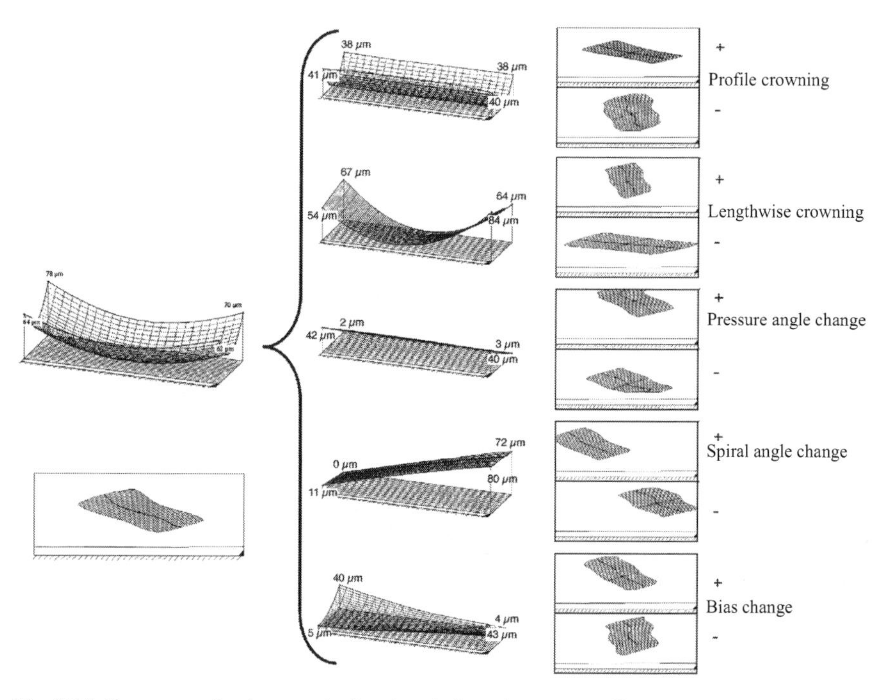

Fig. 3.15 Parameters for the quantitative description of any ease-off topography

3.3.4 Additional Motions

This section describes the effects, on the ease-off, of higher order coefficients in the manufacturing kinematics. Visualization of the contact lines between the tool and the tooth flank during the generating process, Fig. 3.16, illustrates the mode of operation.

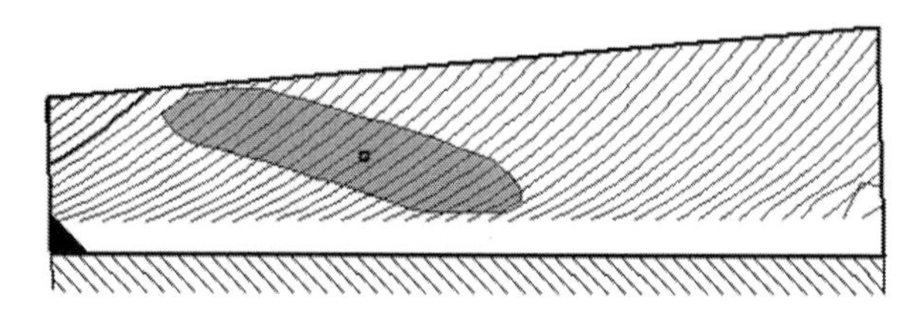

Fig. 3.16 Position of the contact lines between the tool and the tooth flank

Figure 3.16 shows the position of the contact lines between the tool and the tooth flank, termed generatrixes, for a generated pinion. Additional motions dependent on the respective cradle angle α will act on all points of the generatrixes belonging to this α, which are seen advancing diagonally from line to line across the tooth flank. In consequence, additional motions of the type created by higher order coefficients will modify the shape of the flank in the same diagonal direction.

Modifications along each generatrix can be achieved solely through the form of the tool. This principle applies irrespectively of the system of base functions employed in the manufacturing kinematics. For the case of a series function of the tool rotation dependent on the cradle position or roll angle α, the polynomial coefficients of the function produce the flank modifications shown in Fig. 3.17:

$$\beta(\alpha) = a_\beta + b_\beta(\alpha - \alpha_m) + c_\beta(\alpha - \alpha_m)^2 + \ \dots \ + g_\beta(\alpha - \alpha_m)^6 \qquad (3.34)$$

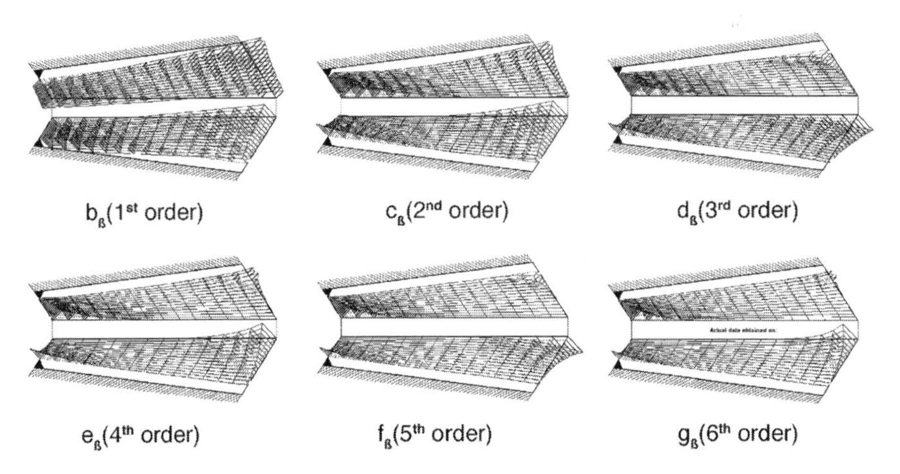

Fig. 3.17 Flank modifications by additional modified roll motion

Tooth flank form modifications are of particular interest for manufacturing methods in which both flanks of a tooth slot are machined in a single cut. In such cases, the additional motions act on both flanks, but in partly different directions.

Modified roll β(α) (see Table 3.3) is an additional rotation of the work piece dependent on the roll angle; modified roll changes material removal on both tooth flanks but in opposing directions. Helical motion χ(α) is a tool feed in the direction of the cradle axis which also depends on the roll angle. Helical motion modifies both tooth flanks in the same direction.

The effect of additional motions in modified roll and helical motion may readily be illustrated using the example of a face milled pinion. Figure 3.18 shows tooth flank modifications due to modified roll, employing coefficient c_β, and flank modifications due to helical motion, employing coefficient c_χ. They are designed such that modifications on one flank compensate those on the other.

In line with this example, it is also possible to make modifications affecting the flank only at one end of the tooth. A suitable combination of odd and even coefficients is used to influence solely the toe or heel of the tooth flank. Many options are therefore available to the user to optimize the tooth flanks to suit a particular application for given conditions.

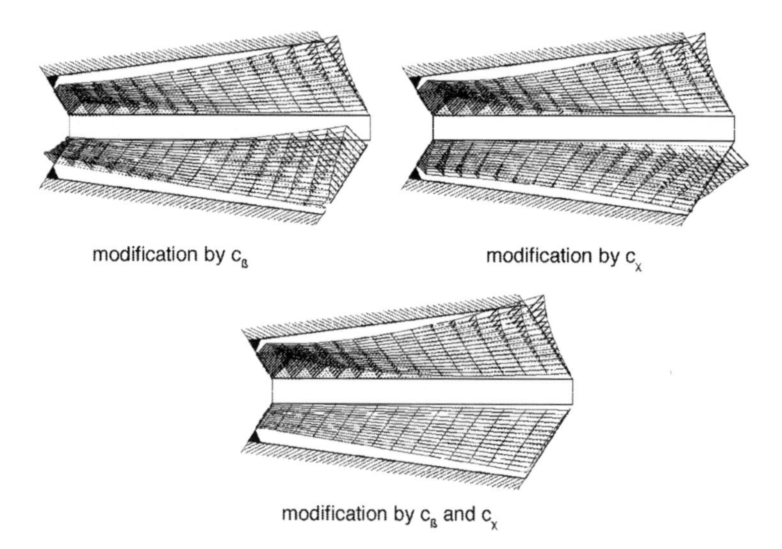

Fig. 3.18 Flank modifications using modified roll and helical motion

These additional motions can only be realized in a generating process. Tooth flank modifications on plunge cut wheels are performed by means of Modified Crowning [SIGM06]. This is described in greater detail in Sect. 3.4.4.

An alternative method for non-generated wheels is the Flared Cup™ method [KREN91]. It employs a flared cup grinding wheel which contacts the tooth flank of the non-generated wheel only along a line which essentially runs perpendicular to the root cone. The lengthwise tooth form is produced by appropriate machine motions to which small additional motions can be superimposed to achieve the desired modification.

3.4 Displacement Behavior

3.4.1 Horizontal and Vertical Displacements

Instead of describing the ease-off characteristics like longitudinal twist or crowning, contact pattern shift is a test which has been in use for decades (Fig. 3.19). The gear set is meshed and the contact pattern is determined at low torque. The contact pattern is shifted on the wheel by altering the pinion mounting distance (H) and the hypoid offset (V) while backlash is maintained constant by adjusting the mounting distance (J) of the wheel. In the US, the preferred symbols are (P), (E), (G).

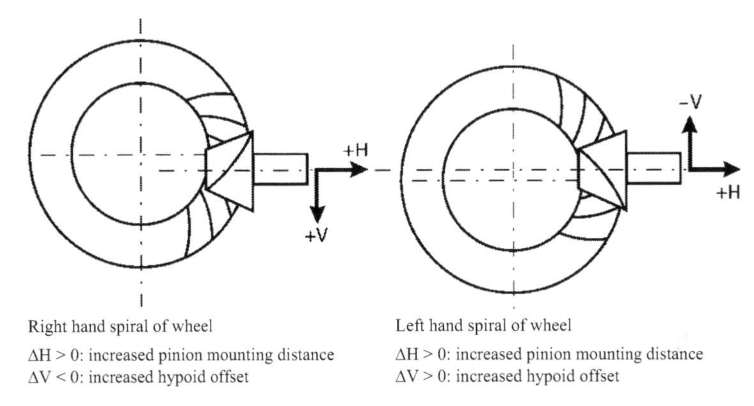

Fig. 3.19 Definition of horizontal and vertical displacements H and V

3.4.2 *Displacements Caused by Tooth Forces*

The relative displacements of a bevel gear set result from manufacturing tolerances on the housing and load-induced deflections. The method used to calculate the tooth forces on the teeth and bearings is described in Sect. 2.5. Beside deflections of the tooth flanks, these tooth forces deform the gear housing, the bearings and the gear bodies. These effects can be represented by relative displacements between the meshing tooth flanks of the pinion and wheel.

A set of bevel gears with positive offset, comprising a right-handed spiral wheel and a left-handed spiral pinion, is shown in Fig. 3.20. The drawing shows the view along the wheel axis. The pinion tooth flank exerts a force F_2 perpendicular to the surface of the tooth flank of the wheel, which counteracts with reactive force F_1 on the pinion.

These forces do not necessarily act in the drawing plane, i.e. only the components which lie in the drawing plane are shown in Fig. 3.20. Force F_1 is split into two components F_H and F_V. In this configuration, the tooth forces reduce the hypoid offset and shift the pinion axially, thereby increasing its mounting distance t_B. A tapered roller bearing in an O-configuration is usually mounted behind the pinion such that the tooth forces press the pinion against the bearing. This mode of operation is termed "drive", its opposite being "coast" (see Sect. 2.2.4.5).

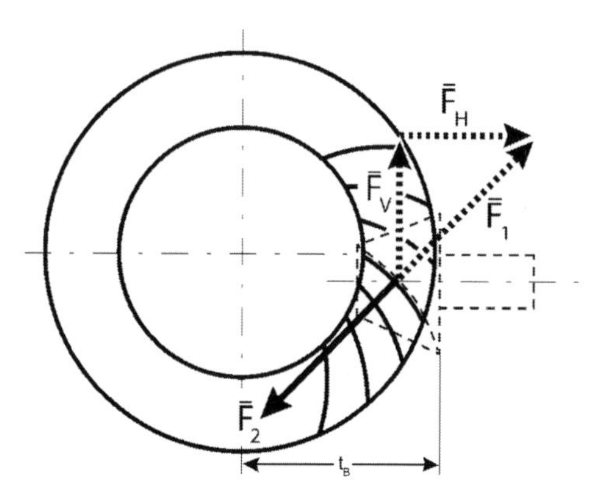

Fig. 3.20 Bevel gear set with positive offset in "drive" mode

If Fig. 3.21, the bevel gear set from Fig. 3.20 is shown in coast mode. In this case, the concave tooth flank of the wheel drives the convex tooth flank of the pinion. In the opposite direction of rotation, the convex pinion tooth flank drives the concave wheel tooth flank. The pinion exerts force F_2 on the tooth of the wheel, which reacts with force F_1 on the pinion. Force F_1 is split into components F_H and F_V. In coast mode, the tooth forces cause an increase in the hypoid offset and a reduction in the mounting distance, i.e. an axial shift of the pinion towards the center of the gearbox. This axial displacement is harder to absorb for a tapered roller bearing in O-configuration behind the pinion, and there is risk that the bevel gear will jam if backlash is insufficient.

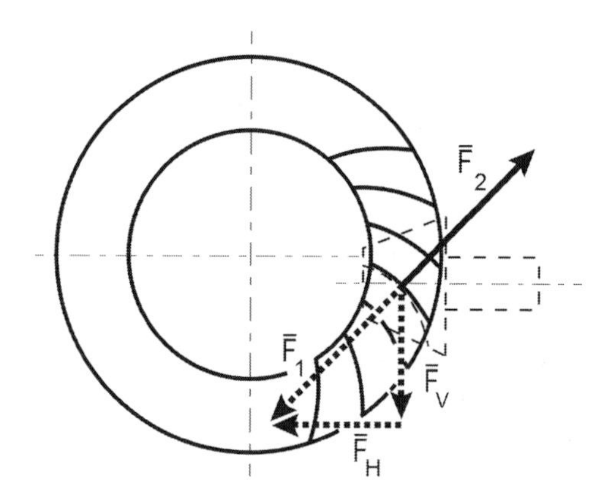

Fig. 3.21 Bevel gear set with positive offset in "coast" mode

In general terms, the displacement behavior for the drive and coast modes may be simplified by stating that in the drive mode the offset is reduced and the mounting distance increased, while in the coast mode the offset is increased and the mounting distance reduced.

3.4.3 Contact Pattern Displacement

Figure 3.22 shows, on the tooth flanks of the wheel, how the center of the contact pattern is displaced when varying values ΔV and ΔH. In each case, the arrows on the lines indicate the positive direction for the parameter change. It will be evident that on the drive side, i.e. the convex tooth flank on the wheel, with displacements $\Delta V < 0$ and $\Delta H > 0$ induced by the reactions on the teeth, the contact pattern moves towards heel and tip. On the coast side, with displacements $\Delta V > 0$ and $\Delta H < 0$, the contact pattern also moves towards heel, but with a tendency towards tooth root. This phenomenon is typical of bevel gears which were cut with a relatively large tool radius r_{c0} (Table 3.1).

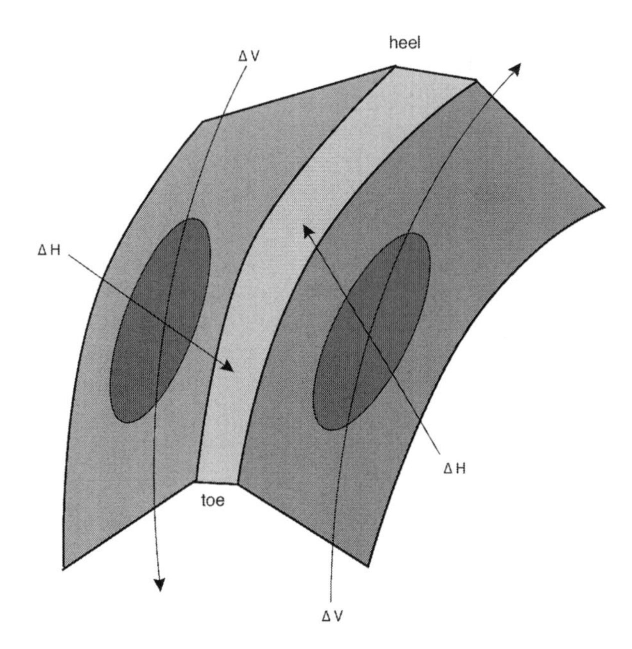

Fig. 3.22 Displacement of the center of the contact pattern on the tooth flanks of the wheel for different ΔV and ΔH values

The fat points in Fig. 3.23 show the way in which the contact pattern is displaced as a result of incremental changes in ΔV and ΔH. The case shown here is a design with a large tool radius. The ΔH trajectory rotates clockwise with a decrease in tool radius. The influence of the tool radius is described in greater detail in Sect. 3.4.4.

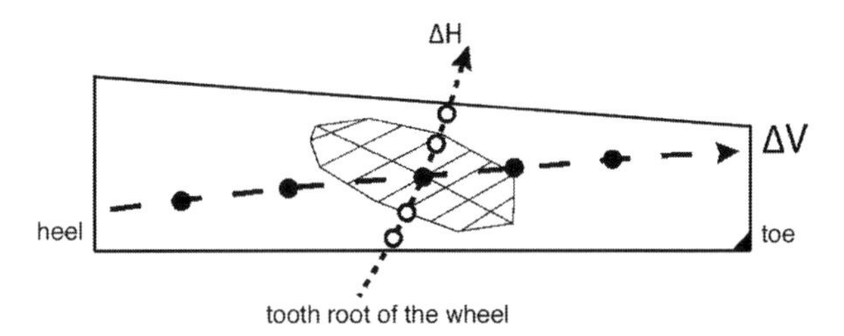

Fig. 3.23 Center of contact pattern displacement in "drive" mode

Despite its inaccuracies, the manual method for contact pattern analysis, known as VH check, is widely used. The contact pattern is displaced to the toe and heel positions for both the drive and coast directions of rotation. Changes in V and H values are noted. The extreme contact pattern positions should be centered in the tooth depth direction of the flank, with the edges just touching the ends of the tooth. For evaluation, the amounts for "V" and "H" are added in each case for the toe (1 or 3) and heel positions (2 or 4; see Fig. 3.24). The sum of the amounts of "V" values provides an estimate of the lengthwise crowning, while the quotient of the sums of V and H yields the bias. However, there are uncertainties involved in the VH check as test load and visual evaluation are dependent on the person performing the test.

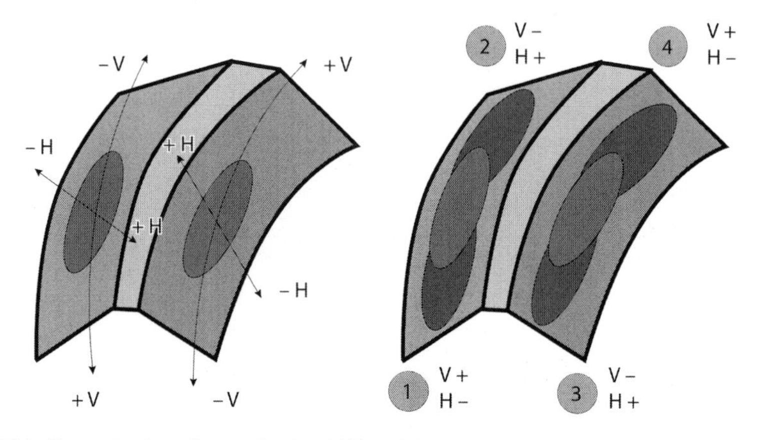

Fig. 3.24 Determination of crowning by shifting the contact pattern

To define bias behavior, a ratio of ΔV to ΔH displacing the centre of the contact pattern towards the heel, at mid-tooth depth, is assumed, as shown in Fig. 3.25.

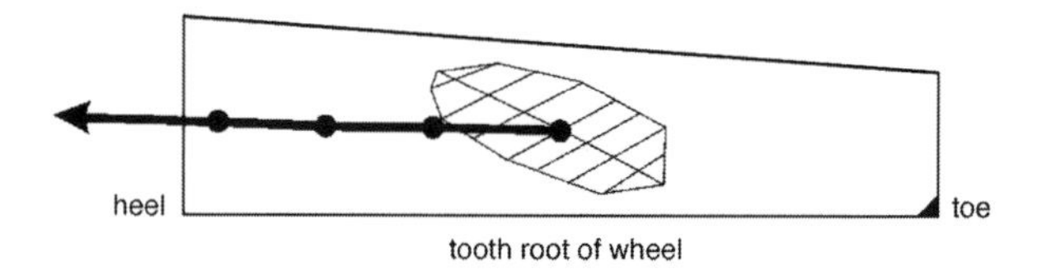

Fig. 3.25 Horizontal shift of the contact pattern for changes in ΔH and ΔV

Bias-in and bias-out are defined in reference to the mean spiral angle of the pinion β_{m1} according to Table 3.6.

Table 3.6 Definition of bias-in and bias-out

Designation	Formula	No.
Bias-in	$\frac{\Delta V}{\Delta H} > \cot(\beta_{m1})$	(3.35)
Bias-out	$\frac{\Delta V}{\Delta H} < \cot(\beta_{m1})$	(3.36)

3.4.4 Influence of the Tool Radius

The curvature in the tooth lengthwise direction is determined mainly by the tool radius r_{c0}. Lengthwise crowning has only a slight effect on the radius of curvature in the lengthwise direction.

Figure 3.23 shows the displacement behavior of the contact pattern on a gear set cut with a large tool radius. A change in ΔV displaces the center of the contact pattern along the thick dashed line, a change in ΔH along the thin dashed line. In simplified terms, it may be stated that a ΔV displacement shifts the contact pattern along the face width, and a ΔH displacement shifts it along the tooth depth. In drive mode, i.e. when $\Delta V < 0$ and $\Delta H > 0$, the center of the contact pattern migrates to the heel and, if there is bias-in, towards tooth tip, provided that ΔV and ΔH are equal in magnitude. Through suitable adjustments in ΔV and ΔH, the contact pattern can reach almost any point on the tooth flank, provided sufficient backlash is available for the displacement. This property is useful for the lapping operation as will be described in Sect. 6.6. Therefore, through V and H combinations, virtually any desired contact pattern position can be obtained and a contact pattern correction can consequently be obtained by lapping in the displaced V and H positions.

The above displacement behavior is detrimental in a gearbox. The lever arm of the bending moment at the point of load application on the tooth of the wheel is larger, and an increase in tooth root stress in the heel zone is expected. Gears designed in this way, without a pre-corrected position for the contact pattern, often show a tendency for the wheel tooth to break at the heel as a result of excessive tooth root stress. To obtain a tooth with the necessary load capacity, it is desirable to have a displacement of the contact pattern which is insensitive to the typical combination of ΔV and ΔH, combined to the largest possible normal module.

The relationships in Table 3.7 are valid for the basic crown gear shown in Fig. 3.26, with a number of teeth z_p, spiral angle β_x at cone distance R_x and distance ρ_{P0} (crown gear center to cutter center).

Table 3.7 Relationships for Fig. 3.26 expressed in the form of equations

Designation	Formula	No.
Transverse module at cone distance R_x	$m_{xt} = \frac{2R_x}{z_p}$	(3.37)
Normal module at R_x	$m_{xn} = m_{xt} \cos \beta_x$	(3.38)
Triangles from Fig. 3.26	$R_x \cos \beta_x = \rho_{P0} \sin \varepsilon_x$	(3.39)
It follows from (3.37) and (3.39) that:	$m_{xn} = \frac{2\rho_{P0} \sin \varepsilon_x}{z_p}$	(3.40)

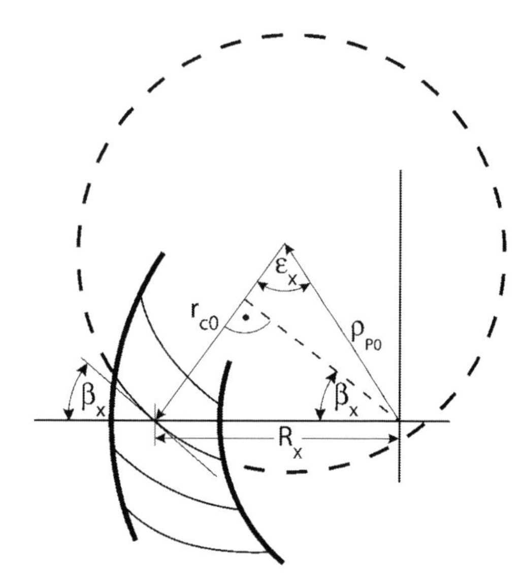

Fig. 3.26 Spiral angle, cone distance and tool radius of the virtual crown gear

According to Equation (3.40), the normal module is maximum when ε is 90°. This point on the pitch cone is referred to as the N point. At this point, the parameters of the lengthwise curvature are equivalent to those of an involute. The largest possible normal module is obtained with the N point in the center of the tooth.

It follows:

$$r_{c0} = R_m \sin \beta_m \tag{3.41}$$

These conditions are termed the "rectangular case". Its particular properties with reference to displacement behavior are shown in Fig. 3.27.

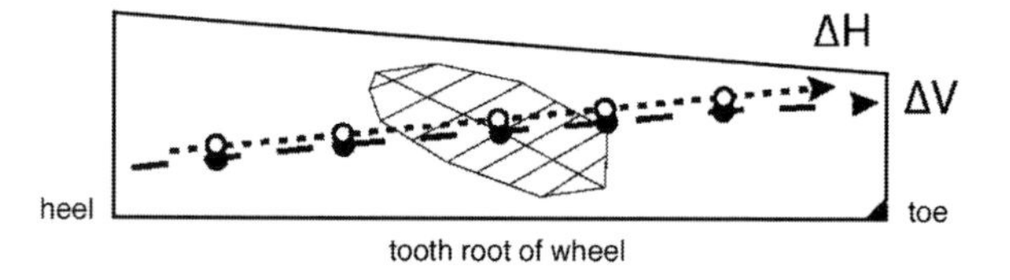

Fig. 3.27 Displacement behavior in the rectangular case

The behavior of the contact pattern for a ΔV displacement is scarcely altered when compared to that in Fig. 3.23, but the direction of the ΔH displacement now runs parallel to the ΔV line such that a typically load-induced total $\Delta V \cong -\Delta H$ displacement does not significantly change the position of the contact pattern.

Because it is necessary to account for both the geometry of the gear and the influences from the environment when considering the load capacity of a bevel gear set, designs with the rectangular case have the highest load capacity. There is barely any displacement of the center of the contact pattern under load, such that the leverage of the exerted force on the tooth root is not increased. And the normal module is at its greatest at the point of maximum stress.

3.4.5 *Ease-Off Design*

After establishing the fundamentals of the behavior of contact pattern displacement, this section examines the possibilities for tooth flank design to obtain a low-noise gear with good load capacity. At a given load, smooth contact of the tooth flanks is desirable in order to minimize the transmission error and hence noise excitation. To obtain a gear with high load capacity, it is necessary to ensure that the contact pattern for a given load is centralized on the tooth flank, and that no pressure peaks occur on the edges of the tooth.

Smooth contact is achieved by means of small crowning, while a limited contact pattern on the tooth flank under load is attained through larger crowning. This contradiction can be resolved by using local crowning which varies along the tooth.

Since in the case of bias-in behavior, the wheel contact pattern moves from toe towards heel and tooth tip as the load increases, it is possible to define various load zones on the tooth flank. Fig. 3.28 shows a typical ease-off with a bias-in characteristic. As the load increases, the contact pattern expands and is displaced in the direction of the heel and tip.

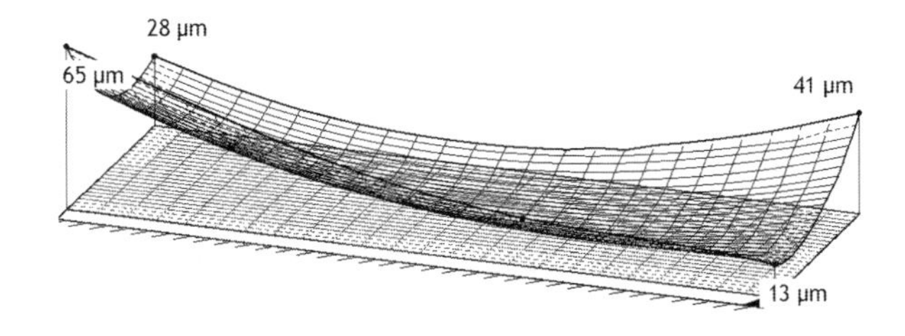

Fig. 3.28 Ease-off with bias-in characteristic for a right-hand spiral bevel wheel

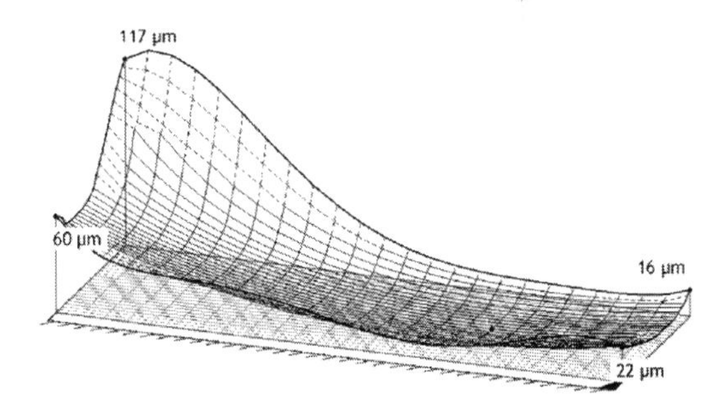

Fig. 3.29 Ease-off topography with local crowning

Figure 3.29 contains an ease-off topography which allows for locally different crowning requirements. No notable displacements are to be expected for the low-load state, suggesting a contact pattern position in the front third of the face width, near the toe. Since, due to the low load, no elliptical contact flats on the tooth are to be expected, local crowning should be kept to a minimum in order to minimize transmission error. Increasing crowning in the toe-root area allows compensation for assembly tolerances of the pinion mounting distance. This will prevent edge contact for negative axial pinion shifts ΔH caused by assembly or manufacturing tolerances. Local ease-off modification for the heel-tip zone applying very large crowning does not always prevent edge contact. The aim is to make sure that no pressure peaks occur at the edge of the tooth. In this area, it is desirable that the center of the contact pattern, and thus the maximum contact pressure, lies in the center of the mean tooth depth. This ensures that tooth root stresses on the wheel and pinion are balanced.

This example is intended to show that it is entirely possible to reconcile the contradictory criteria for low-noise operation and high load capacity on bevel gears using variable crowning along the tooth flank.

Figure 3.30 shows four different ease-off topographies of a gear set, with and without load-induced shaft displacement. The same displacement was entered for all four examples. In example d), the lengthwise crowning was optimized by a correction on the wheel, using the Modified Crowning method.

ease-off description	contact distance without load-induced displacement and deflection	contact distance with load-induced displacement but without deflection
a) gear set with constant lengthwise crowning, contact distance not optimized	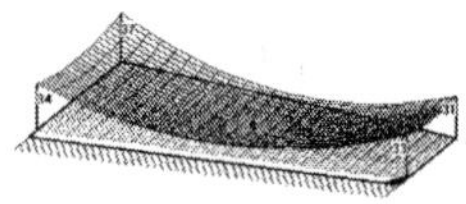	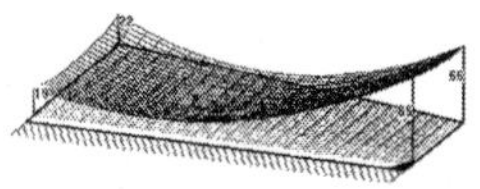
b) gear set with constant lengthwise crowning, contact distance optimized	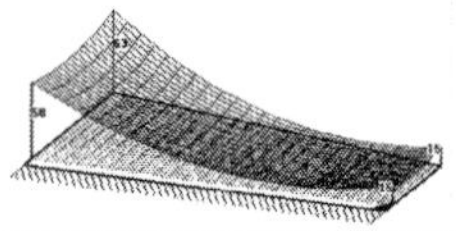	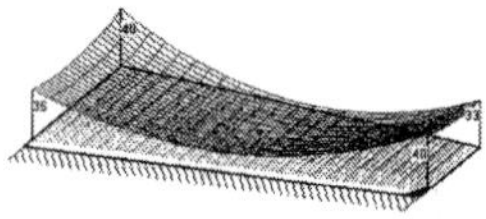
c) gear set with reduced constant lengthwise crowning for low load	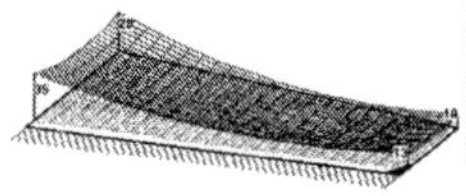	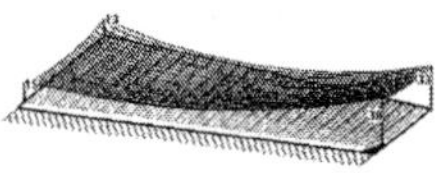
d) gear set with reduced lengthwise crowning and optimized by modified crowning	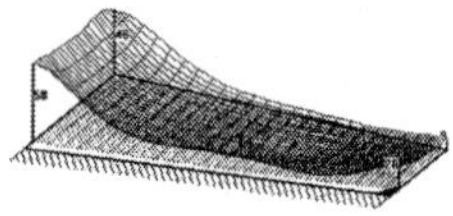	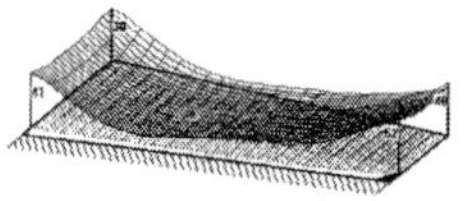

Fig. 3.30 Ease-off topographies with various modifications to compensate load-induced displacements

contact pattern description	contact pattern without load-induced displacement and deflection	contact pattern with load-induced displacement and deflection
a) gear set with contant lengthwise crowning, contact pattern not optimized	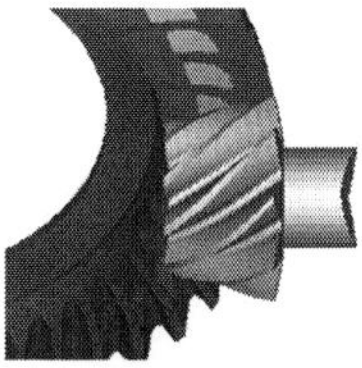	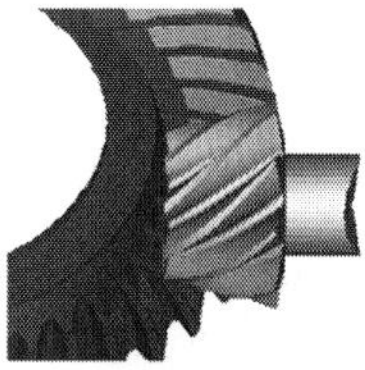
b) gear set with contant lengthwise crowning, contact pattern optimized		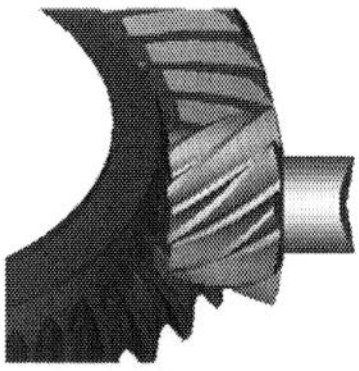
c) gear set with reduced constant lengthwise crowning for a low load	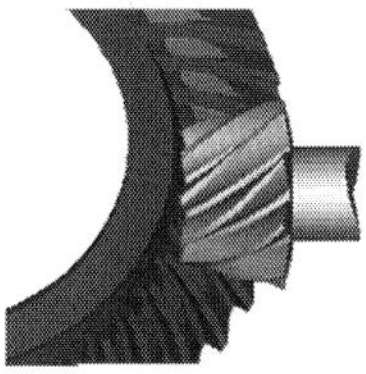	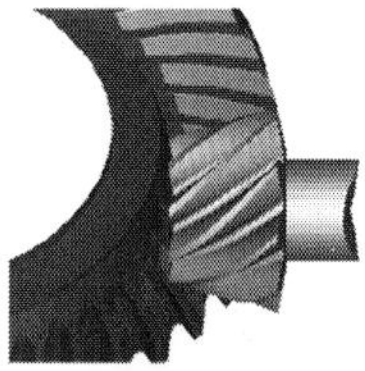
d) gear set with reduced lengthwise crowning and optimized by modified crowning		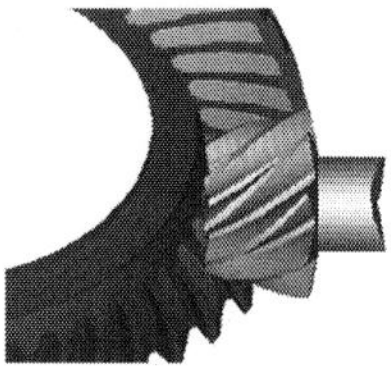

Fig. 3.31 Contact patterns of the gear sets in Fig. 3.30

Modified Crowning Despite the wide range of modifications which can be applied during the generating process of the pinion, so far it was not possible to design variable lengthwise crowning along the face width without any side-effects. With the Modified-Crowning method, it is now possible to change the flank form of non-generated wheels using existing cutting or grinding tools.

The form of the tooth flanks on a non-generated wheel can be altered only via the plunging position which is defined by parameters radial distance φ, cradle angle α, machine root angle γ, sliding base χ, offset η and distance mccp of the axis intersection point of the wheel from the machine center. All values being constant, only the sliding base χ can be used as a feed axis.

If, for example, the radial distance φ is changed, the main effect is to alter the spiral angle of the two tooth flanks, but in opposite directions. If, on the other hand, the machine root angle γ is changed, and the sliding base χ is adjusted so that the same tooth depth is machined in the center of the tooth, the main effect will be to change the spiral angle of the two flanks in the same direction. This approach is used to shift the contact pattern from tooth center towards toe, and so to minimize any load-induced displacement (see Fig. 3.31).

The Modified-Crowning principle employs several different plunging positions which are successively combined in a uniform machine motion. This is accompanied by a transformation of the known setting values of the plunging position. Using the transformed values, it is possible to design the sequence of different plunging positions in such a way that the equations of motion correspond to those of a generating motion, although no generating motion is involved. The reference variable is the transformed cradle angle α, which must proceed exactly from α_1 to α_2. Variables φ, γ and η are then dependent on α.

If the machine motions for a non-generated wheel are compared to those for a wheel with Modified Crowning, it is apparent that at the end of the plunging process, some machine axes have slightly changed in position, involving only a minimal effect on the cutting cycle time. Modified Crowning is particularly advantageous in the continuous indexing method where the additional motions are performed only once for all teeth. Figure 3.30 shows the ease-off and Fig. 3.31 the associated contact pattern for Example d), with and without load, for a gear with Modified Crowning. The modification is a parabolic relief on the heel, with tangentially constant transition.

3.5 Material Specification

3.5.1 Introduction

Plastics, cast materials and steel are used as materials for gears. The criteria in choosing a material are the engineering design parameters, production aspects, costs and anticipated loads. Typical gear stresses are always cyclical and may roughly be divided into bending stress, contact stress, compressive stress and shear stress.

The basic engineering design criteria are the available space and the transmitted power. Other restrictions on material selection are imposed by the production facilities and their cost-effectiveness for the volume required. Material selection is therefore a matter of a balanced comparison between the requirements profile for the bevel gear set and the properties profile of the material. These two profiles are outlined in Table 3.8.

Table 3.8 Requirements and properties profiles

Bevel gear requirements profile	Material properties profile
Cost-effectiveness	Material costs
Process-reliable manufacturing	Alloy composition
Technical properties:	Castability
– Stiffness	Weldability
– Tooth root strength, contact strength	Forgeability
– Fatigue strength	Heat treatability
– Thermal shock resistance	Young's modulus, shear modulus
– Heat resistance	Tensile strength, yield strength
– Wear resistance	Fatigue strength, endurance limit
– Corrosion resistance	Wear resistance
– Health and safety requirements	Mechanical fracturing parameters
– Mass	

Depending on their properties profile, materials may be transformed into different states and their physical properties may then differ according to their state.

3.5.2 *Materials for Bevel Gears*

Plastic For economic reasons, plastics are used in applications with low requirements in terms of tooth root and contact strength. Within certain limits, their state can be further adapted to the requirements profile by means of composite techniques.

Sintered metal More heavily stressed gears, which can no longer be made of plastic, are produced in a sintered process if the number of parts is sufficient to justify it. The higher load capacity, when compared to plastic, must be contrasted with greater mass. Sintered gears are accordingly to be classified between plastic and steel gears in terms of their power-to-weight ratio.

Grey cast iron and cast steel They are less widely-used variants as gear materials. They have a low to medium load capacity compared to gears made from steel. In general, these materials are easy to machine. Depending on the temperature control and cooling conditions in the casting process, it is possible to adjust the failsafe running properties and heat treatability of the material.

Structural and tempering steels If the gear geometry does not allow casting, or if the requirements profile demands higher-quality materials, metal-cutting processes must be applied where structural and tempering steels can be used; they show high machinability and medium load capacity, and are characterized by high toughness and lower costs than those of higher quality steels.

Surface-hardened tempering steels For higher requirements in contact strength, the tooth surface load capacity must be improved by means of heat treatment.

Steels suited for flame or induction hardening are typical representatives of this group of materials.

Steels for thermo chemical heat treatment Nitriding and case hardening steels are alloyed specially for this type of heat treatment. These materials are suitable for the highest requirements in terms of load capacity, geometry and surface quality. The heat treatment processes are complex and in many cases require subsequent hard finishing. With this group of steels, it is possible to achieve the combination of surface hardness and toughness needed to satisfy the highest demands in load capacity and wear resistance. In practice, gear sizes are limited by the size of the heat treatment equipment.

Through-hardening steels A special place is occupied by through-hardening steels, which ensure uniformly high surface hardness irrespective of the amount of stock removed during hard finishing.

3.5.3 Case Hardening Steels

Bevel gears are usually highly loaded and therefore require high quality materials and production. The property profile is characterized by demands for surface hardness, pitting and micropitting resistance, wear and scuffing load capacity, and tooth root and core strength.

Figure 3.32 illustrates the tooth root and flank load capacity as a function of case hardening depth. The case hardening depth is defined as the depth at which the material exhibits a hardness of 550 HV. A large case hardening depth would be ideal for flank load capacity, while the tooth root load capacity requires a smaller case hardening depth.

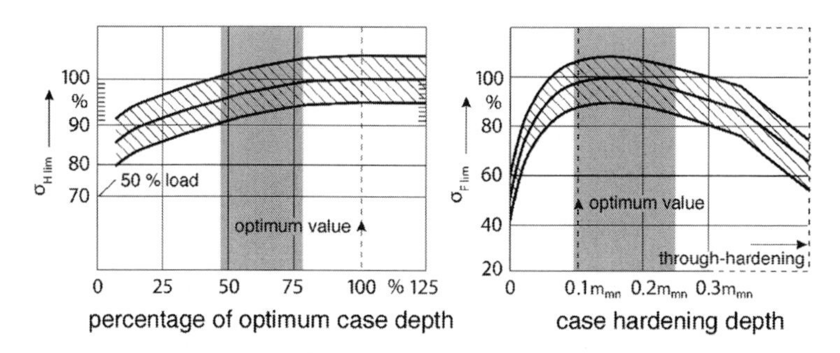

Fig. 3.32 Influence of case hardening depth on tooth root and flank load capacity [NIEM86.2]

Figure 3.33 shows the recommended case hardening depth for the optimum of root and flank load capacity as a function of the gear module.

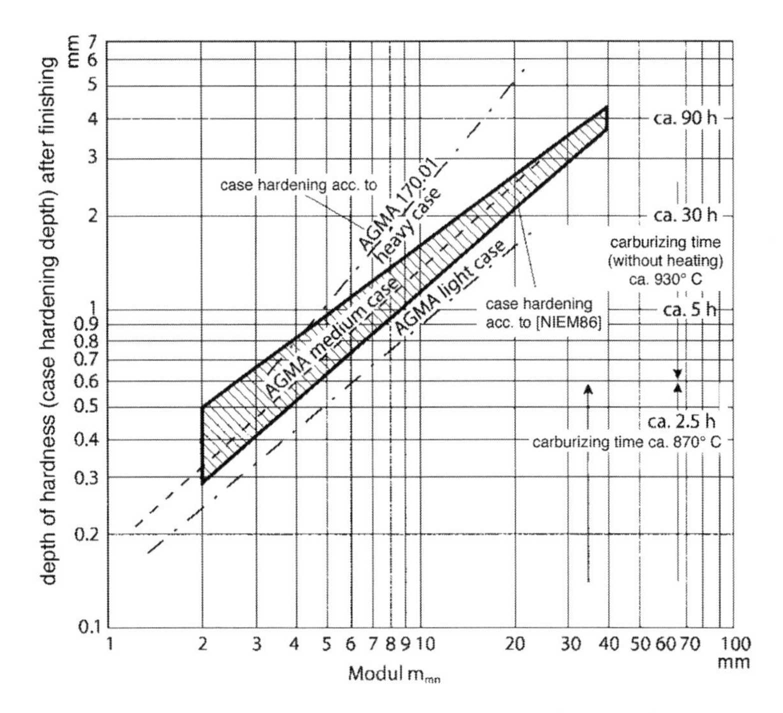

Fig. 3.33 Recommended case hardening depth for the optimum of root and flank load capacity

Generally the hardness profile shown in Fig. 3.34 is desirable. The hardness curve must be adjusted to provide sufficient hardness at the depth of maximum stress.

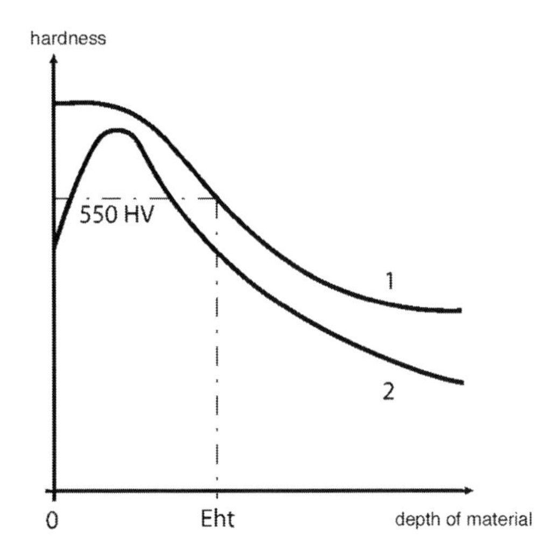

Fig. 3.34 Hardness and stress distribution versus depth of material.
1 hardness distribution,
2 stress distribution

The task is, therefore, to find a material which provides the necessary surface hardness, possesses the required core toughness, has a suitable core hardness and, in addition, allows adjustment of the hardness gradient from the surface to the core. This property spectrum can best be obtained by using a case hardening steel to which a case hardening heat treatment is applied.

Case hardening steels have a low carbon content of less than 0.2 % and defined quantities of alloying elements like manganese, chromium, molybdenum or nickel. This initial state guarantees that the necessary core hardness will be obtained, while remaining relatively easy to machine. However, steels with such low carbon content cannot be hardened directly. To be hardened, they are exposed to carbon diffusion at high temperature in a carbon delivering environment, a process known as carburizing. The desired case hardening depth is, among other factors, set by the duration of the carburizing process. Carburisation is followed by a quenching operation and a subsequent tempering to the desired hardness. The process is detailed in [DIN17022-3] and in Sect. 6.3.

The achievable core hardness depends on the size of the gear as well as on the alloys of the case hardening steel. This relationship is shown in Fig. 3.35 for several types of case hardening steels e.g. 16MnCr5, 15NiCr13, 18CrNiMo7-6, 20NiCrMo2-2 and 25MoCr4. It is apparent that 16MnCr5 is more suitable for small components, while 18CrNiMo7-6 is preferable for larger bevel gears.

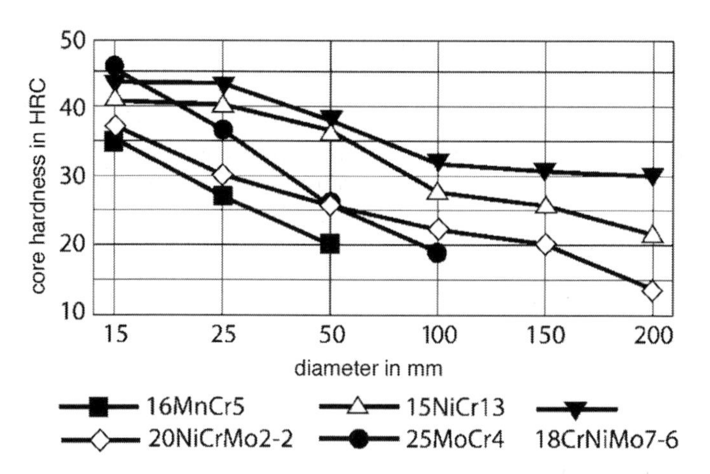

Fig. 3.35 Core hardness of different case hardening steels for various reference diameters

Core hardness for gears is defined at a point in the gear body which lies perpendicular to the 30° tangent in the tooth root, at a distance from the surface equal to five times the case hardening depth. For testing, a cylindrical specimen with a diameter showing a cooling behavior similar to that of the gear at the defined point is used. The core hardness can then be determined in reference to the measured hardness at the centre of the specimen.

Another aspect in material selection is related to the expected distortions during quenching. Pinions are generally quenched without support, while wheels are preferably placed in a quenching press. This is done in order to improve the run-out and flatness of the gear body. A quenching die consists of an expanding mandrel for the bore and a clamping ring holding the wheel flat (see Fig. 6.24). Distortions are relatively small with fixture hardening. If no quenching press is available, the acceptable amount of distortion must be considered in the design.

Apart from specifying the composition of the material, Standard ISO 6336–5 contains three quality classes. Table 3.9 indicates attributes for the standardized qualities ML, MQ and ME. These attributes are the minimum requirements for the material. Especially where large bevel gears are concerned, it is usual to find company-specific material specifications which go far beyond the values listed in Table 3.9.

Table 3.9 Material qualities according to [ISO6336-5]

	Quality requirements for materials		
Parameters	**ML**	**MQ**	**ME**
Steel-making	No specifications	No specifications	HV, ESU, CAB
Degree of purity	DIN 17210 ~K4 < 50 (Oxide)	50 % of DIN 17210 ~K4 < 25 (Oxide)	K1 < 20 K2 < 5 K3 ~ 0 (oxides + sulphides)
Strain	No specifications	>3	3–5
Austenite grain size	No specifications	5–8	5–8
Case hardness	56–64 HRC	58–63 HRC	59–63 HRC
Surface structure	No specifications	Fine acicular martensite	Fine acicular martensite
Core hardness	>20 HRC	>20 HRC >34 HRC (Ni-Leg.)	>40 HRC
Core structure	No specifications	martensite + bainite	martensite + bainite

3.6 Choice of Lubricant

3.6.1 Introduction

The lubricant is intended mainly to reduce friction and wear and to dissipate the heat generated in the gear mesh. In addition, it protects components against corrosion and is the carrier medium for various additives like oxidation inhibitors, pour point depressants, detergents and dispersants, whose role is to improve oil properties, along with wear and scuffing protective additives to prevent gear damage.

3.6.2 Selection of the Lubricant

In the great majority of applications, oil is the preferred lubricant. It can be introduced readily into the gear mesh and can dissipate frictional heat. Lubricating grease and grease type lubricants with adhesive properties of various NLGI consistency grades are also used on slow-running, open or closed gears used mainly in part load or intermittent operation.

Industrial CLP gear oils according to [DIN51517] are typically used for industrial applications. GL4 and GL5 oils according to API (American Petroleum Institute) with enhanced scuffing protection are employed for vehicle and industrial applications.

3.6.3 Choice of Oil Type

Mineral oils are adequate lubricants in most applications. Synthetic oils are employed where enhanced requirements are imposed on the operating temperature range, on the maximum or minimum temperature and on the lubricant service life. Apart from a more favourable viscosity-temperature behavior (high viscosity index VI), they generally also provide better frictional behavior in contact. Polyalphaolefins, similar in chemical structure to mineral oils and, hence, also miscible, are frequently used, providing an average 10 % reduction in contact losses when compared to mineral oils. Polyglycols, which are usually not miscible with mineral oils, can exhibit up to 30 % more favourable frictional behavior. Polyolester-based lubricants are frequently employed for applications on aircraft gears at high circumferential speeds and high temperatures. Rapidly biodegradable lubricants may also be used in especially environmentally critical applications. Only synthetic (trimethylpropane TMP) esters have sufficient high temperature stability to be eligible here and, incidentally, exhibit good lubricating and frictional properties [DIN51517], [HOEN99].

Lubricants based on synthetic oils are used in most cases for applications with extended oil-change intervals. The already good aging stability of synthetic oils when compared to mineral oils is assisted by lower contact friction leading in turn to lower oil temperatures and hence lower thermal stress.

3.6.4 Choice of Oil Properties

3.6.4.1 Choice of Viscosity

Viscosity should vary directly with load and inversely with circumferential speed, which is in line with its influence on lubricant film formation and hence on such forms of damage as wear, scuffing, micro-pitting and pitting. However, a higher viscosity will lead to greater idling losses. A compromise must therefore be found, particularly for multistage gear units which operate with varying loads and speeds in the different stages. Inadequate viscosity for lubricant film formation needs to be compensated by additives which form protective films on the tooth flanks.

Niemann/Winter [NIEM86.2] provides guidance on the choice of viscosity as a function of circumferential speed (Fig. 3.36). These recommendations do not necessarily have to be followed.

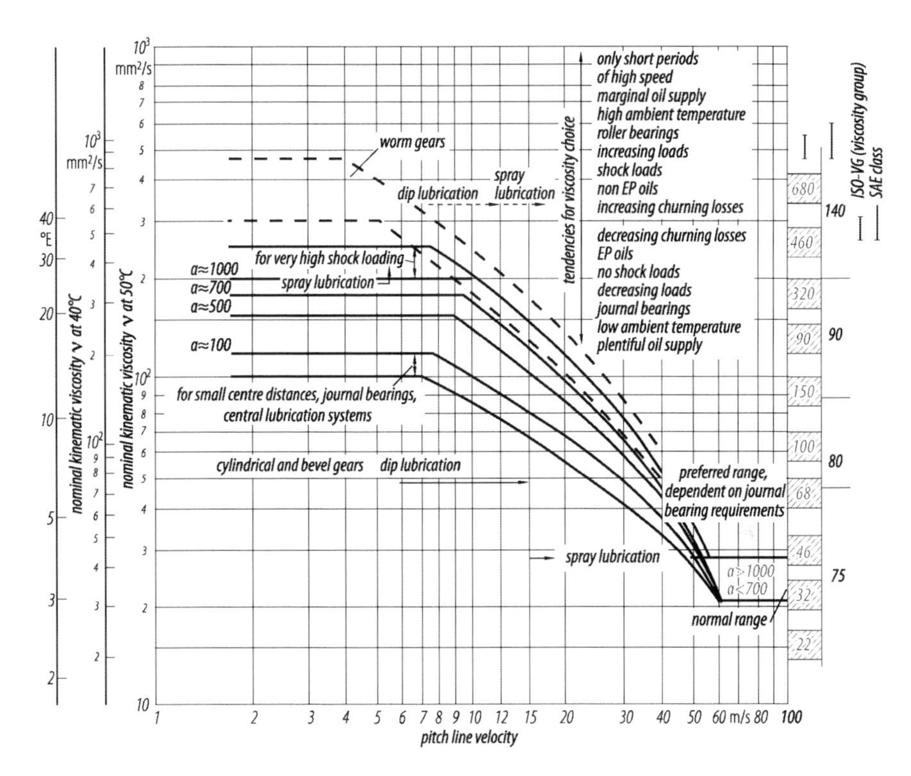

Fig. 3.36 Recommended oil viscosities according to Niemann/Winter [NIEM86.2]

3.6.4.2 Choice of Additives

CLP industrial lubricants according to DIN 51517 will generally suffice for bevel gears with no offset. The greater the offset and hence the longitudinal sliding in the gear mesh, the higher should be the selected scuffing load capacity of the lubricant. Lubricants with high (API GL4) or very high scuffing load capacity (API GL5) are used for automobile axles and other applications. These oils containing extreme pressure (EP) additives are based on organic sulphur-phosphorus compounds in typical concentrations of 4 % (GL4) to 6.5 % (GL5). Unnecessarily high additive concentrations should be avoided because of the aggressiveness of the additives on non-ferrous metals and seal materials, especially at high temperatures.

CLP scuffing load ratings for industrial gear oils can be determined in the A/8.3/90 scuffing test according to [ISO14635.1] A load stage of at least 12 must be demonstrated. For oils according to API GL4, either the A10/16.6R/120 test at 120 °C oil temperature according to [ISO14635.2] or the A10/16.6R/90 test at 90 °C oil temperature according to FVA information sheet 243 [SCHL96] may be used. The scuffing load capacity of oils with maximum load capacities according to API GL5 or above can be examined using the S-A10/16.6R/90 test method according to [SCHL96].

Apart from adequate scuffing load capacity it is important, especially in the case of hypoid gears, to employ a lubricant with sufficiently high micro-pitting load capacity. Micro-pitting capacity may be determined according to FVA information sheet 54/IV [EMME93].

3.6.4.3 Choice of Other Oil Properties

When selecting gear oil, it is necessary to ensure adequate protection of steel and non-ferrous metal materials against corrosion, especially for temporarily decommissioned gear units, for gear systems exposed to large temperature fluctuations with a risk of increased water condensation, and for gear systems in damp locations.

In particular, the compatibility of synthetic lubricants with seals requires careful attention and, for that, reason lubricants and seal materials should be matched individually.

Special attention needs to be paid to foaming behavior with splash lubrication and small quantities of oil of the kind often found in road or rail vehicle gear systems. Apart from construction measures, foam inhibitors may be helpful in minimising oil losses through seals and vents. The foam test according to Flender [GREG06] is a practice-oriented method to assess oil foaming behavior.

The pour point should be at least 5 K below the lowest operational temperature of the lubricant. Synthetic oils are more favourable than mineral oils in this respect.

Depending on the application, attention may also need to be paid to oil water separation ability and air release property.

3.6.5 Oil Feed

Splash lubrication is preferred as a simple, operationally reliable lubricating system for low and medium circumferential speeds up to roughly $v = 20$ m/s. Given suitable internal oil guiding devices in the housing, splash lubrication can still be used at circumferential speeds up to $v = 50$ m/s. An oil level at which the complete face width of the wheel is fully immersed in the sump should be ensured during operation. The oil level while the gear is not running should be correspondingly higher if the oil has to travel some distance, for example to the distant bearings of a pinion shaft. The quantity of oil supplied is often sufficient to provide a lubricant film with enough oil, but the majority of the oil is required to dissipate contact heat. Hypoid gears with large offsets, which generate considerable frictional heat, impose increased demands in this respect.

Forced lubrication is employed for unfavourable shaft positions and continuous circumferential speeds $v > 20$ m/s. This more complex method requires additional elements in the form of pump, filter and cooler, but allows continuous oil management. The injected quantity is chosen such that heat generated in the contact is dissipated completely via the oil, using the designed temperature difference of an external radiator. Additional heat dissipation is also possible via thermal convection and radiation from the housing.

The oil is usually sprayed into the ingoing mesh, producing the lowest gear bulk temperatures at that point. If no-load losses at high circumferential speeds are to be kept low, part of the oil can also be sprayed into the outgoing mesh to cool the gear after contact. Values of 3 to 5 bar have proved effective for the injection pressure. Higher pressures and hence higher oil injection velocities entail a risk of erosion damage at the gear circumference.

3.6.6 Oil Monitoring

Oil change intervals are based partly on the thermal and mechanical stresses in operation and partly on external contamination. At an oil sump temperature of 80 ° C, a service life of about 4000 h operating time may be assumed for mineral oils. The thermal oil life should be halved (doubled) for every 10 K higher (lower) oil temperature. The service life of synthetic lubricants is usually higher by a factor of 2 to 5. The type and quantity of additives play a primary role on oil life.

Multigrade oils, with flatter viscosity-temperature curves than mineral oils, are often used for axle drives in vehicles. All synthetic base oils have a higher natural viscosity index (VI) than mineral oils. Mineral oils with VI-improvers based on high molecular weight additives like, for example, polymethacrylate (PMA) or polyisobutylene (PIB), are also in use. Care must be taken to ensure that these additives possess sufficient shear stability and are not mechanically destroyed in gear and roller bearing contacts, thereby becoming ineffective. The shear stability of VI-improvers can be determined using the taper roller bearing shear test according to [DIN 51350.6].

Oil sensors may be helpful to determine oil change intervals. In the simplest case, the times at which the oil is exposed to different temperature ranges are accumulated. New systems are under development to monitor additional physical oil parameters like viscosity, total acid content, dielectric properties, or chemical oil parameters like selective infra-red spectrum in the region of the wavelengths of additives or degradation products.

References

[BAER91] Bär, G., Liebschner, B.: Fitting Flanks and Contact Properties of Hypoid Gears. 8. World Congress on the Theory of Machines and Mechanisms, Proceedings. vol. 4, Prag (1991)

[DIN17022–3] Verfahren der Wärmebehandlung; Einsatzhärten. Ausgabe: 1989–04

[DIN51517] Schmierstoffe – Schmieröle – Teil 3: Schmieröle CLP; Mindestanforderungen. Ausgabe 2004–01

[DIN51350.6] Prüfung im Shell-Vierkugel-Apparat – Bestimmung der Scherstabilität von polymerhaltigen Schmierstoffen. Ausgabe 1996–08

[DUTS94] Dutschke, R.: Geometrische Probleme bei Herstellung und Eingriff bogenverzahnter Kegelräder. Diss. TU Dresden (1994)

[EMME93] Emmert, S., Schönnenbeck, G.: Testverfahren zur Untersuchung des Schmierstoffeinflusses auf die Entstehung von Grauflecken bei Zahnrädern. Forschungsvereinigung Antriebstechnik, Frankfurt, Informationsblatt Nr. 54/IV (1993)

[GREG06] Gregorius, H.: Schaum im Getriebe – Fluch oder Segen? Erneuerbare Energien 1/2006, S. 40–43

[HOEN99] Höhn, B.-R., Michaelis, K., Döbereiner, R.: Performance of Rapidly Biodegradable Lubricants in Transmissions – Possibilities and Limitations. COST 516 Tribology Symposium Antwerpen, Belgium, 20./21.Mai 1999, pp. 20–30

[ISO6336–5] Calculation of load capacity of spur and helical gears part 5 Second edition (2003)

[ISO14635.1] DIN ISO 14635–1: Zahnräder – FZG-Prüfverfahren – Teil 1: FZG-Prüfverfahren A/8,3/90 zur Bestimmung der relativen Fresstragfähigkeit von Schmierölen (2000)

[ISO14635.2] DIN ISO 146365–2: Zahnräder – FZG-Prüfverfahren – Teil 2: FZG-Stufentest A10/16,6R/120 zur Bestimmung der relativen Fresstragfähigkeit von hoch EP-legierten Schmierölen (2004)

[KN3028] Klingelnberg-Werknorm (2001)

[KREN91] Krenzer, T.: CNC Bevel Gear Generators and Flared Cup Formate Gear Grinding, AGMA Technical Paper 91 FTM 1

[MAAG90] MAAG GEAR BOOK, Fig. 1.02 (January 1990)

[NIEM86.2] Niemann, G., Winter, H.: Maschinenelemente Band II. Springer, Berlin (1986)

[NIEM86.3] Niemann, G., Winter, H.: Maschinenelemente Band III. Springer, Berlin (1986)

[SCHL96] Schlenk, L., Eberspächer, C.: Verfahren zur Bestimmung der Fresstragfähigkeit hochlegierter Schmierstoffe in der FZG-Zahnrad-Verspannungs-Prüfmaschine. Forschungsvereinigung Antriebstechnik, Frankfurt, Informationsblatt Nr. 243 (1996)

[SIGM06] Müller, H., Kirsch, R., Romalis, M.: Modified Crowning, Sigma Report 16 (2006)

Chapter 4
Load Capacity and Efficiency

4.1 Gear Failure Modes

The following sections describe the most frequent failure modes of bevel gears and the most important parameters affecting them.

4.1.1 Classification of Failure Modes

The main failure modes occurring on gears may broadly be sub-divided into tooth breakage and surface failure. Tooth breakage modes are only dependent on the geometry and material of a gear and the conditions under which it operates. Surface failures are additionally dependent on the lubrication conditions in the contact zone (Fig. 4.1) and also on the specific lubricant properties [NIEM86.2]. If the tooth flanks are completely separated by an elastohydrodynamic lubricant film (EHD lubrication), the main type of failure is pitting. With a decreasing lubricant film thickness, more surface contacts occur (mixed lubrication) and the risk of flank damage due to micropitting increases. This, in turn, also affects the pitting load capacity. Wear and scuffing take place only when in boundary lubrication mode, i.e., when the lubricant protective layers fail, potentially causing subsequent failure.

Figure 4.2 qualitatively indicates the stress limits for the most frequent failure modes as a function of the circumferential speed. These test results are based on a case hardened gear set with a design balanced for all failure modes, using a lubricant with EP additives (see Sect. 3.6.4.2).

At low circumferential speeds, wear and micropitting are predominant and are primarily caused by low lubricant film thicknesses. At higher circumferential

---

The original version of this chapter was revised. The erratum to this chapter is available at DOI 10.1007/978-3-662-43893-0_9.

J. Klingelnberg (ed.), *Bevel Gear*, DOI 10.1007/978-3-662-43893-0_4

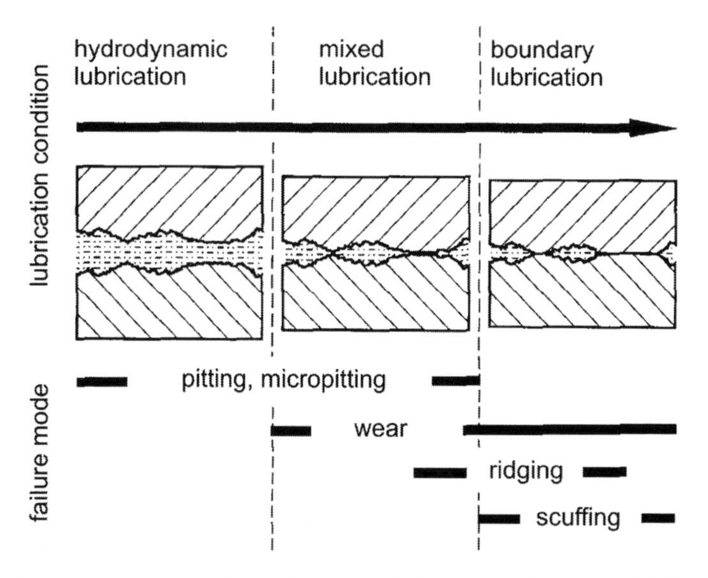

Fig. 4.1 Gear surface failure modes as a function of lubricating conditions [NIEM86.2]

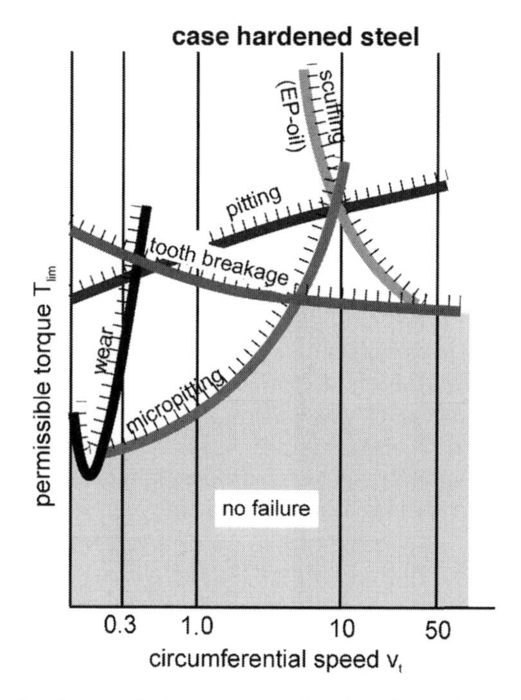

Fig. 4.2 Failure modes of case carburized gears as a function of operating conditions [NIEM86.2]

speeds, the lubricant film grows in thickness, and the permissible torque with respect to these failure modes increases. At medium and high speeds, the pitting load capacity becomes the limiting failure mode for service life. Suitable gear designs will place the tooth root breakage limit far enough beyond the other limits.

4.1.2 Tooth Root Breakage

The typical damage pattern for tooth root breakage is shown in Fig. 4.3.

Fig. 4.3 Tooth root breakage

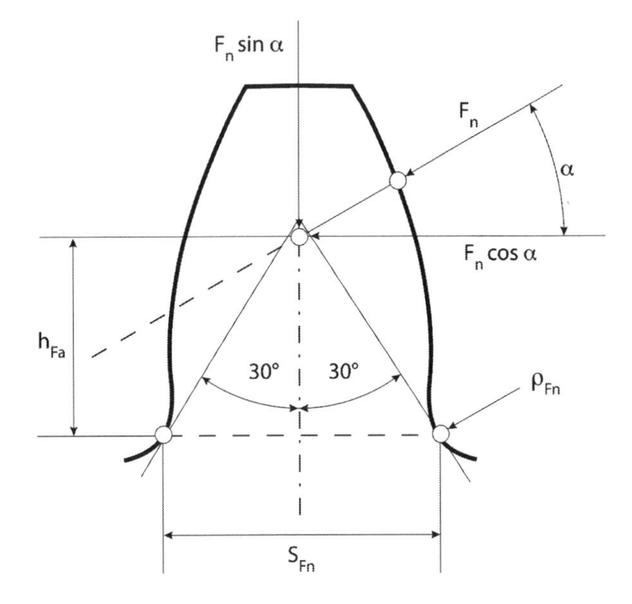

Fig. 4.4 Geometry variables in determining tooth root stress [NIEM86.2]

Apart from compressive and shear stress, the root of a gear tooth is mainly affected by bending stress. The bending stress is determined by the transmitted torque and the geometry of the tooth. The contact point of the 30° tangent in the tooth fillet is regarded as the critical zone (see Fig. 4.4), in which the first cracks are initiated when the stress limit is exceeded [NIEM86.2].

The tooth load capacity can be increased by any measure which enlarges the critical cross-section in the root, such as a bigger module, higher pressure angle or a larger face width. A positive profile shift, which prevents undercut on the pinion (addendum modification), likewise enlarges the tooth root chord. The negative profile shift which is simultaneously necessary on the wheel because of the zero shift rule according to Sect. 3.1 (dedendum modification) has only a slight influence, due to the greater number of teeth on the gear. Different tooth root load capacities on the pinion and wheel can be adjusted by means of a thickness modification.

Another means to increase load capacity is to reduce the notch effect in the tooth root, for example by selecting a larger fillet radius or by using protuberance tools to avoid grinding notches.

It is important to note that measures which affect the geometry may have a negative effect on load capacity with respect to other failure modes, like pitting or scuffing for instance.

For material selection and machinability, the choice of a case hardening steel with sufficient basic strength is recommended. High surface quality and shot peening are also positive in terms of tooth root load capacity.

4.1.3 Flank Breakage

Flank breakage, also termed subsurface fatigue, is a fatigue failure which occurs on the active part of the tooth flank. The damage pattern usually shows large areas of fatigue failure with very small forced fracture planes. The initial cracks start sub-surface in the case depth or in the transition zone from case to core [THOM98]. A typical example of flank breakage is shown in Fig. 4.5.

Flank breakage load capacity can be enhanced by measures which reduce Hertzian stress. These include, for example, a larger pitch diameter and a higher contact ratio due to higher tooth numbers (smaller modules), and larger spiral angles. A positive profile shift on the pinion, causing a larger equivalent curvature radius of the contacting tooth flanks, also results in lower stresses but has a negative effect on scuffing load capacity. The following should also be noted: as the width of the Hertzian contact area increases, the maximum stress in the contact area (Fig. 3.34) moves from the surface of the tooth towards the core where the strength may no longer be sufficient [ANNA03].

In terms of material selection and heat treatment, case hardening steels with adequate core hardness, optimum case hardening depth and fine grained structure should be employed.

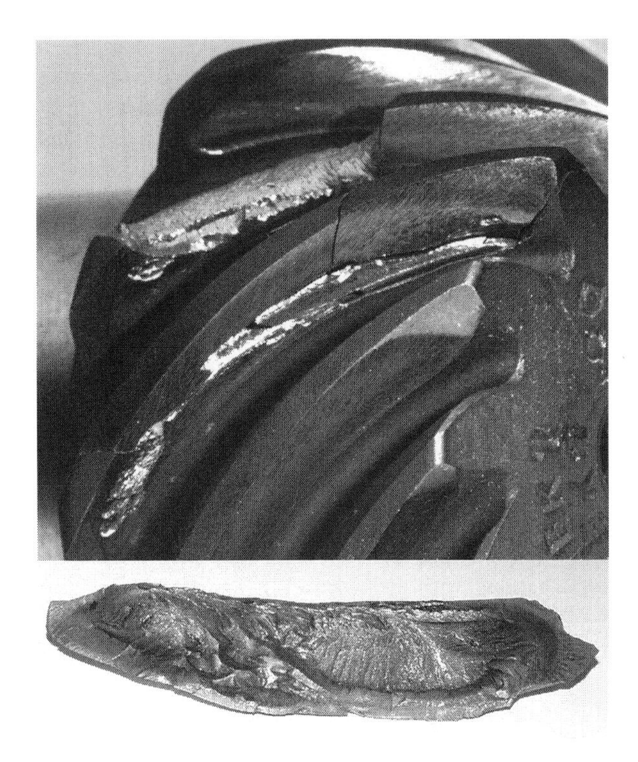

Fig. 4.5 Tooth flank breakage

4.1.4 Pitting

Like tooth flank breakage, pitting is a fatigue failure mode in the active area of the tooth flank [NIEM86.2]. At the surface, the Hertzian stress exceeds the material strength thereby causing shell-shaped craters which, once initiated, continue to grow (Fig. 4.6) and eventually lead to total failure of a gear due to consecutive localized failures.

Fig. 4.6 Pitting

Pitting tends to occur below the pitch circle, in the zone of negative specific sliding (see Sect. 2.4.5). Initially, micro cracks appear in a direction opposed to that of the contact point motion (direction W in Fig. 4.7) [KAES77]. The lubricant, trapped by the rolling motion of the pinion and wheel, fills these micro cracks and, under the contact pressure, causes a bursting effect promoting crack propagation into material depth [KNAU88].

To increase pitting load capacity, it is necessary to reduce the Hertzian contact stress by means of larger radii of curvature. This is achieved by using smaller modules and thus larger contact ratios, positive profile shift for the pinion and larger pressure angles. A higher contact ratio can also be obtained through a larger offset, and hence a larger spiral angle on a hypoid pinion. However, an increase of the offset simultaneously implies an increase in sliding velocity caused by lengthwise sliding and, consequently, an increase in the surface stress. So, the effects of a larger contact ratio and a higher sliding velocity partly neutralize each other. It is therefore necessary to calculate an optimum offset in terms of pitting resistance (see Sect. 4.2.5).

In terms of strength, possible measures are the selection of a case hardening steel, sufficient surface hardness and quality, and use of a higher viscosity lubricant with optimum additives. The choice of a suitable lubricant can also reduce the coefficient of friction of a gear set, again with positive effects on its pitting load capacity. The quality of a lubricant with respect to pitting can be determined in the FZG pitting test [FVA2].

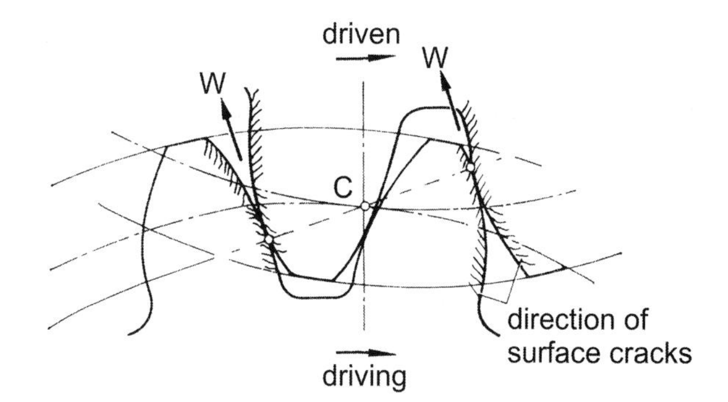

Fig. 4.7 Sliding, rolling and crack orientation on a gear tooth flank [NIEM86.2]

4.1.5 Micropitting

Micropitting is a failure type which occurs when a lubricant film of inadequate thickness coincides with stress peaks on the tooth flank surface. It is characterized by matt grey areas on the surface, induced by fine cracks and micro fractures. As micropitting progresses, particles continually break out of the flank leaving very small craters ("micro" pits). Micropitting thus constitutes a fatigue-induced removal of material from the surface, causing flank form deviations [SCHR00]. This affects the stress distribution on the flanks, leading in turn to an increase in dynamic forces and noise emission. The altered stress distribution may also have a negative effect on pitting load capacity. Typical micropitting on a pinion of a hypoid gear set is shown in Fig. 4.8.

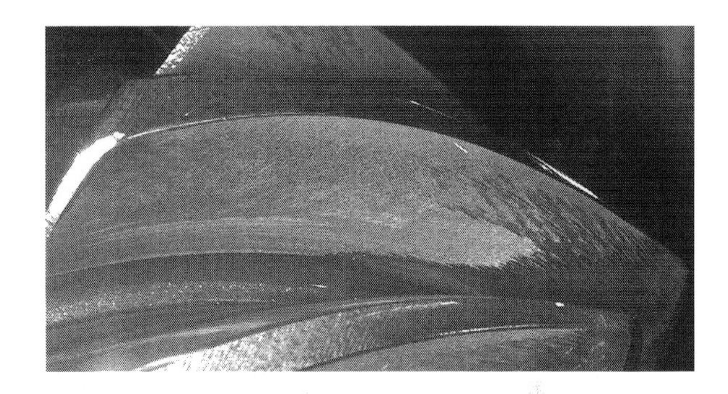

Fig. 4.8 Micropitting

Micropitting is mainly influenced by surface and lubricant properties. The relative lubricant film thickness, defined as the ratio of the minimum lubricant film thickness to the mean tooth flank roughness [SCHR00], is used as a failure criterion. This ratio is compared to a permissible relative lubricant film thickness, the so-called "strength" of the lubricant, derived from the FZG micropitting test [FVA54].

The lubricant film thickness mainly depends on speed conditions and lubricant viscosity rather than on contact stress. The sum of velocities contributes to the formation of the lubricant film while the sliding velocity impedes it. The smallest possible profile shift for the pinion and a low hypoid offset accordingly favour the formation of the lubricant film since less frictional heat is generated in the contact resulting in higher viscosity and a thicker lubricant film.

A higher micropitting load capacity can be obtained by selecting a suitable lubricant with a low coefficient of friction. Additives in the lubricant also have a significant influence on micropitting. In this respect, phosphorus sulphide based additives are usually more advantageous than zinc dithiophosphate based additives [NIEM86.2]. Because smooth flank surfaces are less at risk of micropitting, hard finishing processes or running-in procedures to smooth the active tooth flank surfaces have a positive effect on micropitting load capacity.

4.1.6 Wear

Wear is the continuous destructive removal of material caused by abrasion. It occurs at low circumferential speeds, and hence low film thicknesses [NIEM86.2]. The different forms of wear are described with more details in [ISO10825]. Like micropitting, wear results in tooth flank form deviations occurring in both the addendum and dedendum areas of the teeth. For given tooth flanks of identical hardness, the total weight loss on the pinion and the wheel is roughly equal; therefore, depending on the gear ratio, the change in flank form on the pinion is greater than that on the wheel. And slight differences in hardness lead to increased wear on the softer member [NIEM86.2]. If the case depth is reduced unacceptably by wear, failures like tooth breakage are likely to occur. Figure 4.9 shows a wear pattern accompanied by scuffing.

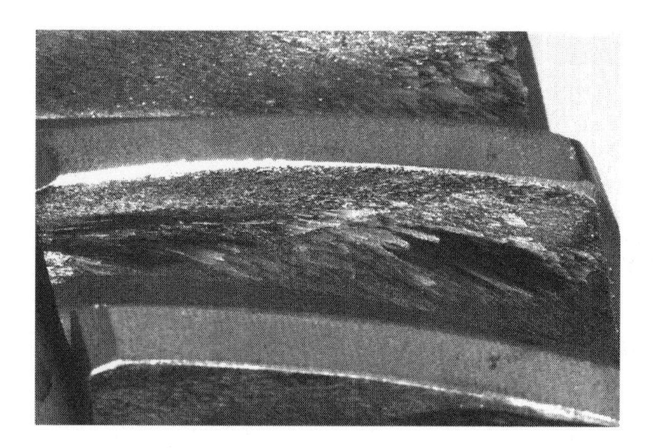

Fig. 4.9 Wear (accompanied by scuffing damage)

All actions which produce a thicker lubricant film, e.g., higher viscosities and circumferential speeds, have a wear-reducing effect. The lubricant type and additives also influence wear load capacity which can be investigated in the FZG wear test [BAYE96]. Tip relief to reduce impact at contact entry and the lowest possible surface roughness are also advantageous.

4.1.7 Ridging and Rippling

When two meshing tooth flanks are not separated by an adequately thick lubricating film, material may be removed continuously from the surfaces, especially on hypoid gears which have a large component in sliding velocity in the lengthwise direction (Fig. 4.10). Material removal creates grooves in the sliding direction, but is not accompanied by scuffing. This failure mode, known as ridging, creates a grooved surface with flank form deviations extending deep into the case depth [FRES81]. This results in increased acoustic excitation of the gear set causing further failures.

Since ridging sits half-way between wear and scuffing, preventive measures appropriate to these two failure modes are recommended.

Another surface failure mode that involves material removal is rippling. If the lubricant film thickness is insufficient, wavy grooves are formed perpendicular to

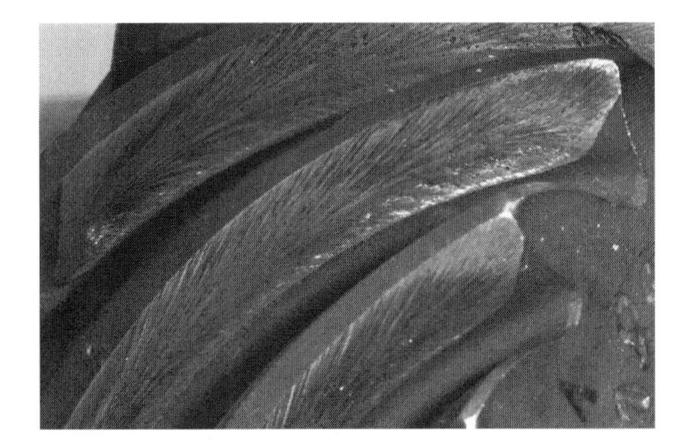

Fig. 4.10 Ridging

Fig. 4.11 Rippling

sliding (Fig. 4.11). These grooves are caused by a friction-induced vibration associated with the stick–slip effect (falling coefficient of friction with increasing sliding velocity). Rippling occurs mainly when EP-lubricants are used where the stress limit depends on the lubricant employed. However, the amount of deviation is only a few micrometres such that rippling cannot be designated as a failure in the true sense. After an extended period of operation, though, rippling generally turns into ridging damage [FRES81].

4.1.8 Scuffing

Scuffing as a failure mode is sub-divided into cold and hot scuffing, neither of which is fatigue failure but is rather caused by short-duration overloads.

Cold scuffing occurs at low circumferential speeds and under unfavourable lubricating conditions.

Hot scuffing occurs at high circumferential speeds and associated high power losses in the contact, which result in severe heating. When the lubricant film thickness is insufficient and the protective lubricant layer fails, the meshing tooth flanks are temporarily welded together and immediately separated again by their relative motion [NIEM86.2]. This results in the characteristic score marks in the sliding direction (Fig. 4.12). These marks occur in the corresponding tooth flank meshing areas on the pinion and wheel. Scuffing causes destructive material removal and tooth flank form deviations. The associated local pressure peaks in turn cause increased temperatures and hence enforce scuffing. The result is often total failure of the gear system.

The most important variables influencing scuffing are sliding velocity and contact pressure. Sliding velocity rises with increasing profile shift and hypoid offset. These variables should therefore be minimized in the interest of optimum scuffing resistance. Suitable crowning can also be helpful when higher sliding velocities are present. Smoothing of the tooth flank surfaces during running-in may also significantly increase scuffing resistance. Phosphating assists the running-in process, and therefore also has a positive effect.

The coefficient of friction, another decisive factor in the amount of heat produced in the contact, should be reduced as much as possible. This can be achieved either through design or through the choice of a suitable lubricant (see Sect. 4.2). Measures such as super finishing (polishing) and lapping (see Sect. 6.6.3.2) can also have a positive effect.

Lubricant and additive properties are determinant factors in scuffing resistance; they can be examined and evaluated by means of one of the FZG scuffing tests [FVA243]. The failure load reached in the test, or the corresponding torque and the gear geometry, are used to determine a permissible temperature which is then compared to the actual temperature occurring during operation.

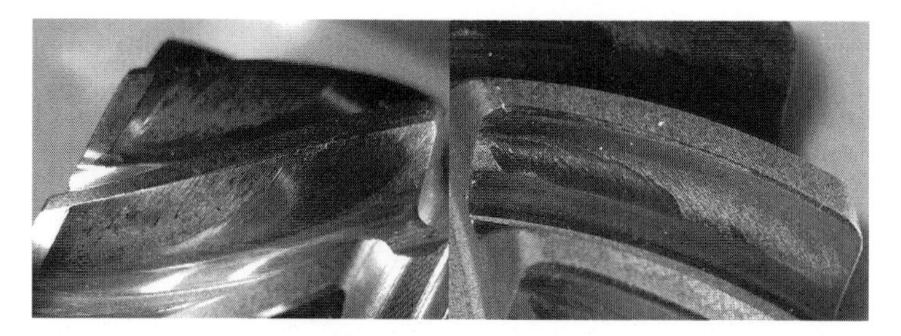

Fig. 4.12 Scuffing on a spiral bevel pinion (*left*) and the mating wheel (*right*)

4.2 Load Capacity Calculation

4.2.1 Standards and Calculation Methods

Various methods are available to calculate the load capacity of bevel gears. Table 4.1 provides an overview of a number of international standards and calculation methods:

Table 4.1 Standards and calculation methods to determine the load capacity of bevel and hypoid gears

	Gear type		**Calculating load capacity**				
Method	**Bevel gears (a = 0)**	**Hypoid gears (a ≠ 0)**	**Tooth root breakage**	**Pitting**	**Scuffing**	**Wear**	**Flank breakage**
[DIN3991]	X		X	X	X		
[FVA411]	X	X	X	X			
[ISO10300]	X	X	X	X			
Annast [ANNA03]	X	X					X
Niemann/Winter	X	X	X	X	X	X	
[ISO/TR13989]	X	X			X		
[AGMA2003]	X		X	X			
Niemann (1965)	X	X	X	X	X		
[DNV41.2]	X		X	X			
German Lloyd	X		X	X			
Lloyd's Register	X		X	X			

The input variables for these standardized calculation methods are the macro geometry, the applied load and rotation speed of the gear set, and the material and lubricant specifications. The results are safety factors representing the ratio of strength to stress, thereby making it possible to state the load capacity of a gear when subjected to a particular load. The required minimum safety factors must be agreed between the supplier and the customer.

The following sections describe the calculation methods to determine the tooth root load capacity (see Sect. 4.2.4) and pitting resistance (see Sect. 4.2.5) according to the FVA-project No. 411 "Hypoid load capacity" [FVA411]. These methods are part of [ISO10300] as "Method B1" since 2014 and allow calculations for hypoid gears whereas only non-offset bevel gears could be calculated with most of older standards like [DIN3991, AGMA2003] and edition 2001 of ISO 10300.

At present there are no internationally standardized methods to calculate the scuffing load capacity of bevel gears. For this reason, Sect. 4.2.6 describes the method of ISO-Technical Report ISO/TR 13989 [ISO/TR13989] for bevel and hypoid gears.

4.2.2 Virtual Cylindrical Gears for Tooth Root and Pitting Load Capacity

The strength values to calculate the tooth root and pitting load capacity of bevel gears according to [ISO10300] (Method B1) and [FVA411] are determined for cylindrical gears. Therefore, the bevel gear geometry is converted to virtual cylindrical gears with representative meshing conditions similar to those of the bevel gear set to be calculated. Virtual cylindrical gears according to [ISO10300] and [FVA411] are determined at the design point P (Fig. 4.13). Note that the fundamental geometry variables of a hypoid gear pair change with the offset, such as the face width and spiral angle of the pinion. Therefore, the helix angle, face width, equivalent radius of curvature, and contact ratio of the virtual cylindrical gear, differ from those according to older standards like [DIN3991]. However, as the

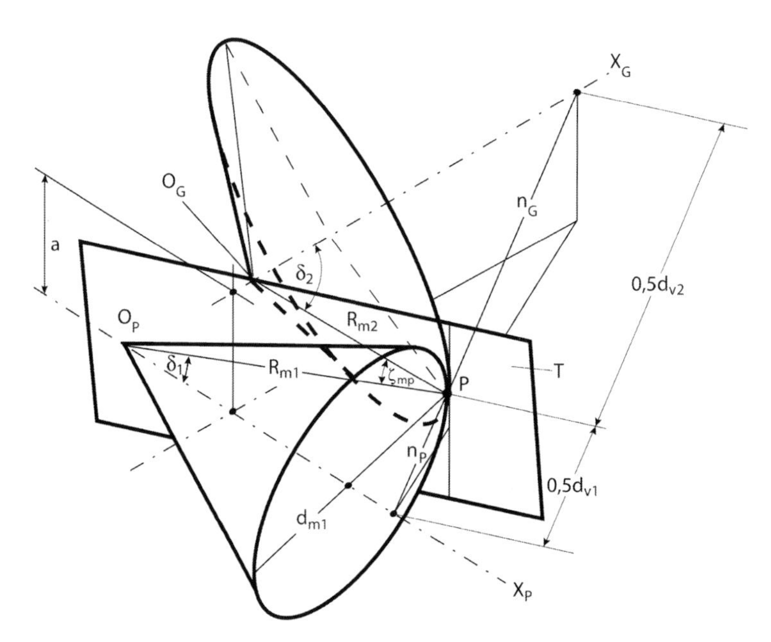

Fig. 4.13 Geometry data to derive the virtual cylindrical gears according to [FVA411] and [ISO10300]

offset value approaches 0, the virtual gears according to ISO 10300 Method B1 approximate those from DIN 3991 (Fig. 4.14).

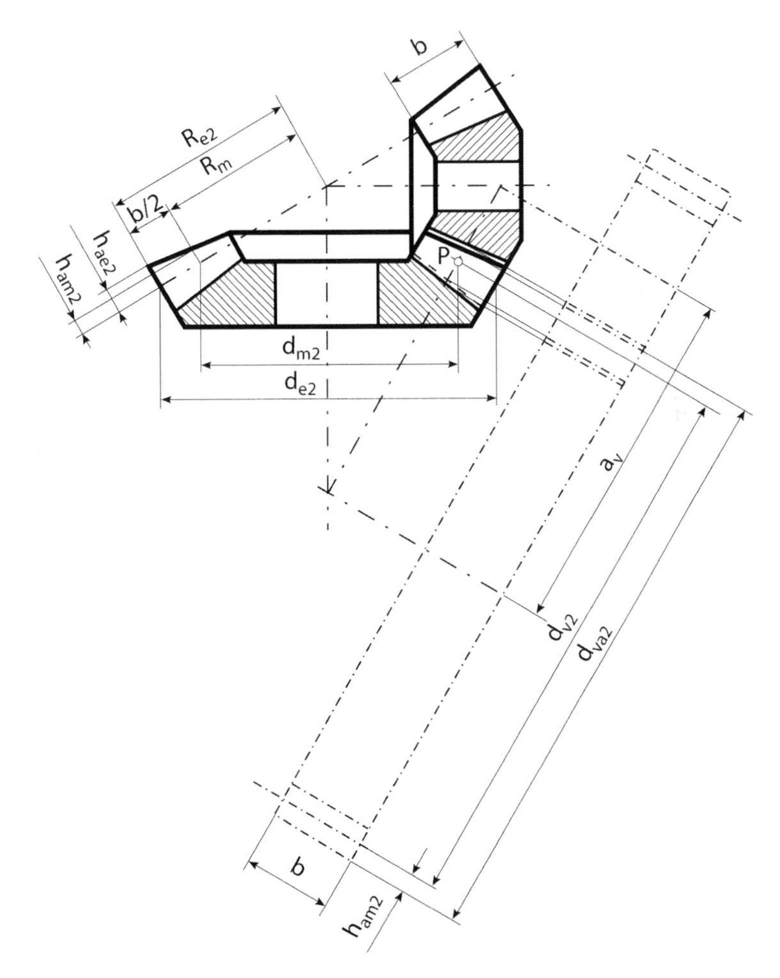

Fig. 4.14 Virtual cylindrical gears according to [DIN3991]

Table 4.2 summarizes the formulae for the determination of the geometry data of the virtual cylindrical gears in the transverse section. Table 4.3 gives the formulae needed for the data in the normal section.

Table 4.2 Geometry of virtual cylindrical gears in the transverse section acc. to [FVA411]

Designation	Formula	No.
Pitch diameter	$d_{v1,2} = \frac{d_{m1,2}}{\cos\delta_{1,2}} = \frac{d_{e1,2}\ R_m}{\cos\delta_{1,2} R_e}$	(4.1)
for a = 0 mm and $\sum = 90°$:	$d_{v1} = d_{m1}\frac{\sqrt{u^2+1}}{u};\quad d_{v2} = d_{v1}u^2$	(4.2)
Center distance	$a_v = (d_{v1} + d_{v2})/2$	(4.3)
Tip diameter	$d_{va1,2} = d_{v1,2} + 2h_{am1,2}$	(4.4)
Root diameter	$d_{vf1,2} = d_{v1,2} - 2h_{fm1,2}$	(4.5)
Helix angle	$\beta_v = (\beta_{m1} + \beta_{m2})/2$	(4.6)
Base helix angle	$\beta_{vb} = \arcsin\ (\sin\beta_v \cos\alpha_n)$	(4.7)
Base diameter	$d_{vb1,2} = d_{v1,2}\cos\alpha_{vet}$	(4.8)
	where: $\alpha_{vet} = \arctan\left(\frac{\tan\alpha_e}{\cos\beta_v}\right)$ $\alpha_e = \alpha_{eD}$ acc. to [ISO23509] on the drive flank $\alpha_e = \alpha_{eC}$ acc. to [ISO23509] on the coast flank	(4.9)
Transverse module	$m_{vt} = m_{mn}/\cos\beta_v$	(4.10)
Number of teeth	$z_{v1,2} = d_{v1,2}/m_{vt}$	(4.11)
Gear ratio	$u_v = z_{v2}/z_{v1}$	(4.12)
for a = 0 mm and $\Sigma = 90°$	$u_v = \left(\frac{z_2}{z_1}\right)^2 = u^2$	(4.13)
Number of teeth for a = 0 mm and $\Sigma = 90°$:	$z_{v1} = z_1\frac{\sqrt{u^2+1}}{u}$	(4.14)
	$z_{v2} = z_2\sqrt{u^2+1}$	(4.15)
Base pitch	$p_{vet} = \frac{m_{mn}\cdot\pi\cdot\cos\alpha_{vet}}{\cos\beta_v}$	(4.16)

Length of the line of contact	$g_{v\alpha} = \frac{1}{2}\left[\sqrt{\left(d_{va1}{}^2 - d_{vb1}{}^2\right)} + \sqrt{\left(d_{va2}{}^2 - d_{vb2}{}^2\right)}\right] - a_v \sin\alpha_{vet}$	(4.17)
Face width	$b_v = b_2 \frac{b_{veff}}{b_{2eff}}$	(4.18)
	where: $b_{veff} = \frac{\left(\frac{b_{2eff}}{\cos\ \zeta_{mP}/2} - g_{v\alpha}\cos\alpha_{vet}\sin\ \zeta_{mP}/2\right)}{1+\tan\gamma'\sin\ \zeta_{mP}/2}$ b_{2eff} = effective contact pattern width on the wheel for a specific load. The width of the contact pattern is estimated, measured, or calculated using a 'loaded tooth contact analysis' software (see Sect. 4.4).	(4.19)
	$\vartheta_{mP} = \arctan(\sin\delta_2 \tan\zeta_m)$	(4.20)
	$\gamma' = \vartheta_{mP} - \zeta_{mP}/2$ ζ_{mP} = offset angle in the crown gear plane ζ_m = offset angle in axial plane according to [ISO23509]	(4.21)
Profile contact ratio	$\varepsilon_{v\alpha} = \frac{g_{v\alpha}}{p_{vet}}$	(4.22)
Overlap ratio	$\varepsilon_{v\beta} = \frac{b_{veff}\sin\beta_v}{m_{mn}\pi}$	(4.23)
Total contact ratio	$\varepsilon_{v\gamma} = \varepsilon_{v\alpha} + \varepsilon_{v\beta}$	(4.24)
Equivalent curvature radius	$\rho_{ers} = \rho_t \cdot \cos^2 w_{Bel}$ where: w_{Bel} = angle between the contact line and pitch cone in the flank tangential plane ρ_t = equivalent radius of curvature in the profile section	(4.25)
	$\rho_t = \left[\frac{\cos\beta_{m1}\cos\beta_{m2}}{[\cos\alpha_n(\tan\alpha_n - \tan\alpha_{\lim}) + \tan\zeta_{mP}\tan w_{Bel}]\cos\zeta_{mP}} \cdot \left(\frac{1}{R_{m2}\tan\delta_2} + \frac{1}{R_{m1}\tan\delta_1}\right)\right]^{-1}$	(4.26)

(continued)

Table 4.2 (continued)

Designation	Formula	No.
	where: $\alpha_n = \alpha_{nD}$ acc. to [ISO23509] on the drive side flank $\alpha_n = -\alpha_{nC}$ acc. to [ISO23509] on the coast side flank	
	$w_{Bel} = \arctan\left(\tan\beta_v \sin\alpha_e\right)$	(4.27)
	where: $\alpha_e = \alpha_{eD}$ acc. to [ISO23509] on the drive side flank $\alpha_e = \alpha_{eC}$ acc. to[ISO23509] on the coast side flank	

Table 4.3 Geometry of virtual cylindrical gears in the normal section acc. to [ISO10300]

Designation	Formula	No.
Number of pinion teeth	$z_{vn1} = \dfrac{z_{v1}}{\cos^2\beta_{vb}\cos\beta_v}$	(4.28)
Number of wheel teeth	$z_{vn2} = z_{vn1}\ u_v$	(4.29)
Pitch diameter	$d_{vn1,2} = \dfrac{d_{v1,2}}{\cos^2\beta_{vb}} = z_{vn1,2}m_{mn}$	(4.30)
Center distance	$a_{vn} = (d_{vn1} + d_{vn2})/2$	(4.31)
Tip diameter	$d_{van1,2} = d_{vn1,2} + 2h_{am1,2}$	(4.32)
Base diameter	$d_{vbn1,2} = d_{vn1,2}\cos\alpha_e$	(4.33)
Profile contact ratio	$\varepsilon_{van} = \dfrac{\varepsilon_{v\alpha}}{\cos^2\beta_{vb}}$	(4.34)

4.2.3 *Virtual Cylindrical Gears for Scuffing Load Capacity*

When calculating scuffing load capacity according to [ISO/TR13989] and efficiency according to [WECH87] (see Sect. 4.3), virtual crossed axes helical gears (Fig. 4.15) are defined with sliding conditions at the calculation point similar to those of hypoid gears (Table 4.4). Depending on the calculation method, the helix angles are used as absolute values or with a positive/negative sign.

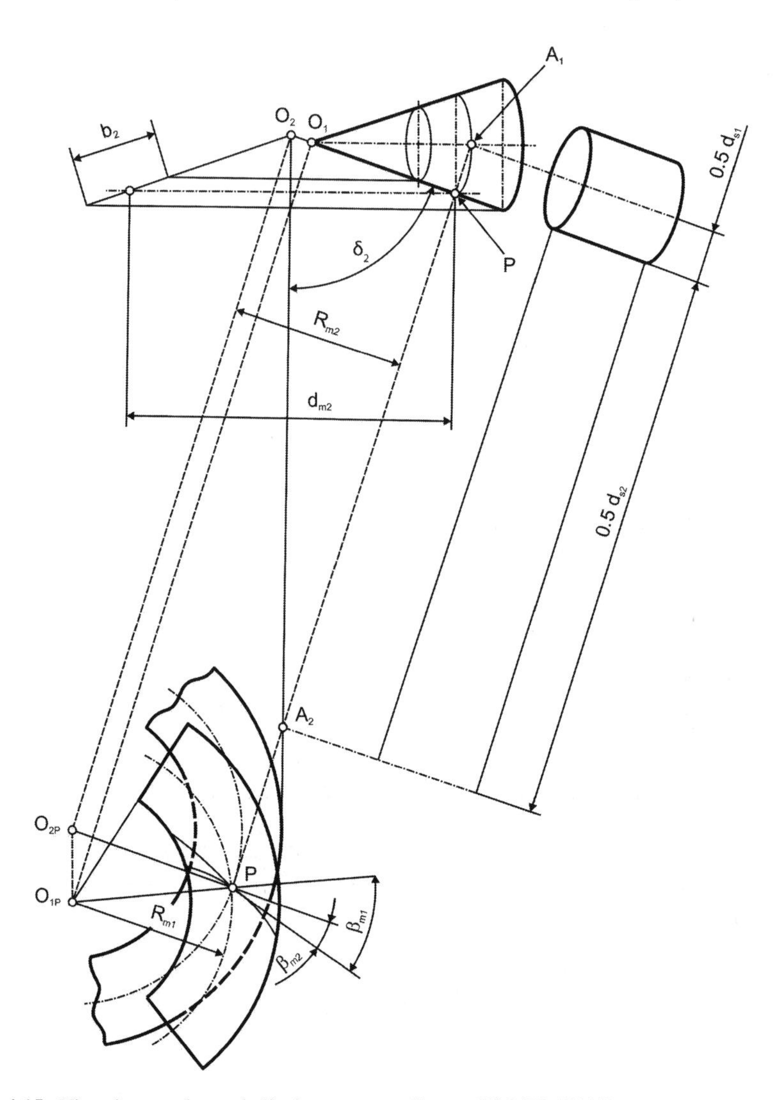

Fig. 4.15 Virtual crossed axes helical gears according to [ISO/TR13989]

Table 4.4 Geometry of virtual crossed axes helical gears according to [ISO/TR13989]

Designation	Formula		No.
Helix angle	$\beta_{s1,2} = \lvert\beta_{m1,2}\rvert$	to calculate scuffing load capacity according to [ISO/TR13989]	(4.35)
	$\beta_{s1,2} = \beta_{m1,2}$	to calculate efficiency according to [WECH87], accounting for hand of spiral (left < 0, right > 0)	
Pressure angle	$\tan\alpha_{st1,2} = \dfrac{\tan\alpha_{sn}}{\cos\beta_{s1,2}}$		(4.36)
with:	$\alpha_{sn} = \alpha_{nD,C}$		
Base helix angle	$\sin\beta_{b1,2} = \dfrac{\sin\beta_{s1,2}}{\cos\alpha_{sn}}$		(4.37)
Shaft angle	$\Sigma = \beta_{m1} - \beta_{m2}$		(4.38)
Pitch diameter	$d_{s1,2} = \dfrac{d_{m1,2}}{\cos\delta_{1,2}}$		(4.39)
Tip diameter	$d_{a1,2} = d_{s1,2} + 2h_{am1,2}$		(4.40)
Base diameter	$d_{b1,2} = d_{s1,2} \cdot \cos\alpha_{st1,2}$		(4.41)
Angle between contact line and tooth trace	$\tan\beta_{B1,2} = \tan\beta_{s1,2}\sin\alpha_{sn}$		(4.42)
Angle between the contact lines	$\varphi = \beta_{B1} + \beta_{B2}$		(4.43)
Module	$m_{sn} = m_{mn}$		(4.44)
Normal base pitch	$p_{en} = m_{sn}\pi\cos\alpha_{sn}$		(4.45)
Equivalent curvature radius	$\rho_{Cn} = \dfrac{\rho_{n1}\rho_{n2}}{\rho_{n1} + \rho_{n2}}$		(4.46)
	where: $\rho_{n1,2} = 0.5 \cdot d_{s1,2}\dfrac{\sin^2\alpha_{st1,2}}{\sin\alpha_{sn}}$		(4.47)
Length of path of contact	$\overline{AE} = g_{an1} + g_{an2}$		(4.48)
	where: g_{an1} = pinion tip path of contact/wheel root path of contact; g_{an2} = wheel tip path of contact/pinion root path of contact		
	$g_{an1} = g_{fn2} = \dfrac{0.5 \cdot \left(\sqrt{d_{a1}^2 - d_{b1}^2} - \sqrt{d_{s1}^2 - d_{b1}^2}\right)}{\cos\beta_{b1}}$		(4.49)
	$g_{an2} = g_{fn1} = \dfrac{0.5 \cdot \left(\sqrt{d_{a2}^2 - d_{b2}^2} - \sqrt{d_{s2}^2 - d_{b2}^2}\right)}{\cos\beta_{b2}}$		(4.50)
Total contact ratio in the normal section	$\varepsilon_n = \dfrac{\overline{AE}}{p_{en}}$		(4.51)
Tip/root contact ratio in the normal section	$\varepsilon_{n1,2} = \dfrac{g_{an1,2}}{p_{en}}$		(4.52)

4.2.4 *Tooth Root Load Capacity*

In order to calculate the tooth root load capacity of a bevel gear, the maximum bending stress which occurs in the tooth root is determined. This stress value is compared to a permissible stress derived from standardized cylindrical test gears (standard strength values in [ISO6336]). Bending stress σ_{F0} is calculated at the contact point of the 30° tangent in the fillet of the loaded flank since tooth root breakage usually starts at this point (Fig. 4.16).

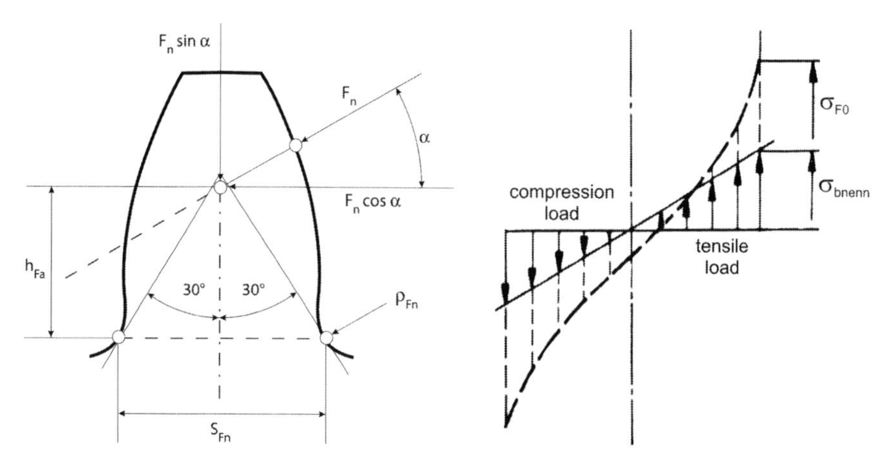

Fig. 4.16 Geometry data to calculate tooth root stress (*left*) and gradient of local tooth root stress (*right*)

4.2.4.1 Tooth Root Stress

The tooth root stress σ_F is calculated from the local tooth root stress σ_{F0} multiplied by load factors and load distribution factors which consider overloading and real load distribution in actual operation of the gears. The local tooth root stress σ_{F0} results from a nominal stress and a number of corrective factors Y for specific influences on tooth root stress, e.g., the notch effect or the complex stress condition in the tooth root regarding compressive stress from the radial load $F_n \cdot \sin\alpha$ and shear stress from the tangential load $F_n \cdot \cos\alpha$ (Table 4.5).

Table 4.5 Calculating the tooth root stress σ_F

Designation	Formula	No.
Tooth root stress	$\sigma_{F1,2} = \sigma_{F0\,1,2}\, K_A K_v K_{F\beta} K_{F\alpha}$ $= \frac{F_{mtv}}{b_v \cdot m_{mn}} Y_{Fa1,2} Y_{Sa1,2} Y_\varepsilon Y_{BS} Y_{LS} \cdot K_A K_v K_{F\beta} K_{F\alpha}$	(4.53) (4.54)
	$F_{mtv} = F_{mt1,2} \frac{\cos\beta_v}{\cos\beta_{m1,2}} = \frac{2000 \cdot T_{1,2}}{d_{m1,2}} \frac{\cos\beta_v}{\cos\beta_{m1,2}}$	(4.55)
	$Y_{Fa1,2}$ = factor considering the influence of the tooth form	
	$Y_{Sa1,2}$ = stress correction factor considering stress increase due to the notch effect and the complex stress condition in the tooth root	
	Y_ε = contact ratio factor (tooth root) to convert the local stress determined for load application at tip to the desired load position	
	Y_{BS} = bevel spiral angle factor considering shorter, inclined contact lines compared to the total face width	
	Y_{LS} = load sharing factor considering load sharing between two or more tooth pairs in contact	
	K_A = application factor considering additional external loads resulting from operation	
	K_v = dynamic factor considering additional dynamic loads	
	$K_{F\beta}$ = face load factor considering a non-uniform load distribution along the face width	
	$K_{F\alpha}$ = transverse load factor considering a non-uniform load distribution between the meshing tooth pairs	

Tooth form factor $Y_{Fa1,2}$ This factor considers the influence of the tooth form, and is calculated separately for the pinion and the wheel, but with the virtual cylindrical gears as described in Sect. 4.2.2 (Table 4.6 for generated gears, Table 4.7 for plunge cut wheels or non-generated wheels). By calculating the tooth root chord s_{Fn} for both tooth flanks, the normal pressure angles on the drive and coast sides are considered. Tooth root chords s_{FnD} and s_{FnC} are calculated with the corresponding geometry data for the drive and coast tooth flanks. The sum of $0.5s_{FnD}$ and $0.5s_{FnC}$ is considered as the tooth root cross-section s_{Fn}. All other variables are calculated with the effective normal pressure angle α_e for the drive or coast tooth flank, respectively. The calculations are iterative using the auxiliary variables E, G, H and ϑ. Iteration starts at $\vartheta = \pi/6$ and is usually concluded with sufficient accuracy after a few steps.

Table 4.6 Calculation of tooth form factors $Y_{Fa1,2}$ for generated gears

Designation	Formula	No.
Tooth form factor	$Y_{Fa1,2} = \frac{6 \frac{h_{FaD,C}}{m_{mn}} \cos\alpha_{FanD,C}}{\left(\frac{s_{FnD,C}}{m_{mn}}\right)^2 \cos\alpha_{nD,C}}$	(4.56)
	where: $\alpha_{FanD,C} = \alpha_{anD,C} - \gamma_{aD,C}$	(4.57)
	$\alpha_{anD,C} = \arccos\left(\frac{d_{vbnD,C}}{d_{vanD,C}}\right)$	(4.58)
	$\gamma_{aD,C} = \frac{1}{z_{vn}}\left[\frac{\pi}{2} + 2(x_{hm}\tan\alpha_{eD,C} + x_{sm})\right] + inv\alpha_{eD,C} - inv\alpha_{an}$	(4.59)

(continued)

Table 4.6 (continued)

Designation	Formula	No.
Individual tooth root chords for the drive and coast tooth flanks	$\frac{s_{FnD,C}}{m_{mn}} = z_{vn} \sin\left(\frac{\pi}{3} - \vartheta_{D,C}\right) + \sqrt{3}\left(\frac{G_{D,C}}{\cos\vartheta} - \frac{\rho_{a0D,C}}{m_{mn}}\right)$	(4.60)
	s_{FnD} (calculate all variables with $\alpha_e = \alpha_{eD}$)	
	s_{FnC} (calculate all variables with $\alpha_e = \alpha_{eC}$)	
Total tooth root chord at the 30° tangent	$s_{Fn} = (s_{FnD} + s_{FnC})/2$	(4.61)
Fillet radius at the 30° tangent	$\frac{\rho_{F\,D,C}}{m_{mn}} = \frac{\rho_{a0D,C}}{m_{mn}} + \frac{2G_{D,C}^2}{\cos\vartheta(z_{vn}\cos^2\vartheta_{D,C} - 2G_{D,C})}$	(4.62)
Bending lever arm (load application at tooth tip)	$\frac{h_{FaD,C}}{m_{mn}} = \frac{1}{2}\left[\left(\cos\gamma_{aD,C} - \sin\gamma_{aD,C}\tan\alpha_{FanD,C}\right)\frac{d_{vanD,C}}{m_{mn}} - z_{vn}\cos\left(\frac{\pi}{3} - \vartheta_{D,C}\right) - \frac{G_{D,C}}{\cos\vartheta} + \frac{\rho_{a0D,C}}{m_{mn}}\right]$	(4.63)
Auxiliary variables	$E_{D,C} = \left(\frac{\pi}{4} - x_{sm}\right) m_{mn} - h_{a0D,C}\tan\alpha_{eD,C} - \frac{\rho_{a0D,C}(1 - \sin\alpha_{eD,C}) - s_{prD,C}}{\cos\alpha_{eD,C}}$	(4.64)
	$G_{D,C} = \frac{\rho_{a0D,C}}{m_{mn}} - \frac{h_{a0D,C}}{m_{mn}} + x_{hm}$	(4.65)
	$H_{D,C} = \frac{2}{z_{vn}}\left(\frac{\pi}{4} - \frac{E_{D,C}}{m_{mn}}\right) - \frac{\pi}{3}$	(4.66)
	$\vartheta_{D,C} = \frac{2G_{D,C}}{z_{vn}}\tan\vartheta_{D,C} - H_{D,C}$	(4.67)

Table 4.7 Calculation of tooth form factor Y_{Fa2} for non-generated wheels

Designation	Formula	No.
Tooth form factor	$Y_{Fa2} = \frac{6\frac{h_{FaD,C}}{m_{mn}}}{\left(\frac{s_{Fn}}{m_{mn}}\right)^2}$	(4.68)
	where: $s_{FnD,C} = \pi \cdot m_{mn} - 2E_{D,C} - 2\rho_{aD,C}\cos 30°$	(4.69)
	$s_{Fn} = \frac{s_{FnD}}{2} + \frac{s_{FnC}}{2}$	(4.70)
	E_2 according to (4.64) with $\alpha_{nD,C}$ instead of $\alpha_{eD,C}$ for the considered tooth flank	
	$\rho_{FD,C} = \rho_{a0D,C}$	
	$h_{FaD,C} = h_{a0D,C} - \frac{\rho_{a0D,C}}{2} + m_{mn} - \left(\frac{\pi}{4} + x_{sm} - \tan\alpha_{nD,C}\right) m_{mn}\tan\alpha_{nD,C}$	(4.71)

Stress correction factor $Y_{Sa1,2}$ The stress correction factor $Y_{Sa1,2}$ accounts for the complex stress condition in the tooth root. It is calculated using the values for the tooth root chord s_{Fn}, the bending lever arm h_{Fa} and the fillet radius ρ_F, and is determined according to FVA411 (Table 4.8).

Table 4.8 Calculation of stress correction factor Y_{Sa}

Designation	Formula	No.
Stress correction factor	$Y_{SaD,C} = (1.2 + 0.13L_{aD,C}) \cdot q_{sD,C}^{\left(1/\left(1.21+2.3/L_{aD,C}\right)\right)}$	(4.72)
	where: $L_{aD,C} = \frac{s_{Fn}}{h_{FaD,C}}$ and $q_{sD,C} = \frac{s_{Fn}}{2\rho_{FD,C}}$	(4.73)

Contact ratio factor Y_ε The contact ratio factor Y_ε is used to convert the tooth form factor Y_{Fa} and stress correction factor Y_{Sa}, valid for load application at the tip, to the considered point of load application. It is determined as a function of the profile contact ratio and the overlap ratio (Table 4.9).

Table 4.9 Calculation of contact ratio factor Y_ε

Designation	Formula	No.
Contact ratio factor	For $\varepsilon_{v\beta} = 0$: $Y_\varepsilon = 0.25 + \frac{0.75}{\varepsilon_{v\alpha}} \geq 0.625$	(4.74)
	For $0 < \varepsilon_{v\beta} < 1$: $Y_\varepsilon = 0.25 + \frac{0.75}{\varepsilon_{v\alpha}} - \varepsilon_{v\beta}\left(\frac{0.75}{\varepsilon_{v\alpha}} - 0.375\right) \geq 0.625$	(4.75)
	For $\varepsilon_{v\beta} > 1$: $Y_\varepsilon = 0.625$	(4.76)

Load sharing factor Y_{LS} The load sharing factor Y_{LS} considers the number of tooth pairs in mesh as a function of the shape of the zone of action. Older standards assumed an elliptical zone of action, whereas FVA 411 bases its calculation of load sharing on a zone in the shape of a parallelogram (Fig. 4.17). In loaded TCAs, the parallelogram has proved to be a better approximation of the contact pattern projected into the plane of action. Therefore, only the calculation method of Y_{LS} according to [FVA411] is presented in Tables 4.10 and 4.11).

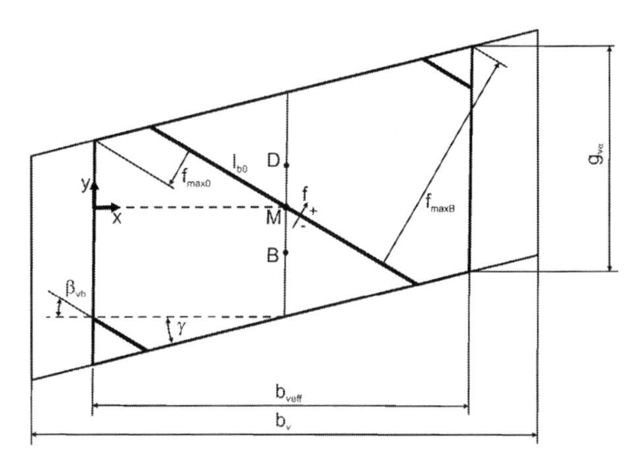

Fig. 4.17 Zone of action to calculate load sharing according to [FVA411]

In accordance with ISO10300 Method B1, a parabolic peak load distribution **a** is assumed to prevail along the line of action. The load distribution **b** along a contact line **c** corresponds to a semi-ellipse (Fig. 4.18). The major axis of the ellipse of Hertzian pressure is taken to be the contact line.

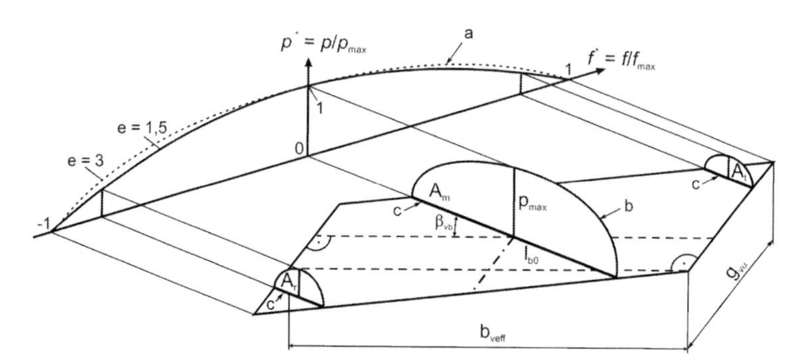

Fig. 4.18 Peak load distribution over the zone of action and load distribution over the contact lines

Table 4.10 Calculation of load sharing factor Y_{LS} for a zone of action in the shape of a parallelogram

Designation	Formula	No.
Face load factor	$Y_{LS} = \dfrac{A_m}{A_t + A_m + A_r}$	(4.77)
	where: A = area of the semi-ellipse over a contact line A_t, A_m, A_r calculated using f_t, f_m, f_r acc. to Table 4.11	
Area of semi-ellipse over a contact line	$A = \dfrac{1}{4} p^* l_b \pi$	(4.78)
Peak load distribution along the path of contact	$p^* = \dfrac{p}{p_{\max}} = 1 - \left(\dfrac{\lvert f \rvert}{\lvert f_{\max} \rvert}\right)^e$ where: e = 1.5 for high crowning (e.g., industry applications) e = 3.0 for low crowning (e.g., automotive applications)	(4.79)
	where: f_{max} = the larger value of f_{max0} and f_{maxb}	(4.80)
	$f_{\max 0} = \dfrac{1}{2}\left(g_{va} - b_{veff}(\tan\gamma + \tan\beta_{vb})\right) \cdot \cos\beta_{vb}$	(4.81)
	$f_{\max b} = \dfrac{1}{2}\left(g_{va} + b_{veff}(\tan\gamma + \tan\beta_{vb})\right) \cdot \cos\beta_{vb}$	(4.82)
	$\gamma = \arctan\left(\dfrac{\tan\gamma'}{\cos\alpha_{vet}}\right)$	(4.83)
	b_{veff} and γ' according to Table 4.4	
Length of the contact line	$l_b = l_{b0}\sqrt{\left(1 - \left(\dfrac{f}{f_{\max}}\right)^2\right)\left(1 - \sqrt{\dfrac{b_{veff}}{b_v}}\right)^2}$	(4.84)

(continued)

Table 4.10 (continued)

Designation	Formula	No.
Theoretical length of the contact line	$l_{b0} = \sqrt{(x_1 - x_2)^2 + (y_1 - y_2)^2}$	(4.85)
Parameter x	$x_1 = \dfrac{\left[f \cos\beta_{vb} + \tan\beta_{vb}\left(f \sin\beta_{vb} + \dfrac{b_{veff}}{2}\right) + \dfrac{1}{2}(g_{v\alpha} + b_{veff}\tan\gamma)\right]}{\tan\gamma + \tan\beta_{vb}}$	(4.86)
	$x_2 = \dfrac{\left[f \cos\beta_{vb} + \tan\beta_{vb}\left(f \sin\beta_{vb} + \dfrac{b_{veff}}{2}\right) - \dfrac{1}{2}(g_{v\alpha} - b_{veff}\tan\gamma)\right]}{\tan\gamma + \tan\beta_{vb}}$	(4.87)
	where: f according to Table 4.11 if $x_{1,2} < 0$, then $x_{1,2} = 0$ if $x_{1,2} > b_{veff}$, then $x_{1,2} = b_{veff}$	
Parameter y	$y_{1,2} = y(x_{1,2}) = -x_{1,2}\tan\beta_{vb} + f\cos\beta_{vb} + \tan\beta_{vb}\left(f \sin\beta_{vb} + \dfrac{b_{veff}}{2}\right)$	(4.88)

Table 4.11 Variable f to calculate the distances of the contact lines from mid-point M for tooth root load capacity

Designation	Formula	No.
$\varepsilon_{v\beta} = 0$	$f_t = (p_{vet} - 0.5p_{vet}\varepsilon_{v\alpha})\cos\beta_{vb} + p_{vet}\cos\beta_{vb}$ $f_m = (p_{vet} - 0.5p_{vet}\varepsilon_{v\alpha})\cos\beta_{vb}$ $f_r = (p_{vet} - 0.5p_{vet}\varepsilon_{v\alpha})\cos\beta_{vb} - p_{vet}\cos\beta_{vb}$	(4.89)
$0 < \varepsilon_{v\beta} < 1$	$f_t = (p_{vet} - 0.5p_{vet}\varepsilon_{v\alpha})\cos\beta_{vb}(1 - \varepsilon_{v\beta}) + p_{vet}\cos\beta_{vb}$ $f_m = (p_{vet} - 0.5p_{vet}\varepsilon_{v\alpha})\cos\beta_{vb}(1 - \varepsilon_{v\beta})$ $f_r = (p_{vet} - 0.5p_{vet}\varepsilon_{v\alpha})\cos\beta_{vb}(1 - \varepsilon_{v\beta}) - p_{vet}\cos\beta_{vb}$	(4.90)
$\varepsilon_{v\beta} \geq 1$	$f_t = +p_{vet}\cos\beta_{vb}$ $f_m = 0$ $f_r = -p_{vet}\cos\beta_{vb}$	(4.91)

Bevel spiral angle factor Y_{BS} The bevel spiral angle factor Y_{BS} for bending stress considers contact lines shorter than the face width because of their inclination (Table 4.12).

Table 4.12 Calculation of bevel spiral angle factor Y_{BS}

Designation	Formula	No.
Bevel spiral angle factor	$Y_{BS} = \dfrac{a_{BS}}{c_{BS}}\left(\dfrac{l_{bb}}{b_a} - 1.05 \cdot b_{BS}\right)^2 + 1$	(4.92)
Auxiliary values	$a_{BS} = -0.0182\left(\dfrac{b_a}{h}\right)^2 + 0.4736\left(\dfrac{b_a}{h}\right) - 0.32$	(4.93)
	$b_{BS} = -0.0032\left(\dfrac{b_a}{h}\right)^2 + 0.0526\left(\dfrac{b_a}{h}\right) + 0.712$	(4.94)
	$c_{BS} = -0.0050\left(\dfrac{b_a}{h}\right)^2 + 0.0850\left(\dfrac{b_a}{h}\right) + 0.54$	(4.95)

(continued)

Table 4.12 (continued)

Designation	Formula	No.
Developed length of one tooth	$b_a = \frac{b_v}{\cos\beta_v}$	(4.96)
Part of face width covered by the contact line	$l_{bb} = l_{bm}\frac{\cos\beta_{vb}}{\cos\beta_v}$	(4.97)
Average tooth depth	$h = 0.5 \cdot (h_{m1} + h_{m2})$	(4.98)

Load factors K_A and K_v The load factors K_A and K_v deal with loads exceeding the nominal torque, and resulting from external overloads and internal excitation during the actual operation of the gears. In the ideal case, both values are determined by experimental measurements or comprehensive system analyses (see [ISO10300] Part 1). If no comparable information is available, the approximate values or equations in Tables 4.13, 4.14, and 4.15 may be employed. In a load spectrum calculation (see Sect. 4.2.7), application factor K_A is set to 1 since the influence of additional external loads on load capacity is taken into account directly in the defined stress classes.

Table 4.13 Approximate values for application factor K_A

Mode of operation of the driving machine	Mode of operation of the driven machine			
	Uniform	Slight shocks	Medium shocks	Heavy shocks
Uniform	1.00	1.25	1.50	1.75 or higher
Slight shocks	1.10	1.35	1.60	1.85 or higher
Medium shocks	1.25	1.50	1.75	2.00 or higher
Heavy shocks	1.50	1.75	2.00	2.25 or higher

Table 4.14 Approximation equations for dynamic factor K_v

Designation	Formula	No.
Dynamic factor	$K_v = K_v{}^* - \frac{K_v{}^* - 1}{0.1} \cdot a_{rel} \geq 1$	(4.99)
	where: $K_v{}^* = N \cdot K + 1 \quad for\ N \leq 0.75$	(4.100)
	$K_v{}^* = \frac{b_v f_{p,eff} c'}{F_{mtv} K_A} c_{v1,2} + c_{v4} + 1 \quad for\ 0.75 < N \leq 1.5$	(4.101)
	$K_v{}^* = \frac{b_v f_{p,eff} c'}{F_{mtv} K_A} c_{v5,6} + c_{v7} \quad for\ N \geq 1.5$	(4.102)
	$a_{rel} = \frac{2\lvert a\rvert}{d_{m2}}$	(4.103)
	$K = \frac{b_v f_{p,eff} c'}{F_{mtv} K_A} c_{v1,2} + c_{v3}$ all c_v factors see Table 4.15	(4.104)

(continued)

Table 4.14 (continued)

Designation	Formula	No.
	For case hardened and nitrided gears: $f_{p,eff} = f_{pt} - y_\alpha$ where $y_\alpha = 0.075\ f_{pt} \leq 3\mu m$	(4.105)
Reference speed	$N = \frac{n_1}{n_{E1}}$	(4.106)
Resonance speed	$n_{E1} = \frac{30 \cdot 10^3}{\pi \cdot z_1} \sqrt{\frac{c_\gamma}{m_{red}}}$	(4.107)
Mesh stiffness	$c_\gamma = c_{\gamma 0} C_F$	(4.108)
	where: $c_{\gamma 0} = 20N/(mm \cdot \mu m)$	(4.109)
	$C_F = \left(F_{mtv} K_A / b_{veff}\right) / (100N/mm) \leq 1$	(4.110)
Reduced mass moment of inertia	$m_{red} = \frac{m_1{}^* m_2{}^*}{m_1{}^* + m_2{}^*}$	(4.111)
Gear mass per unit face width	$m_{1,2}{}^* = \frac{1}{8} \rho_M \pi \frac{d_{m1,2}^2}{\cos^2 \alpha_n}$ For steel $\rho_M = 7.86° 10^{-6}$ kg/mm^3	(4.112)

Table 4.15 c_v factors to calculate the dynamic factor

Influence factor	$1 < \varepsilon_{v\gamma} \leq 2$	$\varepsilon_{v\gamma} > 2$	
c_{v1}	0.32	0.32	$c_{v1,2} = c_{v1} + c_{v2}$
c_{v2}	0.34	$\frac{0.57}{\varepsilon_{v\gamma} - 0.3}$	
c_{v3}	0.23	$\frac{0.096}{\varepsilon_{v\gamma} - 1.56}$	
c_{v4}	0.90	$\frac{0.57 - 0.05\varepsilon_{v\gamma}}{\varepsilon_{v\gamma} - 1.44}$	
c_{v5}	0.47	0.47	$c_{v5,6} = c_{v5} + c_{v6}$
c_{v6}	0.47	$\frac{0.12}{\varepsilon_{v\gamma} - 1.74}$	

Influence factor	$1 < \varepsilon_{v\gamma} \leq 1.5$	$1.5 < \varepsilon_{v\gamma} \leq 2.5$	$\varepsilon_{v\gamma} > 2.5$
c_{v7}	0.75	$0.125 \sin\left[\pi\left(\varepsilon_{v\gamma} - 2\right)\right] + 0.875$	1.0

Face load factors $K_{H\beta}$ and $K_{F\beta}$ The influence of non-uniform load distribution over crowned bevel gear tooth flanks is considered by face load factors $K_{H\beta}$ and $K_{F\beta}$. The tooth root factor $K_{F\beta}$ is calculated as a function of the face load factor for the tooth flank $K_{H\beta}$ and the lengthwise curvature factor K_{F0} (Tables 4.16 and 4.17).

Table 4.16 Calculation of face load factor $K_{F\beta}$

Designation	Formula		No.
Face load factor for the root	$K_{F\beta} = K_{H\beta}/K_{F0}$		(4.113)
Face load factor for the flank	$K_{H\beta} = 1,5K_{H\beta-be}$ $K_{H\beta-be}$ see Table 4.17		(4.114)
Lengthwise curvature factor	$K_{F0} = 0{,}211\left(\frac{r_{c0}}{R_{m2}}\right)^q + 0{,}789$	For spiral bevel gears (face milled)	(4.115)
	$K_{F0} = 1.0$	For straight and Zerol bevel gears	(4.116)
	where: $q = \frac{0.279}{\log_{10}(\sin\beta_{m2})}$		(4.117)

Table 4.17 Approximate values for mounting factor $K_{H\beta\text{-}be}$

	Mounting conditions of the pinion and wheel		
Contact pattern check for:	**Neither member cantilever mounted**	**One member cantilever mounted**	**Both members cantilever mounted**
Each gear pair in its housing under full load	1.00	1.00	1.00
Each gear pair under light test load	1.05	1.10	1.25
A sample gear pair estimated for full load	1.20	1.32	1.50

Transverse load factors $K_{H\alpha}$ and $K_{F\alpha}$ The distribution of the total tangential force on a number of tooth pairs in mesh depends on gear accuracy and tangential force, and is considered by transverse load factor $K_{F\alpha}$ ($K_{H\alpha}$) which is calculated as a function of the total contact ratio, tooth stiffness, pitch error and total circumferential force (Table 4.18).

Table 4.18 Approximate values for transverse load factors $K_{F\alpha}$ and $K_{H\alpha}$

Designation	Formula		No.
Transverse load factor	$K_{F\alpha} = K_{H\alpha} = K_{H\alpha}{}^* - \frac{K_{H\alpha}{}^* - 1}{0.1} a_{rel} \quad \geq 1$		(4.118)
	where: a_{rel} according to Table 4.14		
	For $\varepsilon_{v\gamma} \leq 2$: $K_{H\alpha}{}^* = \frac{\varepsilon_{v\gamma}}{2}\left(0.9 + 0.4\frac{c_\gamma\left(f_{pt} - y_\alpha\right)}{F_{mtH}/b_v}\right)$		(4.119)
	For $\varepsilon_{v\gamma} > 2$: $K_{H\alpha}{}^* = 0.9 + 0.4\sqrt{\frac{2\left(\varepsilon_{v\gamma} - 1\right)}{\varepsilon_{v\gamma}} \cdot \frac{c_\gamma\left(f_{pt} - y_\alpha\right)}{F_{mtH}/b_v}}$		(4.120)
	With: $F_{mtH} = F_{mtv}K_A K_v K_{H\beta}$		(4.121)
Boundary condition	$1 < K_{F\alpha}{}^* < \frac{\varepsilon_{v\gamma}}{\varepsilon_{v\alpha}Y_{LS}}$	Limits for $K_{F\alpha}$	(4.122)
	$1 < K_{H\alpha}{}^* < \frac{\varepsilon_{v\gamma}}{\varepsilon_{v\alpha}Z_{LS}{}^2}$	Limits for $K_{H\alpha}$	(4.123)

4.2.4.2 Permissible Tooth Root Stress

The permissible tooth root stress is calculated separately for the pinion and wheel on the basis of a strength value determined on a cylindrical test gear (Table 4.19). Strength values for different materials can be found in [ISO6336] Part 5.

Table 4.19 Calculation of permissible tooth root stress $\sigma_{FP\,1,2}$

Designation	Formula	No.
Permissible tooth root stress	$\sigma_{FP1,2} = \frac{\sigma_{FE1,2} Y_{NT1,2}}{S_{F\min 1,2}} Y_{\delta relT1,2} Y_{RrelT1,2} Y_{X1,2}$	(4.124)
	$\sigma_{FP1,2} = \frac{\sigma_{F\lim 1,2} Y_{ST1,2} Y_{NT1,2}}{S_{F\min 1,2}} Y_{\delta relT1,2} Y_{RrelT1,2} Y_{X1,2}$	(4.125)
	$\sigma_{FE1,2}$ = Allowable bending stress number (fatigue strength of the un-notched specimen $\sigma_{FE} = \sigma_{Flim} \cdot Y_{ST} = \sigma_{Flim} \cdot 2.0$)	(4.126)
	$\sigma_{F\lim 1,2}$ = Nominal bending stress number considering material, heat treatment and surface influence of the test gears	
	$Y_{ST1,2}$ = stress correction factor for test gear dimensions, $Y_{ST1,2} = 2.0$ for test gears acc. [ISO6336] Part 5	(4.127)
	$S_{F\min 1,2}$ = required minimum safety factor	
	$Y_{\delta relT1,2}$ = Relative notch sensitivity factor in relation to test gear conditions, to consider the notch sensitivity of the material	
	$Y_{RrelT1,2}$ = Relative surface condition factor in relation to test gear conditions, to consider the surface influence in the actual fillet	
	$Y_{X1,2}$ = Size factor, to consider the influence of the module on the permissible tooth root stress	
	$Y_{NT1,2}$ = Life factor, to consider the influence of the required life time (number of load cycles)	

The nominal bending stress number σ_{Flim} may be taken from ISO6336 Part 5. Figure 4.19 shows the graph for case hardened steels and flame and induction hardened heat treatable steels as an example.

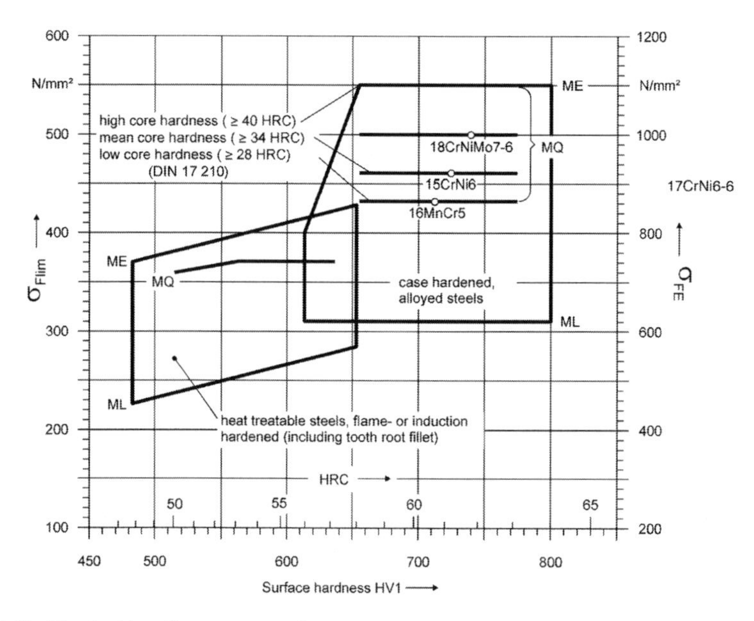

Fig. 4.19 Nominal bending stress number σ_{Flim}

Relative notch sensitivity factor $Y_{\delta relT1,2}$ Factor $Y_{\delta relT}$ is the ratio of the notch sensitivity factor Y_{δ} of the gear under design compared to that of the standard test gear $Y_{\delta T}$, which specifies the amount by which the theoretical stress at fatigue breakage exceeds the fatigue strength. $Y_{\delta relT}$ is a function of the material and of the stress gradient concerned, and can be approximated using the following values (Table 4.20):

Table 4.20 Approximate values for relative notch sensitivity factor $Y_{\delta relT1,2}$

Designation	Formula	No.
Relative notch sensitivity factor	$Y_{\delta relT1,2} = 1.0 \quad \text{for } q_{s1,2} \geq 1.5$	(4.128)
	$Y_{\delta relT1,2} = 0.95 \quad \text{for } q_{s1,2} < 1.5$	(4.129)
	with $q_{s1,2}$ according to Table 4.8	

Relative surface condition factor $Y_{RrelT1,2}$ The relative surface factor Y_{RrelT} considers the influence of the surface quality, especially roughness in the tooth root, on the permissible tooth root stress. It is a function of the material and may be determined as follows (Table 4.21):

Table 4.21 Calculation of the relative surface condition factor $Y_{RrelT1,2}$

Designation	Formula	No.
Relative surface condition factors for $R_z < 1\ \mu m$	$Y_{RrelT1,2} = 1.12$ for heat-treated and case hardened steels $Y_{RrelT1,2} = 1.07$ for constructional steels $Y_{RrelT1,2} = 1.025$ for grey cast iron and nitrocarburized steels	(4.130)
Relative surface condition factors for $1\ \mu m < R_z < 40\ \mu m$	$Y_{RrelT1,2} = 1.674 - 0.529(R_{z1,2} + 1)^{1/10}$ for heat treated and case hardened steels	(4.131)
	$Y_{RrelT1,2} = 5.306 - 4.203(R_{z1,2} + 1)^{1/100}$ for constructional steels	(4.132)
	$Y_{RrelT1,2} = 4.299 - 3.259(R_{z1,2} + 1)^{1/200}$ for grey cast iron and nitrocarburized steels	(4.133)

Size factor $Y_{X1,2}$ Tooth root strength decreases as tooth size increases. The size factor Y_X considers this influence. It is calculated as a function of the mean normal module and the material (Table 4.22):

Table 4.22 Calculation of size factor $Y_{X1,2}$

Designation	Formula		No.
Size factor	$Y_{X1,2} = 1.03 - 0.006\,m_{mn}$ with $0.85 \leq Y_{X1,2} \leq 1.0$	For construction, heat treatable steel, spheroidal cast iron, and black malleable cast iron	(4.134)
	$Y_{X1,2} = 1.05 - 0.01\,m_{mn}$ with $0.80 \leq Y_{X1,2} \leq 1.0$	For flame, induction and case hardened, nitrided and nitrocarburized steels	(4.135)
	$Y_{X1,2} = 1.075 - 0.015\,m_{mn}$ with $0.70 \leq Y_{X1,2} \leq 1.0$	For grey cast iron	(4.136)

Life factor $Y_{NT1,2}$ For a required life time (number of load cycles N_L) which is below the endurance limit for bending fatigue at $N_L = 3 \cdot 10^6$ load cycles, the permissible tooth root stress increases in accordance with the SN-curve. ($N_L = 1$: usually means one revolution of the gear concerned). This effect is considered by life factor Y_{NT}. Y_{NT} depends principally on the material employed and on the heat treatment (Fig. 4.20).

Under optimum conditions with regard to material and manufacturing quality, $Y_{NT} = 1$ may be expected for load cycles numbers $N_L > 3 \cdot 10^6$. In the static stress range ($N_L < 10^3$), the life factor is set at 2.5. Interpolations are made between these two values for static and endurance strength.

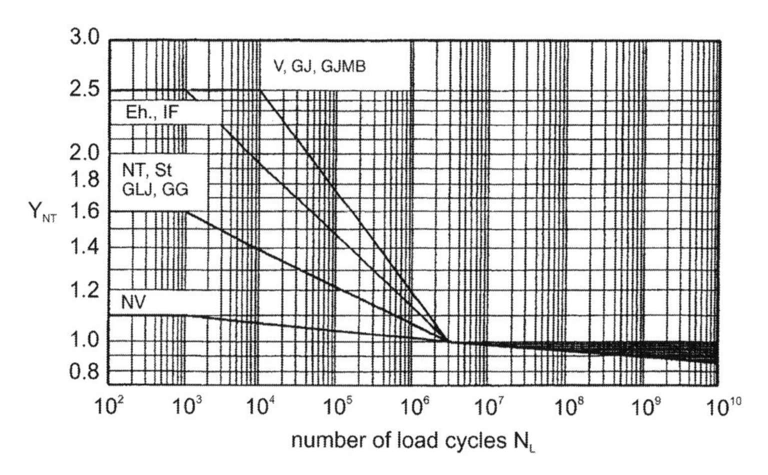

Fig. 4.20 Life factor Y_{NT} for tooth bending. GJL, grey cast iron; GJS, spheroidal cast iron; GJMB, black malleable cast iron; St, normalized low carbon steel; V, through hardened wrought steel; Eh, case hardened steel; IF, induction or flame hardened steel; NV(nitr.), nitrided heat treatable or case hardening steel; NT(nitr.), nitrided nitriding steel; NV(nitrocar.), nitrocarburized heat treatable or case hardening steel

4.2.4.3 Safety Factor Against Tooth Root Breakage

The safety factor for tooth root breakage is calculated using the values for tooth root stress and permissible tooth root stress determined in Sects. 4.2.4.1 and 4.2.4.2 (Table 4.23):

Table 4.23 Calculating the safety factor against tooth root breakage

Designation	Formula	No.
Safety factor (tooth root breakage)	$S_{F1,2} = \frac{\sigma_{FP\,1,2}(S_{F,\min 1,2} = 1)}{\sigma_{F1,2}} > S_{F,\min 1,2}$	(4.137)

The required minimum safety factor should be agreed between supplier and customer. [ISO10300] gives reference values of $S_{F,min} = 1.3$ for spiral bevel gears and $S_{F,min} = 1.5$ for straight and helical bevel gears.

4.2.5 Pitting Load Capacity

In order to calculate the pitting load capacity of a bevel gear set, the decisive contact stress is determined and is compared to the strength values derived from standard test gears (strength values in [ISO6336]).

4.2.5.1 Contact Stress

The contact stress σ_H depends mainly on the tooth geometry, accuracy of production, and operating conditions. The calculation method in [ISO10300] is based on the Hertzian stress theory. The determinant point of load application depends on the overlap ratio $\varepsilon_{v\beta}$ (Table 4.24):

- for $\varepsilon_{v\beta} = 0$: calculation at the inner point of single contact B
- for $\varepsilon_{v\beta} > 1$: calculation at the mid-point of the zone of action M
- for $0 < \varepsilon_{v\beta} < 1$: interpolation between the inner point B and the mid-point M

Table 4.24 Calculation of contact stress σ_H

Designation	Formula	No.
Contact stress	$\sigma_H = \sigma_{H0}\sqrt{K_A K_v K_{H\beta} K_{H\alpha}} =$	(4.138)
	$= \sqrt{\frac{F_n}{l_{bm}\rho_{ers}}} Z_{M-B} Z_E Z_{LS} Z_K \sqrt{K_A K_v K_{H\beta} K_{H\alpha}}$	(4.139)
	where: σ_{H0} = nominal stress	
	$F_n = \frac{F_{mt1}}{\cos\alpha_n \cos\beta_{m1}}$ = nominal normal load at mean point P	(4.140)
	l_{bm} = mean length of contact line according to Table 4.10 with f_m according to Table 4.11	
	ρ_{ers} = decisive equivalent curvature radius according to Table 4.2	
	Z_{M-B} = mid-zone factor; converts curvature parameters to the decisive point of load application	
	Z_{LS} = load sharing factor, which accounts for the number of simultaneously meshing teeth	
	Z_E = elasticity factor, which accounts for material properties Z_K = bevel gear factor: $Z_K = 0.85$ using [ISO10300] Method B1, whereas $Z_K = 1$ using [FVA411] (see Sect. 4.2.5.2, Bevel slip factor $Z_{S1,2}$)	
	K_A = application factor, which accounts for additional external loads resulting from operating conditions	
	K_v = dynamic factor, which accounts for additional internal dynamic loads	
	$K_{H\beta}$ = face load factor, which accounts for non-uniform load distribution over the face width	
	$K_{H\alpha}$ = load sharing factor, which accounts for non-uniform load sharing between meshing tooth pairs	

Load factors K_A, K_v, $K_{H\beta}$ and $K_{H\alpha}$ are determined as in Tables 4.13, 4.14, 4.15, 4.16, 4.17, and 4.18

Mid-zone factor $Z_{M\text{-}B}$ The mid-zone factor $Z_{M\text{-}B}$ (Tables 4.25 and 4.26) is used to modify the curvature parameters of the pitch point to the determinant point of load application as a function of the overlap ratio $\varepsilon_{v\beta}$. Assuming an involute tooth profile, the following may be stated for gears without profile shift and for zero shift gears.

Table 4.25 Calculation of mid-zone factor Z_{M-B}

Designation	Formula	No.
Mid-zone factor	$Z_{M-B} = \frac{\tan\alpha_{vet}}{\sqrt{\left[\sqrt{\left(\frac{d_{va1}}{d_{vb1}}\right)^2 - 1} - F_1\frac{\pi}{z_{v1}}\right]\cdot\left[\sqrt{\left(\frac{d_{va2}}{d_{vb2}}\right)^2 - 1} - F_2\frac{\pi}{z_{v2}}\right]}}$ see Table 4.26 for auxiliary factors F_1 and F_2	(4.141)

Table 4.26 Factors to calculate mid-zone factor Z_{M-B}

Factor	F_1	F_2
$\varepsilon_{v\beta} = 0$	2	$2\,(\varepsilon_{v\alpha} - 1)$
$0 < \varepsilon_{v\beta} < 1$	$2 + (\varepsilon_{v\alpha} - 2)\,\varepsilon_{v\beta}$	$2\varepsilon_{v\alpha} - 2 + (2 - \varepsilon_{v\alpha})\,\varepsilon_{v\beta}$
$\varepsilon_{v\beta} \geq 1$	$\varepsilon_{v\alpha}$	$\varepsilon_{v\alpha}$

Elasticity factor Z_E Material-specific influences on Hertzian contact stress, such as the modulus of elasticity and Poisson's ratio, are considered by using the elasticity factor Z_E (Table 4.27):

Table 4.27 Calculation of the elasticity factor Z_E

Designation	Formula		No.
Elasticity factor	$Z_E = \sqrt{\dfrac{1}{\pi\left(\dfrac{1-\nu_1{}^2}{E_1} + \dfrac{1-\nu_2{}^2}{E_2}\right)}}$		(4.142)
For $E_1 = E_2 = E$ and $\nu_1 = \nu_2 = \nu$:	$Z_E = \sqrt{\dfrac{E}{2\pi\,(1-\nu^2)}}$	For steel/steel pair: $Z_E = 189.8$	(4.143)

Load sharing factor Z_{LS} The load sharing over several meshing tooth pairs is considered by the factor Z_{LS} as calculated in Table 4.28. The distances *f* of contact lines for pitting load capacity are calculated in Table 4.29.

Table 4.28 Calculation of load sharing factor Z_{LS}

Designation	Formula		No.
Load sharing factor	$Z_{LS} = \sqrt{Y_{LS}}$	use Y_{LS} formulae according to Table 4.10, but with *f* values according to Table 4.29	(4.144)

Table 4.29 Variable *f* to calculate the distances of contact lines from mid-point M for pitting load capacity

Designation	Formula	No.
$\varepsilon_{v\beta} = 0$	$f_t = -(p_{vet} - 0.5p_{vet}\varepsilon_{v\alpha})\cos\beta_{vb} + p_{vet}\cos\beta_{vb}$ $f_m = -(p_{vet} - 0.5p_{vet}\varepsilon_{v\alpha})\cos\beta_{vb}$ $f_r = -(p_{vet} - 0.5p_{vet}\varepsilon_{v\alpha})\cos\beta_{vb} - p_{vet}\cos\beta_{vb}$	(4.145)
$0 < \varepsilon_{v\beta} < 1$	$f_t = -(p_{vet} - 0.5p_{vet}\varepsilon_{v\alpha})\cos\beta_{vb}(1 - \varepsilon_{v\beta}) + p_{vet}\cos\beta_{vb}$ $f_m = -(p_{vet} - 0.5p_{vet}\varepsilon_{v\alpha})\cos\beta_{vb}(1 - \varepsilon_{v\beta})$ $f_r = -(p_{vet} - 0.5p_{vet}\varepsilon_{v\alpha})\cos\beta_{vb}(1 - \varepsilon_{v\beta}) - p_{vet}\cos\beta_{vb}$	(4.146)
$\varepsilon_{v\beta} \geq 1$	$f_t = +p_{vet}\cos\beta_{vb}$ $f_m = 0$ $f_r = -p_{vet}\cos\beta_{vb}$	(4.147)

4.2.5.2 Permissible Contact Stress

The permissible loading of the tooth flanks depends not only on the value of the contact stress, but also on the lubricating conditions. Thus, the influences of lubricant viscosity, rotation speed and tooth flank roughness are considered by estimation factors. However, the scatter of these parameters demonstrates that not yet all influences are accounted for. The specific behavior of a lubricant for pitting load capacity may therefore be examined in a FZG pitting test [FVA2]. The calculation method according to [FVA411] additionally employs hypoid factor Z_{Hyp} to consider the negative effect of lengthwise sliding on the formation of the lubricant film. Unlike [ISO10300], FVA 411 as well employs slip factor Z_S to consider the effect of positive specific sliding in hypoid gears (Table 4.30).

Table 4.30 Calculation of the permissible contact stress σ_{HP}

Designation	Formula	No.
Permissible contact stress	$\sigma_{HP1,2} = \frac{\sigma_{H\lim 1,2} Z_{NT1,2}}{S_{H\min 1,2}} Z_{X1,2} Z_L Z_R Z_v Z_W Z_{S1,2} Z_{Hyp}$	(4.148)
	$\sigma_{H\lim 1,2}$ = allowable stress number for contact stress	
	$S_{H\min 1,2}$ = required minimum safety factor	
	$Z_{X1,2} = 1$ = size factor	
	Z_L = lubricant factor	
	Z_R = roughness factor	
	Z_v = speed factor	
	Z_W = work hardening factor	
	$Z_{NT1,2}$ = life factor	
	$Z_{S1,2}$ = bevel slip factor	
	Z_{Hyp} = hypoid factor	

The allowable stress number for contact stress $\sigma_{H\lim}$ can be taken from [ISO6336] Part 5. The diagram for case hardened alloyed steels and flame and induction hardened heat treatable steels is shown as an example in Fig. 4.21.

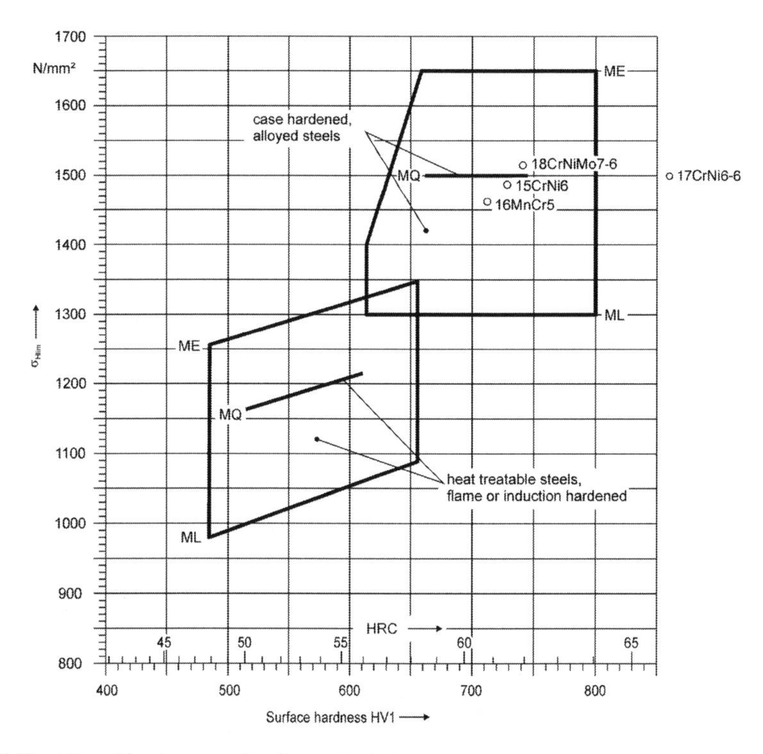

Fig. 4.21 Allowable stress number for contact stress σ_{Hlim}

Size factor Z_X The influence of gear size on the permissible contact stress is considered by size factor Z_X. The influence of gear size stems from statistical features like the distribution of imperfections in the microstructure of the material, from strength theory like smaller stress gradients with larger dimensions, and from technological aspects in the homogeneity of the microstructure. Z_X must be determined separately for the pinion and the wheel. In the absence of detailed information, the size factor is set at $Z_X = 1$.

Influence factors for lubricant film formation Z_L, Z_R, Z_v The influence factors for the lubricant film formation can be determined according to Tables 4.31, 4.32, and 4.33. If different materials are used for the pinion and the wheel, the calculation is made for the softer member.

Table 4.31 Calculation of lubricant factor Z_L

Designation	Formula	No.
Lubricant factor	$Z_L = C_{ZL} + \dfrac{4 \cdot (1.0 - C_{ZL})}{\left(1.2 + \dfrac{134}{v_{40}}\right)^2}$	(4.149)
	where: $C_{ZL} = 0.08 \dfrac{\sigma_{H\lim} - 850}{350} + 0.83$ $850\ \mathrm{N/mm^2} < \sigma_{\mathrm{Hlim}} < 1200\ \mathrm{N/mm^2}$	(4.150)

Table 4.32 Calculation of roughness factor Z_R

Designation	Formula	No.
Roughness factor	$Z_R = \left(\frac{3}{Rz_{10}}\right)^{C_{ZR}}$	(4.151)
	where: $Rz_{10} = \frac{Rz_1 + Rz_2}{2} \sqrt[3]{\frac{10}{\rho_{red}}}$	(4.152)
	$\rho_{red} = \frac{a_v \sin \alpha_{vt}}{\cos \beta_{vb}} \frac{u_v}{(1+u_v)^2}$	(4.153)
	$C_{ZR} = 0.12 + \frac{1000 - \sigma_{H\lim}}{5000}$ $850\ \text{N/mm}^2 < \sigma_{H\lim} < 1200\ \text{N/mm}^2$	(4.154)

Table 4.33 Calculation of speed factor Z_v

Designation	Formula	No.
Speed factor	$Z_v = C_{ZV} + \frac{2 \cdot (1.0 - C_{ZV})}{\sqrt{0.8 + \frac{32}{v_{mt2}}}}$	(4.155)
	where: $C_{ZV} = 0.08 \frac{\sigma_{H\lim} - 850}{350} + 0.85$ $850\ \text{N/mm}^2 < \sigma_{H\lim} < 1200\ \text{N/mm}^2$	(4.156)
	$v_{mt2} = \frac{d_{m2} n_2}{19098}$	(4.157)

As an approximation, the following values for the product of Z_L, Z_R and Z_v may be used:	
for heat-treated, milled gears	$Z_L \cdot Z_R \cdot Z_v = 0.85$
for milled and lapped gears	$Z_L \cdot Z_R \cdot Z_v = 0.92$
for hardened and ground or hard skived gears	$Z_L \cdot Z_R \cdot Z_v = 1.0$ ($Rz_{10} < 4$ μm) $Z_L \cdot Z_R \cdot Z_v = 0.92$ ($Rz_{10} > 4$ μm)

Work hardening factor Z_W When different materials are used for a pinion and a wheel which are of different hardnesses, pitting resistance of the softer member tends to rise due to work hardening, smoothing, and other influences. Provision is made that the harder member has a sufficiently smooth surface ($Rz < 6$ μm) to minimize the risk of wear. This effect is considered by the work hardening factor Z_W (Table 4.34).

Table 4.34 Calculation of work hardening factor Z_W

Designation	Formula	No.
Work hardening factor	$Z_W = 1.2 - \frac{HB - 130}{1700}$ where: HB = Brinell hardness of the softer member only $Z_W = 1$ for the harder member $Z_W = 1$ if pinion and wheel are of the same hardness $Z_W = 1.2$ for $HB < 130$; $Z_W = 1.0$ for $HB > 470$	(4.158)

Bevel slip factor $Z_{S1,2}$ The slip factor considers the influence of specific sliding of the tooth flanks on the permissible contact stress. It is determined as a function of the mid-zone factor $Z_{M\text{-}B}$. Linear interpolation of Z_S is applied for $0.98 < Z_{M\text{-}B} < 1.0$. In [ISO10300] the bevel slip factor is set to 1.0 whereas a bevel gear factor $Z_K = 0.85$ is introduced in the calculation of the contact stress (Table 4.24). This usually leads to similar safety factors for the pinion, but higher safety factors for the wheel (Table 4.35).

Table 4.35 Calculation of bevel slip factor Z_S

Designation	Formula	No.
Slip factor	$Z_{S1} = 1.175; \quad Z_{S2} = 1.0 \quad \text{for } Z_{M-B} < 0.98$	(4.159)
	$Z_{S1} = 1.0; \quad Z_{S2} = 1.175 \quad \text{for } Z_{M-B} > 1.0$	(4.160)

Life factor $Z_{NT1,2}$ The curves for life factors Z_{NT} are shown for various materials in Fig. 4.22.

For a required life time (in load cycles N_L) below the endurance limit for contact stress, set at $N_L = 5 \cdot 10^7$ load cycles for case hardened gears, the permissible contact stress increases in accordance with the SN-curve. This is considered by life factor Z_{NT} which depends mainly on the material employed and on the heat treatment. $Z_{NT} = 1$ may be expected for load cycles $N_L > 5 \cdot 10^7$, provided optimum conditions in terms of material and manufacturing quality are met. In the static stress range ($N_L < 10^3$), the life factor is set at 1.6. Interpolations are made between these two values for the static and endurance strengths.

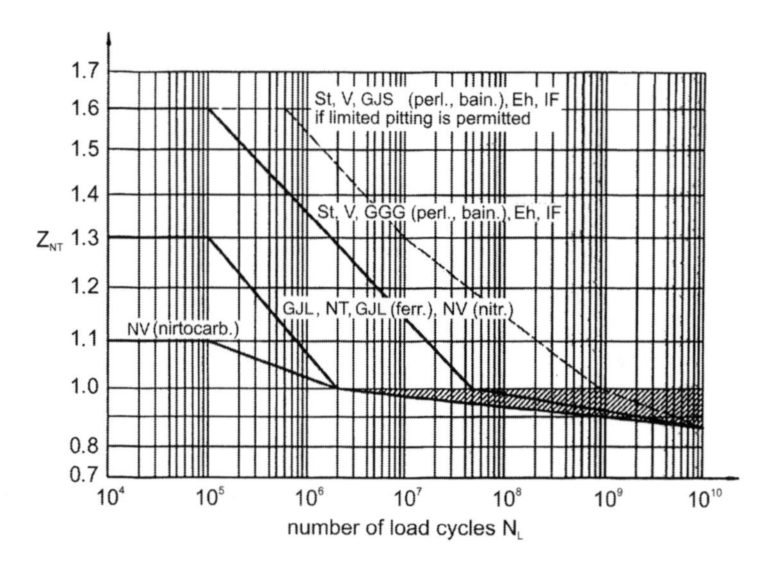

Fig. 4.22 Life factor Z_{NT} for pitting (see Fig. 4.20 for abbreviations)

Hypoid factor Z_{Hyp} The hypoid factor Z_{Hyp} considers the effects of the offset-induced lengthwise sliding component in hypoid gears, when compared to bevel gears, in reducing load capacity. A value of $Z_{Hyp} = 1$ applies for bevel gears (Table 4.36).

Table 4.36 Calculation of hypoid factor Z_{Hyp}

Designation	Formula	No.
Hypoid factor	$Z_{Hyp} = 1 - 0.3\left(\frac{v_{g,par}}{v_{\Sigma,senk}} - 0.15\right)$ $v_{g,par}$ = sliding velocity parallel to the contact line $v_{\Sigma,senk}$ = sum of velocities perpendicular to the contact line (all variables are calculated at the design point)	(4.161)
Sum of velocities perpendicular to the contact line	$v_{\Sigma senk} = v_{\Sigma m} \sin\left(v_{Bel} + \lvert w_{Bel} \rvert\right)$	(4.162)
	where: $v_{\Sigma m} = \sqrt{{v_{\Sigma h}}^2 + {v_{\Sigma s}}^2}$	(4.163)
	$v_{\Sigma h} = \lvert 2 v_{mt1} \cos\beta_{m1} \sin\alpha_n \rvert$	(4.164)
	$v_{\Sigma s} = \left\lvert v_{mt1}\left(\sin\beta_{m1} + \frac{\sin\beta_{m2}\cos\beta_{m1}}{\cos\beta_{m2}}\right)\right\rvert$	(4.165)
	$v_{Bel} = \lvert \arctan\left(v_{\Sigma h} \;/\; v_{\Sigma s}\right) \rvert$ w_{Bel} according to Table 4.2	(4.166)
Sliding velocity parallel to the contact line	$v_{g,par} = v_g \cos \lvert w_{Bel} \rvert$	(4.167)
(For $\Sigma = 90°$)	$v_g = \sqrt{\left(v_{mt1}\sin\phi_1\right)^2 + \left(v_{mt2}\sin\phi_2\right)^2 + \left(v_{mt1}\cos\phi_1 - v_{mt2}\cos\phi_2\right)^2}$	(4.168)
Auxiliary quantities to determine the sliding velocity	$\phi_1 = \arcsin\left(2h_1/d_{m1}\right)$ $\quad h_1 = \frac{d_{m1}/\cos\delta_1}{d_{m1}/\cos\delta_1 + d_{m2}/\cos\delta_2} \cdot a$	(4.169)
	$\phi_2 = \arcsin\left(2h_2/d_{m2}\right)$ $\quad h_2 = a - h_1$	(4.170)

4.2.5.3 Safety Factor Against Pitting

The safety factor against pitting is calculated using the values for contact stress and permissible contact stress as determined in Sects. 4.2.5.1 and 4.2.5.2. The ratio of permissible contact stress to calculated contact stress represents the safety factor in terms of acceptable contact pressure. To quantify the permissible torque, the square of the safety factor is to be used. The required minimum safety factor should be agreed between supplier and customer. [ISO10300] gives $S_{H,min} = 1.0$ as a reference value (Table 4.37).

Table 4.37 Calculation of resistance to pitting

Designation	Formula	No.
Safety factor against pitting	$S_{H1,2} = \frac{\sigma_{HP1,2}(for\, S_{H,\min 1,2} = 1)}{\sigma_H} > S_{H,\min 1,2}$	(4.171)

4.2.6 *Scuffing Load Capacity*

Two methods are available to calculate the scuffing load capacity of gears. The contact temperature method is based on the flash temperature, which varies along the path of contact. The integral temperature method calculates a weighted mean contact temperature based on the flash temperature. In both cases, the calculated temperature is compared to a permissible temperature determined in a scuffing test with standard test gears.

4.2.6.1 Contact Temperature Method for Non-offset Bevel Gears

For the calculation of scuffing load capacity, non-offset bevel gears are approximated by virtual cylindrical gears as described in Sect. 4.2.2.

Operational contact temperature The maximum contact temperature in the mesh, $\vartheta_{\mathrm{Bmax}}$, consists of the bulk temperature ϑ_{M} and the maximum flash temperature $\vartheta_{\mathrm{flmax}}$ (the maximum value of the flash temperature ϑ_{fl} along the path of contact). Determination of the flash temperature is based on [BLOK67] (Fig. 4.23 and Tables 4.38 and 4.39).

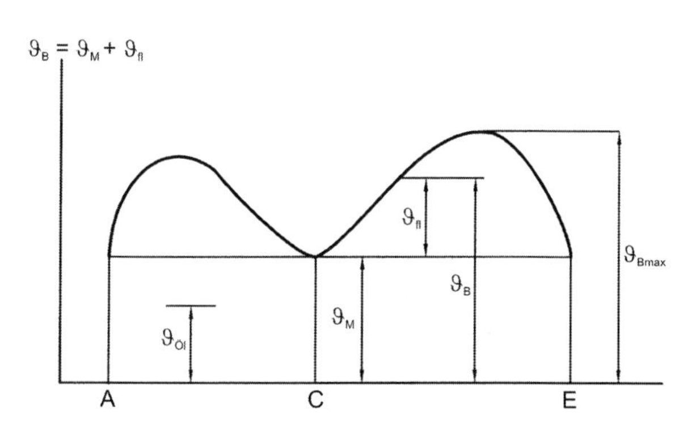

Fig. 4.23 Progression of the contact temperature ϑ_{B} along the path of contact

Table 4.38 Calculation of the maximum contact temperature $\vartheta_{\mathrm{Bmax}}$

Designation	Formula	No.
Maximum contact temperature	$\vartheta_{\mathrm{Bmax}} = \vartheta_M + \vartheta_{fl\mathrm{max}}$	(4.172)

Table 4.39 Calculation of the flash temperature ϑ_{fl} for bevel gears without offset

Designation	Formula	No.
Flash temperature	$\vartheta_{fl} = \mu_m \cdot X_M \cdot X_J \cdot X_G \cdot (X_\Gamma \cdot w_{Bt})^{3/4} \cdot \frac{v_{mt}^{1/2}}{R_m^{1/4}}$	(4.173)
	w_{Bt} = load per unit face width (unit load) including load factors	
	X_Γ = load sharing factor	
	$v_{mt} = v_{mt1,2} = \frac{\pi d_{m1,2} n_{1,2}}{1000 \cdot 60}$	(4.174)
	R_m = pitch cone distance	
	μ_m = mean coefficient of friction	
	X_M = flash factor	
	X_J = mesh approach factor (contact temperature method)	
	X_G = geometry factor	

Load per unit face width w_{Bt} The unit load w_{Bt} represents the tangential force in relation to the effective face width, considering the load factors according to [ISO10300] (Table 4.40).

Table 4.40 Calculation of the unit load w_{Bt}

Designation	Formula	No.
Unit load including load factors	$w_{Bt} = K_A K_v K_{B\beta} K_{B\alpha} K_{mp} F_{mt} / b_{2eff}$	(4.175)
	where: K_A, K_v from Table 4.13 to 4.15 $K_{B\beta}$ and $K_{B\alpha}$ are the load factors for scuffing, if no better knowledge is available use $K_{H\beta}$ and $K_{H\alpha}$ from Table 4.16 to 4.18 K_{mp} = 1 number of mates meshing with the actual gear	
	$F_{mt} = F_{mt1} = F_{mt2} = \frac{2000 \cdot T_{1,2}}{d_{m1,2}}$ b_{2eff} = effective contact pattern width on the wheel (see Table 4.2, deviating from [ISO/TR13989])	(4.176)

Load sharing factor X_Γ The load sharing factor considers the distribution of the total tangential load between the meshing tooth pairs. It is calculated according to Table 4.41 as a function of dimensionless parameter Γ_y on the path of contact.

Table 4.41 Calculation of load sharing factor X_Γ (scuffing)

Designation	Formula	No.
Load sharing factor	$X_\Gamma = \frac{1.5}{\varepsilon_\alpha} - \frac{(\Gamma_Y - \Gamma_M)^2}{(\Gamma_E - \Gamma_A)^2} \frac{6}{\varepsilon_{\alpha v}}$	(4.177)
	where: Γ_Y = dimensionless parameter on the path of contact at the considered contact point Y $\Gamma_Y = \left(\frac{\tan \alpha_{y1}}{\tan \alpha_t} - 1\right)$	(4.178)

(continued)

Table 4.41 (continued)

Designation	Formula	No.
	$\Gamma_A = -\frac{\tan \delta_2}{\tan \delta_1}\left(\frac{\tan \alpha_{a2}}{\tan \alpha_t} - 1\right)$	(4.179)
	$\Gamma_B = \frac{\tan \alpha_{a1}}{\tan \alpha_t} - 1 - \frac{2\pi \cos \delta_1}{z_1 \tan \alpha_t}$	(4.180)
	$\Gamma_D = -\frac{\tan \delta_2}{\tan \delta_1}\left(\frac{\tan \alpha_{a2}}{\tan \alpha_t} - 1\right) + \frac{2\pi \cos \delta_1}{z_1 \tan \alpha_t} a_v$	(4.181)
	$\Gamma_E = \left(\frac{\tan \alpha_{a1}}{\tan \alpha_{wt}} - 1\right)$	(4.182)
	$\Gamma_M = \frac{\Gamma_A + \Gamma_E}{2}$	(4.183)
	$\tan \alpha_{a1} = \sqrt{\left(\frac{\cos \alpha_t}{1 + 2h_{am1} \cos \delta_1 / d_{m1}}\right)^2 - 1}$	(4.184)
	$\tan \alpha_{a2} = \sqrt{\left(\frac{\cos \alpha_t}{1 + 2h_{am2} \cos \delta_2 / d_{m2}}\right)^2 - 1}$	(4.185)

Mean coefficient of friction of the gear μ_m The mean coefficient of friction, μ_m, relevant to scuffing load capacity, is calculated according to Table 4.42 as a function of the unit load, the sum of velocities and the equivalent curvature radius combined with the lubricant properties and tooth surface roughness.

Table 4.42 Calculation of mean coefficient of friction μ_m

Designation	Formula	No.
Mean coefficient of friction	$\mu_m = 0.060 \cdot \left(\frac{w_{Bt}}{v_{\Sigma C} \rho_{redC}}\right)^{0.2} \cdot \eta_{oil}{}^{-0.05} X_L X_R$	(4.186)
	where: w_{Bt} = unit load including load factors according to Table 4.40	
	$v_{\Sigma C} = 2v_{mt} \sin \alpha_{vt}$ = sum of velocities at pitch point C	(4.187)
Equivalent curvature radius at point C	$\rho_{redC} = \frac{\rho_{C1} \rho_{C2}}{\rho_{C1} + \rho_{C2}}$	(4.188)
	$\rho_{C1} = R_m \tan \delta_1 \sin \alpha_t$	(4.189)
	$\rho_{C2} = R_m \cdot u \cdot \tan \delta_1 \sin \alpha_t$	(4.190)
	η_{oil} = viscosity at oil temperature	
	$X_R = \left(\frac{Ra_1 + Ra_2}{2}\right)^{0.25}$ = roughness factor	(4.191)
	X_L = lubricant factor 1.0 for mineral oils 0.6 for water-soluble polyglycols 0.7 for non-water-soluble polyglycols 0.8 for polyalphaolefins 1.3 for phosphate esters 1.5 for traction fluids	

Flash factor X_M The flash factor X_M considers the influence of the pinion and wheel materials on the flash temperature. It contains the modulus of elasticity, Poisson's ratio and the thermal properties of the materials (Table 4.43).

Table 4.43 Calculation of flash factor X_M

Designation	Formula		No.
Flash factor	$X_M = \frac{E^{\frac{1}{4}}}{(1-\nu^2)^{\frac{1}{4}} B_M}$	If $E_1 = E_2$ and $\nu_1 = \nu_2$	(4.192)
	where: $B_M = \sqrt{0.001 \cdot \lambda_M \ \rho_M \ c_M}$ λ_M = thermal conductivity ρ_M = density c_M = specific thermal capacity per unit of mass		(4.193)
	For common steels:	$X_M = 50.0 \frac{K\, s^{\frac{1}{2}}\, mm}{N^{\frac{3}{4}}\, m^{\frac{1}{2}}}$	(4.194)

Mesh approach factor X_J This factor X_J considers the negative effect at contact entry, where no oil film is built in a zone of high sliding at the tip of the driven gear (Table 4.44).

Table 4.44 Calculation of mesh approach factor X_J

Designation	Formula		No.
Mesh approach factor for "pinion driving wheel"	$X_J = 1$	at $\Gamma_Y \geq 0$	(4.195)
	$X_J = 1 + \frac{C_{eff} - C_{a2}}{50} \left(\frac{-\Gamma_Y}{\Gamma_E - \Gamma_A} \right)^3$	at $\Gamma_Y < 0$	(4.196)
Mesh approach factor for "wheel driving pinion"	$X_J = 1$	at $\Gamma_Y \leq 0$	(4.197)
	$X_J = 1 + \frac{C_{eff} - C_{a1}}{50} \left(\frac{\Gamma_Y}{\Gamma_E - \Gamma_A} \right)^3$	at $\Gamma_Y > 0$	(4.198)
	$C_{eff} = \frac{K_A K_{mp} F_t}{b \cos \alpha_t \ c_\gamma}$		(4.199)
	where: $K_{mp} = 1$ number of mates meshing with the actual gear		
	$c_\gamma = 20$ N/(mm · μm) = approximate value for mesh stiffness according to [ISO10300] Part 1		
	$C_{a\,1,2}$ = amount of tip relief pinion/wheel		

Geometry factor X_G The geometry factor calculates the Hertzian stress and sliding velocity at the tip of the pinion tooth (Table 4.45).

Table 4.45 Calculation of geometry factor X_G

Designation	Formula	No.
Geometry factor	$X_G = 0.51 \cdot X_{\alpha\beta} \cdot (\cot\delta_1 + \cot\delta_2)^{1/4} \cdot \frac{\left|\sqrt{1+\Gamma_y} - \sqrt{1+\Gamma_y \cdot \tan\delta_1/\tan\delta_2}\right|}{(1+\Gamma_y)^{1/4} \cdot (1-\Gamma_y \cdot \tan\delta_1/\tan\delta_2)^{1/4}}$ where: $X_{\alpha\beta}$ = influence factor of pressure and spiral angle; $X_{\alpha\beta} = 1$ approximate value for gears with a normal pressure angle of $\alpha_n = 20°$	(4.200)

Bulk temperature ϑ_M The maximum contact temperature is given by the maximum flash temperature ϑ_{flmax} and the bulk temperature ϑ_M, which can be calculated as described in Table 4.46.

Table 4.46 Calculation of bulk temperature ϑ_M

Designation	Formula	No.
Bulk temperature	$\vartheta_M = \vartheta_{oil} + 0.47 \cdot X_S X_{mp} \vartheta_{flm}$	(4.201)
	where: X_S = lubrication factor 1.0 for dip lubrication; 1.2 for injection lubrication; 0.2 for dip lubrication under optimum cooling conditions	
	$X_{mp} = \frac{1+n_p}{2}$ (n_p = number of wheels meshing with one pinion)	(4.202)
	$\vartheta_{flm} = \frac{1}{(\Gamma_E - \Gamma_A)} \int_A^E \vartheta_{fl} d\Gamma_Y$ = mean flash temperature	(4.203)

Permissible contact temperature The calculated maximum contact temperature ϑ_{Bmax} is compared to the permissible contact temperature ϑ_S. This can be determined in a scuffing test, e.g., the FZG-A/8.3/90 test [DIN51354] for lubricants with no or few additives or other test procedures [FVA243] for EP-lubricants. For the FZG-A/8.3/90 test, the permissible contact temperature is calculated according to the following relationship (Table 4.47):

Table 4.47 Calculation of permissible contact temperature ϑ_S

Designation	Formula	No.
Permissible contact temperature	$\vartheta_S = 80 + (0.85 + 1.4X_W)X_L(S_{FZG})^2$	(4.204)
	where: X_W = structure factor = 1.00 for steels with usual austenite content (20–30 %) = 1.15 for steels with below-average austenite content = 0.85 for steels with above-average austenite content = 0.45 for stainless steel = 1.25 for phosphatized steels = 1.50 for nitrated and copper-plated steels X_L = lubricant factor according to Table 4.42 S_{FZG} = load stage in FZG-A/8.3/90 test	

Safety factor for scuffing according to the contact temperature method The ratio of the permissible contact temperature to the actual contact temperature, minus the oil temperature in each case, is the safety factor S_B according to the contact temperature method. It gives an approximate value for the factor by which the load on a gear set can be increased while maintaining a safety factor equal to 1 (Table 4.48).

Table 4.48 Calculation of scuffing safety factor S_B

Designation	Formula	No.
Scuffing safety factor	$S_B = \dfrac{\vartheta_S - \vartheta_{oil}}{\vartheta_{B\max} - \vartheta_{oil}}$	(4.205)

4.2.6.2 Integral Temperature Method for Non-offset Bevel Gears

Integral temperature J_{int} This temperature consists of the bulk temperature ϑ_M and the weighted mean tooth flank temperature ϑ_{flaint}. The mean flank temperature ϑ_{flaint} is in turn calculated from the flash temperature at the pinion tip without load sharing ϑ_{flaE}, and is converted to a mean flank temperature using the contact ratio factor X_ε. (Fig. 4.24 and Tables 4.49 and 4.50)

Table 4.49 Calculation of the integral temperature ϑ_{int}

Designation	Formula	No.
Integral temperature	$\vartheta_{int} = \vartheta_M + C_2 \cdot \vartheta_{flaint} = \vartheta_M + C_2 \cdot \vartheta_{flaE} \cdot X_\varepsilon$	(4.206)
	where: $C_2 = 1.5$	

Table 4.50 Calculation of the weighted mean flank temperature for bevel gears $\vartheta_{\text{flaint}}$

Designation	Formula		No.
Weighted mean flank temperature	$\vartheta_{flaint} = \vartheta_{flaE} \cdot X_\varepsilon = \mu_{mC} X_M X_{BE} X_{\alpha\beta} \dfrac{w_{Bt}{}^{0.75} v_{mt}{}^{0.5}}{\lvert a_v \rvert^{0,25}} \dfrac{X_E}{X_Q X_{Ca}} \cdot X_\varepsilon$		(4.207)
	w_{Bt} = unit load according to Table 4.40		
	$v_{mt} = v_{mt1,2} = \dfrac{d_{m1,2}\ n_{1,2}}{19098}$	Circumferential speed on the pitch cone at mid-face width	(4.208)
	a_v = centre distance of the virtual cylindrical gear		
	μ_{mC} = mean coefficient of friction at pitch point C		
	X_M = flash factor to consider material properties according to Table 4.43		
	X_{BE} = geometry factor for the pinion tooth tip considering the radii of curvature and sliding velocity		
	$X_{\alpha\beta} \approx 1$ = influence factor of the pressure and spiral angles		
	X_E = running-in factor to consider the negative influence of inadequately run-in tooth flanks		
	X_Q = mesh approach factor to consider the meshing impact from the tip of the driven gear at contact entry		
	X_{Ca} = tip relief factor to consider the positive influence of a tip relief (profile crowning)		
	X_ε = contact ratio factor to convert the contact temperature at pinion tip to an applicable mean tooth flank temperature		

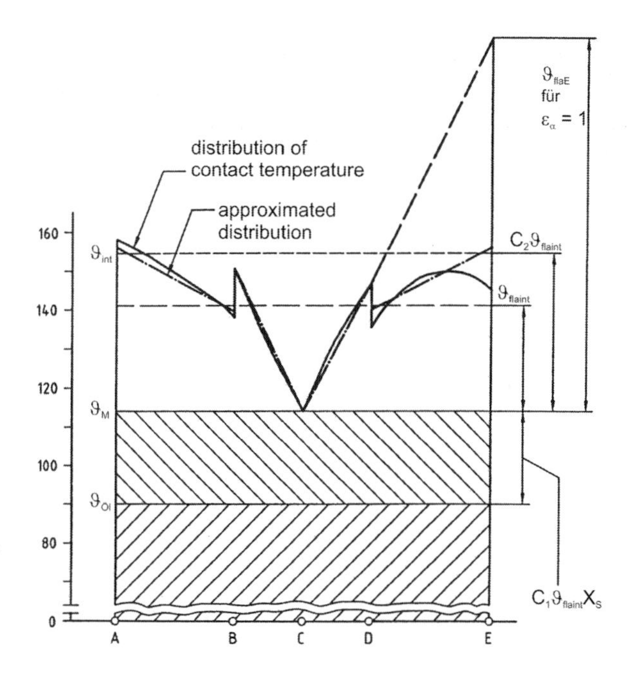

Fig. 4.24 Contact temperature over the path of contact for a cylindrical gear with profile contact ratio of $1.0 \leq \varepsilon_{\alpha v} < 2.0$

Mean coefficient of friction at the pitch point μ_{mC} The mean coefficient of friction μ_{mC} can be calculated with a good precision according to Table 4.51, using the values for load per unit face width, sum of velocities and equivalent radius of curvature at the pitch point C. The influences of the lubricant and tooth surface roughness are also taken into account.

Table 4.51 Calculation of the mean coefficient of gear friction μ_{mC}

Designation	Formula	No.
Mean coefficient of gear friction	$\mu_{mC} = 0.048 \cdot \left(\frac{w_{Bt}}{v_{\Sigma C} \rho_{redC}}\right)^{0.2} \eta_{oil}^{-0.05} X_L X_R$	(4.209)
	w_{Bt} = load per unit face width (unit load) including load factors according to Table 4.40	
	$v_{\Sigma C} = 2 v_{mt} \sin \alpha_{vt}$ = sum of velocities at C	(4.210)
	$\rho_{redC} = \frac{u}{(1+u)^2} a_v \frac{\sin \alpha_{vt}}{\cos \beta_b}$ = equivalent curvature radius at C	(4.211)
	η_{oil} = viscosity at oil temperature	
	X_L = lubricant factor according to Table 4.42	
	X_R = roughness factor according to Table 4.42	

Geometry factor X_{BE} The geometry factor X_{BE} considers the radii of curvature and sliding velocity at pinion tip when calculating the local flash temperature ϑ_{flaE} (Table 4.52).

Table 4.52 Calculation of geometry factor X_{BE}

Designation	Formula	No.
Geometry factor	$X_{BE} = 0.51 \cdot \sqrt{(u_v + 1)} \cdot \frac{\sqrt{\rho_{E1}} - \sqrt{\rho_{E1}/u_v}}{(\rho_{E1} \cdot \rho_{E2})^{0,25}}$	(4.212)
Radii of curvature at tooth tip	where: $\rho_{E1} = 0.5\sqrt{d_{va1}^2 - d_{vb1}^2}$	(4.213)
	$\rho_{E2} = a_v \sin \alpha_{vt} - \rho_{E1}$	(4.214)

Running-in factor X_E Scuffing load capacity of inadequately run-in tooth flanks can decrease to a quarter of the value of adequately run-in teeth. The running-in factor X_E considers this negative influence as a function of the surface roughness value Ra of the tooth flanks (Table 4.53).

Table 4.53 Calculation of running-in factor X_E

Designation	Formula		No.
Running-in factor	$X_E = 1 + (1 - \Phi_E) \cdot \frac{30 \cdot Ra}{\rho_{redC}}$	ρ_{redC} acc. to Table 4.51	(4.215)
	where: $\Phi_E = 0$ for non-run-in gears $\Phi_E = 1$ for fully run-in gears ($Ra \approx 0.6\ Ra_{original}$)		

Approach factor X_Q The approach factor X_Q considers the negative influence of meshing impact at the tip of the driven gear (Table 4.54).

Table 4.54 Calculation of approach factor X_Q

Designation	Formula	No.
Approach factor	$X_Q = 1.00$ for $\varepsilon_f/\varepsilon_a \le 1.5$	(4.216)
	$X_Q = 1.40 - \frac{4\,\varepsilon_f}{15\varepsilon_a}$ for $1.5 < \varepsilon_f/\varepsilon_a \le 3$	(4.217)
	$X_Q = 0.60$ for $3 \le \varepsilon_f/\varepsilon_a$	(4.218)
	where: $\left.\begin{array}{l}\varepsilon_f = \varepsilon_{v2}\\ \varepsilon_a = \varepsilon_{v1}\end{array}\right\}$ "pinion drives wheel"	(4.219)
	$\left.\begin{array}{l}\varepsilon_f = \varepsilon_{v1}\\ \varepsilon_a = \varepsilon_{v2}\end{array}\right\}$ "wheel drives pinion"	(4.220)
	$\varepsilon_{v1,2} = \frac{z_{v1,2}}{2\pi} \cdot \left[\sqrt{\left(\frac{d_{va1,2}}{d_{vb1,2}}\right)^2 - 1} - \tan\alpha_{vt}\right]$	(4.221)

Tip relief factor X_{Ca} The tip relief factor X_{Ca} considers the positive influence of tip relief (profile crowning). A gear with an optimum crowning for the maximum load ($C_a = C_{eff}$) is assumed for the calculation (Table 4.55).

Table 4.55 Calculation of tip relief factor X_{Ca}

Designation	Formula	No.
Tip relief factor	$X_{Ca} = 1 + \left[0.06 + 0.18 \cdot \left(\frac{C_a}{C_{eff}}\right)\right] \cdot \varepsilon_{v\max} + \left[0.02 + 0.69 \cdot \left(\frac{C_a}{C_{eff}}\right)\right] \cdot \varepsilon_{v\max}^2$	(4.222)
	where: $\varepsilon_{v\max} = \varepsilon_{v1}$ or ε_{v2} (see Table 4.54) whichever is greater; $C_a = C_{eff}$	

Contact ratio factor X_ε The contact ratio factor X_ε, used to convert the contact temperature at pinion tip to a mean tooth flank temperature, is calculated according to Table 4.56, as a function of the contact ratios.

Table 4.56 Calculation of contact ratio factor X_ε

Designation	Formula	No.
Conditions	Contact ratio factor	(4.223)
$\varepsilon_{v\alpha} < 1$, $\varepsilon_{v1} < 1$, $\varepsilon_{v2} < 1$	$X_\varepsilon = \frac{1}{2 \cdot \varepsilon_{v\alpha} \cdot \varepsilon_{v1}} \cdot \left(\varepsilon_{v1}^2 + \varepsilon_{v2}^2\right)$	
$1 \leq \varepsilon_{v\alpha} < 2$, $\varepsilon_{v1} < 1$, $\varepsilon_{v2} < 1$	$X_\varepsilon = \frac{1}{2 \cdot \varepsilon_{v\alpha} \cdot \varepsilon_{v1}} \cdot \left[0.70 \cdot \left(\varepsilon_{v1}^2 + \varepsilon_{v2}^2\right) - 0.22 \cdot \varepsilon_{v\alpha} + 0.52 - 0.60 \cdot \varepsilon_{v1} \cdot \varepsilon_{v2}\right]$	
$1 \leq \varepsilon_{v\alpha} < 2$, $\varepsilon_{v1} \geq 1$, $\varepsilon_{v2} < 1$	$X_\varepsilon = \frac{1}{2 \cdot \varepsilon_{v\alpha} \cdot \varepsilon_{v1}} \cdot \left(0.18 \cdot \varepsilon_{v1}^2 + 0.70 \cdot \varepsilon_{v2}^2 + 0.82 \cdot \varepsilon_{v1} - 0.52 \cdot \varepsilon_{v2} - 0.30 \cdot \varepsilon_{v1} \cdot \varepsilon_{v2}\right)$	
$1 \leq \varepsilon_{v\alpha} < 2$, $\varepsilon_{v1} < 1$, $\varepsilon_{v2} \geq 1$	$X_\varepsilon = \frac{1}{2 \cdot \varepsilon_{v\alpha} \cdot \varepsilon_{v1}} \cdot \left(0.70 \cdot \varepsilon_{v1}^2 + 0.18 \cdot \varepsilon_{v2}^2 - 0.52 \cdot \varepsilon_{v1} + 0.82 \cdot \varepsilon_{v2} - 0.30 \cdot \varepsilon_{v1} \cdot \varepsilon_{v2}\right)$	
$2 \leq \varepsilon_{v\alpha} < 3$, $\varepsilon_{v1} \geq \varepsilon_{v2}$	$X_\varepsilon = \frac{1}{2 \cdot \varepsilon_{\nu\alpha} \cdot \varepsilon_{\nu1}} \cdot \left(0.44 \cdot \varepsilon_{\nu1}^2 + 0.59 \cdot \varepsilon_{\nu2}^2 + 0.30 \cdot \varepsilon_{\nu1} - 0.30 \cdot \varepsilon_{\nu2} - 0.15 \cdot \varepsilon_{\nu1} \cdot \varepsilon_{\nu2}\right)$	
$2 \leq \varepsilon_{v\alpha} < 3$, $\varepsilon_{v1} < \varepsilon_{v2}$	$X_\varepsilon = \frac{1}{2 \cdot \varepsilon_{v\alpha} \cdot \varepsilon_{v1}} \cdot \left(0.59 \cdot \varepsilon_{v1}^2 + 0.44 \cdot \varepsilon_{v2}^2 - 0.30 \cdot \varepsilon_{v1} + 0.30 \cdot \varepsilon_{v2} - 0.15 \cdot \varepsilon_{v1} \cdot \varepsilon_{v2}\right)$	
	where: $\varepsilon_{v\alpha} = \varepsilon_{v1} + \varepsilon_{v2}$	

Bulk temperature ϑ_M The bulk temperature consists of the oil temperature and a part of the integral temperature weighted with constant C_1 which represents the thermal conductivity parameters. Lubrication factor X_S is affected by the lubricating conditions (Table 4.57).

Table 4.57 Calculation of the bulk temperature ϑ_M

Designation	Formula	No.
Bulk temperature	$\vartheta_M = \vartheta_{oil} + C_1 X_{mp} \vartheta_{flaint} \cdot X_S$	(4.224)
	where: $C_1 = 0.7$	
	$X_{mp} = \frac{1 + n_p}{2}$ = multiple mating factor	(4.225)
	n_p = number of mates meshing with the actual gear	
	X_S = lubrication factor 1.0 for dip lubrication 1.2 for injection lubrication 0.2 for complete immersion of the gears	

Permissible integral temperature The permissible integral temperature is calculated from the load stage determined in a scuffing test. The following is applicable to the FZG-A/8.3/90 test [DIN51354] (Table 4.58):

Table 4.58 Calculation of the permissible integral temperature $\vartheta_{\mathrm{intS}}$

Designation	Formula	No.
Permissible integral temperature	$\vartheta_{\mathrm{int}S} = \vartheta_{MT} + C_2 X_{wrelT}\,\vartheta_{flaintT}$	(4.226)
	where: $C_2 = 1.5$	
	$\vartheta_{MT} = 80 + 0.23 \cdot T_{1T} \cdot X_L$	(4.227)
	$\vartheta_{flaintT} = 0.2 \cdot T_{1T} \cdot \left(\frac{100}{\upsilon_{40}}\right)^{0.02} \cdot X_L$	(4.228)
	$T_{1T} = 3.726 \cdot (S_{FZG})^2$ X_L according to Table 4.42	(4.229)

Scuffing safety factor according to the integral temperature method In the integral temperature method, the scuffing safety factor S_{intS} is represented by the ratio between the allowable and the actual integral temperature. This ratio between the allowable and actual integral temperature, minus the oil temperature in each case, may be used as an approximate indicator of the load safety factor S_{SI} (Table 4.59).

Table 4.59 Calculation of scuffing safety factor $S_{\mathrm{int\ S}}$

Designation	Formula		No.
Scuffing safety factor	$S_{\mathrm{int}S} = \frac{\vartheta_{\mathrm{int}S}}{\vartheta_{\mathrm{int}}} > S_{S\min}$		(4.230)
	$S_{SI} = \frac{w_{Bt\max}}{w_{Bt\,eff}} \approx \frac{\vartheta_{\mathrm{int}S} - \vartheta_{oil}}{\vartheta_{\mathrm{int}} - \vartheta_{oil}}$		(4.231)
Minimum safety factors	$S_{S\min} < 1$	high risk of scuffing	
	$1 \leq S_{S\min} \leq 2$	scuffing risk dependent on exact knowledge of tooth flank roughness, run-in state, overloads, etc.	
	$S_{S\min} > 2$	low scuffing risk	

4.2.6.3 Contact Temperature Method for Hypoid Gears

As for bevel gears, the scuffing load capacity of hypoid gears according to [ISO/TR13989] is calculated with the contact or integral temperature approach, but using the virtual crossed-axes helical gears described in Sect. 4.2.3.

Contact temperature during operation Analogously to Table 4.38, the actual maximum contact temperature consists of the bulk temperature and the maximum flash temperature. The latter is calculated according to Table 4.60, considering the hypoid-specific property of differently oriented tangential velocities v_{F1} and v_{F2} on the tooth flanks.

Table 4.60 Calculation of the flash temperature for hypoid gears ϑ_{fl}

Designation	Formula	No.
Flash temperature	$\vartheta_{fl} = 1.11 \cdot \mu_m \cdot X_J \cdot X_\Gamma \cdot \dfrac{w_{Bn} \cdot \lvert v_{F1} - v_{F2} \rvert}{(2b_H)^{1/2} \cdot B_{M1} \cdot (v_{F1} \sin\gamma_1)^{1/2} + B_{M2} \cdot (v_{F2} \sin\gamma_2)^{1/2}}$	(4.232)
	μ_m = mean coefficient of friction according to Table 4.42	
	X_J = mesh approach factor according to Table 4.44	
	X_Γ = load sharing factor according to Table 4.41	
	$w_{Bn} = \dfrac{w_{Bt}}{\cos\alpha_{wn} \cos\beta_w}$ = unit load in normal section	(4.233)
	where: w_{Bt} = unit load according to Table 4.40 $\alpha_{wn} = \arcsin(\cos\alpha_{wt} \cos\beta_b)$	(4.234)
	$\beta_w = \arctan\left(\dfrac{\tan\beta_b}{\cos\alpha_{wt}}\right)$	(4.235)
	$v_{F1,2}$ = tooth flank tangential velocities according to Table 4.61	
	$\gamma_{1,2} = \arctan\left(\dfrac{\sin\alpha_{sn}}{\tan\beta_{s1,2}}\right)$ angle between the tooth flank tangential velocities $v_{F1,2}$ and the contact line	(4.236)
	B_M = thermal contact coefficient according to Table 4.43	
	b_H = semi-width of Hertzian contact	

Tooth flank tangential velocities $v_{F1,2}$ In calculating the flash temperature for hypoid gears, it is necessary to consider the local tooth flank tangential velocities $v_{F1,2}$ and the associated angle relative to the contact line $\gamma_{1,2}$ of the virtual crossed-axes helical gears. These values can be determined using the calculation method according to [WECH87]. The circumferential speeds of the pinion and wheel, v_{t1} and v_{t2}, at the relevant contact point are projected on the tooth flank tangential plane T (component w) and on the plane of action E (component c) (Fig. 4.25).

Components w and c are again divided into their components perpendicular to (w_α, c_α) and parallel to (w_β, c_β) the tooth trace. The tooth flank tangential velocities $v_{F1,2}$ for the pinion and wheel are then obtained by vector addition (Fig. 4.26 and Table 4.61).

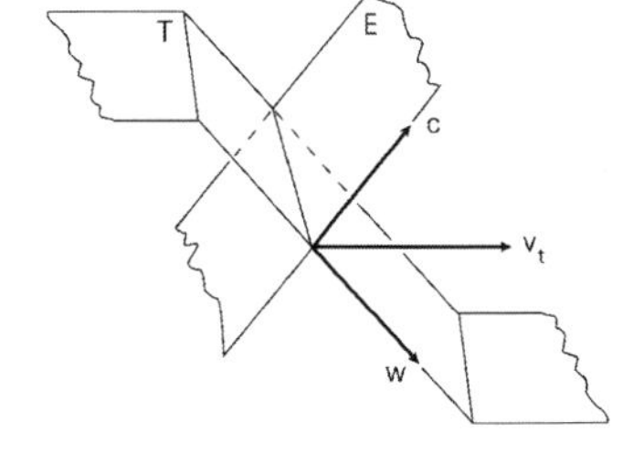

Fig. 4.25 Projection of the circumferential speed v_t at the relevant contact point on the tooth flank tangential plane T and the plane of action E

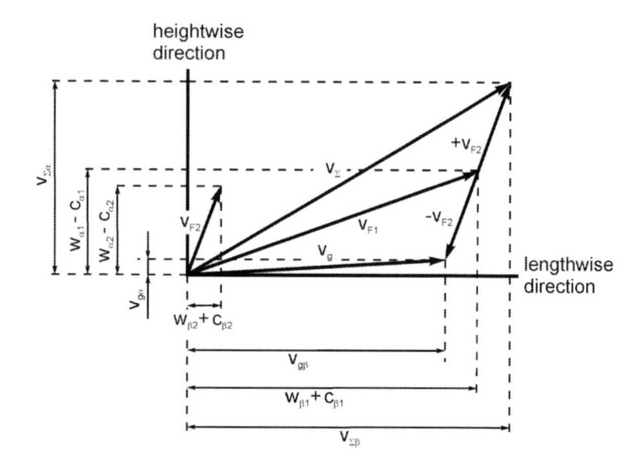

Fig. 4.26 Tooth flank tangential velocities v_{F1} and v_{F2} as a vector addition of the velocity components in the heightwise (α) and lengthwise (β) directions

Table 4.61 Tooth flank tangential velocities according to [WECH87]

Designation	Formula	No.
Speed components w and c on the pinion	$w_1 = v_{t1} \cdot \left(\sin \alpha_{st1} + \frac{g_{t1}(X)}{r_1} \right) = v_{mt1} \cdot \left(\sin \alpha_{st1} + \frac{g_{t1}(X)}{r_1} \right) \cdot \frac{r_1}{r_{s1}}$	(4.237)
	$c_1 = v_{t1} \cdot \cos \alpha_{st1} = v_{mt1} \cdot \cos \alpha_{st1} \cdot \frac{r_1}{r_{s1}}$	(4.238)
Speed components w and c on the wheel	$w_2 = v_{t2} \cdot \left(\sin \alpha_{st2} - \frac{g_{t2}(X)}{r_2} \right) = v_{mt2} \cdot \left(\sin \alpha_{st2} - \frac{g_{t2}(X)}{r_2} \right) \cdot \frac{r_2}{r_{s2}}$	(4.239)
	$c_2 = v_{t2} \cdot \cos \alpha_{st2} = v_{mt2} \cdot \cos \alpha_{st2} \cdot \frac{r_2}{r_{s2}}$	(4.240)
Speed components w and c in the profile direction	$w_{\alpha 1,2} = w_{1,2} \cdot \cos \beta_{B1,2}$	(4.241)
	$c_{\alpha 1,2} = c_{1,2} \cdot \sin \beta_{b1,2} \cdot \sin \beta_{B1,2}$	(4.242)
Speed components w and c in the lengthwise direction	$w_{\beta 1,2} = w_{1,2} \cdot \sin \beta_{B1,2}$	(4.243)
	$c_{\beta 1,2} = c_{1,2} \cdot \sin \beta_{b1,2} \cdot \cos \beta_{B1,2}$	(4.244)
Tooth flank tangential velocity: pinion	$v_{F1} = \sqrt{(w_{\alpha 1} - c_{\alpha 1})^2 + (w_{\beta 1} + c_{\beta 1})^2}$	(4.245)
Tooth flank tangential velocity: wheel	$v_{F2} = \sqrt{(w_{\alpha 2} - c_{\alpha 2})^2 + (w_{\beta 2} + c_{\beta 2})^2}$	(4.246)
	where: $r_{s1,2}$ = pitch radii of the crossed-axes helical gears $r_{1,2}$ = distance of the relevant contact point from the axis of the crossed-axes helical gear	
Distance of the relevant contact point X from the pitch point along the path of contact	$g_{t1,2}(X) = g_n(X) \cdot \frac{\sin \alpha_n}{\sin a_{t1,2}}$ with g_n (X) according to Table 4.62	(4.247)

Table 4.62 Calculation of sections $g_n(X)$ of the contact points A, B, D and E

Designation	Formula	No.
	$g_n(A) = -g_{an2}$	(4.248)
	$g_n(B) = g_{an1} - p_{en}$	(4.249)
	$g_n(D) = -g_{an2} + p_{en}$	(4.250)
	$g_n(E) = g_{an1}$	(4.251)

Width of the contact band b_H According to [ISO/TR13989], the width of the Hertzian contact band can be approximated by the minor half-axis b of the contact ellipse. For the relevant contact point X this can be calculated according to Niemann/Winter [NIEM86.2] (see Table 4.63).

Table 4.63 Hertzian contact width or minor half-axis b of the contact ellipse

Designation	Formula	No.
Width of Hertzian contact band	$2b_H \approx 2b = 2 \cdot Z_F\ \eta \cdot \sqrt[3]{K_A F_n \cdot \rho_n}$	(4.252)
	where: $Z_F = \sqrt[3]{1.5 \cdot \left(\frac{1-\nu_1^2}{E_1} + \frac{1-\nu_2^2}{E_2}\right)}$ = material factor	(4.253)
	η = auxiliary half-axis value according to Table 4.67 K_A = application factor according to Table 4.13	
	$F_n = \frac{F_{t1}}{\cos\alpha_{sn}\cos\beta_{s1}} = \frac{2000 \cdot T_1}{d_{m1} \cdot \cos\alpha_{sn}\cos\beta_{s1}}$ = tooth normal force	(4.254)
	$\rho_n = \frac{\rho_{n1}\rho_{n2}}{\rho_{n1}+\rho_{n2}}$ equivalent curvature radius in normal section	(4.255)
Equivalent radius of curvature in normal section	$\rho_{Yn} = \frac{\rho_{Yn1}\rho_{Yn2}}{\rho_{Yn1}+\rho_{Yn2}}$	(4.256)
	where: $\rho_{Yn1} = \left(0.5\sqrt{d_{s1}{}^2 - d_{b1}{}^2} + g_n(X)\right) / \cos\beta_{b1}$	(4.257)
	$\rho_{Yn2} = \left(0.5\sqrt{d_{s2}{}^2 - d_{b2}{}^2} - g_n(X)\right) / \cos\beta_{b2}$	(4.258)

The bulk temperature according to Table 4.46 is then determined using the calculated flash temperature.

Permissible contact temperature The permissible contact temperature is calculated in the same way as for non-offset bevel gears, according to Table 4.47.

Scuffing safety factor for the contact temperature method Analogously to Table 4.48, the ratio of the permissible contact temperature (Table 4.47) to the maximum contact temperature (sum of the maximum contact and bulk temperatures) represents the scuffing safety factor for hypoid gears.

4.2.6.4 Integral Temperature Method for Hypoid Gears

Integral temperature As in Table 4.49, the integral temperature during operation is a combination of the bulk and the mean tooth flank temperatures (Tables 4.64, 4.65, 4.66, 4.67, and 4.68).

Table 4.64 Calculation of the integral temperature ϑ_{int} for hypoid gears

Designation	Formula	No.
Integral temperature	$\vartheta_{\text{int}} = \vartheta_M + C_{2H} \cdot \vartheta_{flaint,h}$	(4.259)
	where: $C_{2H} = 1.8$	

Table 4.65 Calculation of the mean flank temperature $\vartheta_{\text{flaint}}$ for hypoid gears

Designation	Formula		No.
Mean flank temperature	$\vartheta_{flaint,h} = 110 \cdot \sqrt{F_n K_A K_{B\beta} v_{t1}} \cdot \mu_{mC} \cdot \dfrac{X_E X_G X_\varepsilon}{X_Q X_{Ca}}$		(4.260)
	$F_n = \dfrac{2000 \cdot T_1}{\cos \alpha_n \cos \beta_{m1} d_{m1}}$	= tooth normal force	(4.261)
	K_A = application factor according to Table 4.13		
	$K_{B\beta} = 1.5 \cdot K_{B\beta be}$ with $K_{B\beta be}$ according to Table 4.17		
	$v_{t1} = \dfrac{d_{m1} n_1}{19098}$	Pinion circumferential velocity on the pitch cone at mid face width	(4.262)
	μ_{mC} = mean coefficient of friction at pitch point C		
	X_E = running-in factor according to Table 4.53, but with ρ_{Cn} according to Table 4.4 instead of ρ_{redC}		
	X_G = geometry factor		
	X_Q = mesh approach factor according to Table 4.54		
	X_{Ca} = tip relief factor according to Table 4.55 with ε_{nmax} instead of ε_{vmax} ε_{nmax} = value of ε_{n1} or ε_{n2}, whichever is greater (Table 4.4)		
	X_ε = contact ratio factor		

Table 4.66 Calculation of the mean coefficient of friction μ_{mC} for hypoid gears

Designation	Formula	No.
Mean coefficient of friction	$\mu_{mC} = 0.045 \cdot \left(\dfrac{w_{Bt}}{v_{\Sigma C} \rho_{Cn}} \right)^{0.2} \eta_{oil}^{-0.05} X_L X_R$	(4.263)
	w_{Bt} = load per unit face width according to Table 4.40	
	with: $b_{eff}/\cos \beta_{b2}$ instead of b_{eff} F_n according to Table 4.65 instead of F_{mt}	
	$v_{\Sigma C}$ = sum of velocities at C according to Table 4.51	
	ρ_{Cn} = equivalent radius of curvature according to Table 4.4	
	η_{oil} = viscosity at oil temperature	
	X_L = lubricant factor according to Table 4.42	
	X_R = roughness factor with ρ_{Cn} according to Table 4.4 instead of ρ_{redC}	

Table 4.67 Calculation of the geometry factor X_G for hypoid gears

Designation	Formula	No.
Geometry factor	$X_G = \dfrac{(\sin\Sigma/\cos\beta_{s2})\sqrt{1/\rho_{Cn}}}{\sqrt{L\sin\beta_{s1}} + \sqrt{L\cos\beta_{s1}\tan\beta_{s2}}}$	(4.264)
	where: $L = \frac{2}{3}\xi^2\eta$	(4.265)
	for $0 \leq \cos\vartheta < 0.949$: $\ln\xi = \dfrac{\ln(1-\cos\vartheta)}{\left(-1.53 + 0.333\ \ln(1-\cos\vartheta) + 0.0467\ [\ln(1-\cos\vartheta)]^2\right)}$	(4.266)
	$\ln\eta = \dfrac{\ln\ (1-\cos\vartheta)}{\left(1.525 - 0.86\cdot\ln\ (1-\cos\vartheta) - 0.0993\cdot[\ln(1-\cos\vartheta)]^2\right)}$	(4.267)
	for $0.949 \leq \cos\vartheta < 1$: $\ln\ \xi = (-0.4567 - 0.4446\ \ln\ (1-\cos\vartheta) + 0.1238\ [\ln\ (1-\cos\vartheta)]^2)^{1/2}$	(4.268)
	$\ln\eta = -0.333 + 0.2037\ \ln\ (1-\cos\vartheta) + 0.0012\ [\ln\ (1-\cos\vartheta)]^2$	(4.269)
	where: $\cos\vartheta = \rho_{Cn}\sqrt{\dfrac{1}{\rho_{n1}^2} + \dfrac{1}{\rho_{n2}^2} + \dfrac{2\cos 2\varphi}{\rho_{n1}\rho_{n2}}}$ ρ_{Cn}, $\rho_{n1,2}$, φ according to Table 4.4	(4.270)

Table 4.68 Calculation of the contact ratio factor X_ε for hypoid gears

Designation	Formula	No.
Contact ratio factor	$X_\varepsilon = \dfrac{1}{\sqrt{\varepsilon_n}}\cdot\left[1 + 0.5\cdot g^*\cdot\left(\dfrac{v_{g\gamma1}}{v_{gs}-1}\right)\right]$	(4.271)
	$g^* = \dfrac{g_{an1}^2 + g_{an2}^2}{g_{an1}^2 + g_{an1}\cdot g_{an2}}$ ($g_{an1,2}$ according to Table 4.4)	(4.272)
Sliding velocity at pinion tip	$v_{g\gamma1} = \sqrt{v_{g\alpha1}^2 - v_{g\beta1}^2}$	(4.273)
	$v_{g\alpha1} = v_{g1}\cos\gamma_1 + v_{g2}\cos\gamma_2$	(4.274)
	$v_{g\beta1} = v_{gs} + v_{g1}\sin\gamma_1 - v_{g2}\sin\gamma_2$	(4.275)
	where: $v_{gs} = v_{t1}\dfrac{\sin\Sigma}{\cos\beta_{s2}}$	(4.276)
	$v_{g1} = 2v_{t1}g_{an1}\dfrac{\cos\beta_{b1}}{d_{s1}}$	(4.277)
	$v_{g2} = 2v_{t2}g_{fn2}\dfrac{\cos\beta_{b2}}{d_{s2}}$	(4.278)

The bulk temperature according to Table 4.57 is then determined using the calculated mean tooth flank temperature.

Permissible integral temperature The permissible integral temperature is calculated as per Table 4.58.

Scuffing safety factor for the integral temperature method Analogously to Table 4.59, the ratio of the maximum integral temperature to the permissible integral temperature represents the scuffing safety factor for hypoid gears.

4.2.7 Calculation of the Load Capacity for a Load Spectrum

In general, time-varying loads are applied to gear sets. In the rating calculation, these load variations are approximated with application factor K_A. In the more precise calculation of load capacity, a spectrum of loads is used while application factor K_A is set to 1. The load spectrum indicates the amplitudes and numbers of load cycles to which the gear is subjected. The evolution of stress levels over time may either be determined on real gear sets taken from similar applications, or calculated by numerical simulation. The loads are divided into classes and their respective numbers of cycles are added. Figure 4.27 shows an example of a load spectrum with 10 classes.

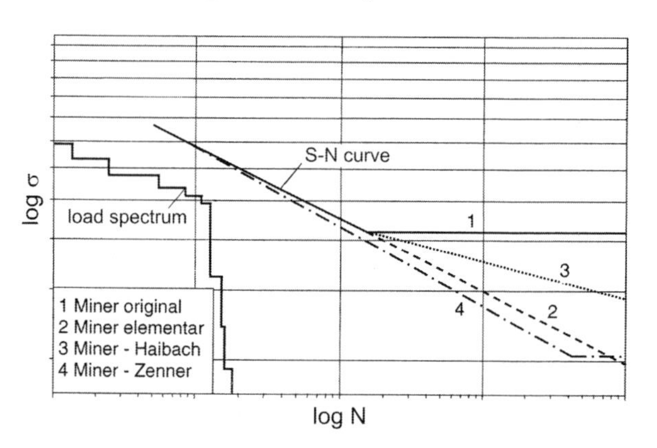

Fig. 4.27 Example of load spectrum and hypotheses of damage accumulation

The calculated load spectrum is compared to a characteristic strength function (e.g., the S-N curve). The damage condition of a gear is calculated for an assumed damage accumulation. Figure 4.27 shows the most familiar hypotheses: Miner original, Miner elementary, Miner-Haibach and Miner-Zenner. These hypotheses presume that all stresses above a certain stress level have an equally damaging effect on a loaded component, irrespective of the timing or the order in which the stresses occur. The hypotheses differ chiefly according to the way in which they consider stresses below the original endurance limit. The Standards [DIN3990] and [ISO6336] for cylindrical gears provide rules to calculate load spectra. The method according to [DIN3990] uses damage accumulation according to "Miner original", with a fixed S-N curve. Standard strength functions for static and endurance strength are given via fixed inflexion points for various materials. The endurance strength value determines the static strength level via the fixed factors Y_N and Z_N. For the gradient of the curve representing the endurance strength refer to [ISO6336].

The ratio of the number of load cycles of a stress class n_I to the corresponding permissible stress number N_I at this stress level represents the damage inflicted on the component in the form of a "lifetime consumption". The sum of these individual damage cases n_I/N_I results in the sum of damage D. If this value reaches 1, failure may theoretically be expected. In practice, however, it is apparent that damage may also occur at a sum of damage $D \neq 1$, and even at $D < 1$. This suggests employing, for the service strength calculation, a specific critical sum of damage dependent on the type of damage and on the type and level of the load spectrum (relative Miner's rule), e.g., $D = 0.2 \ldots 0.3$ for tooth root breakage [FVA188] or $D = 0.85 \ldots 1.0$ for pitting [FVA125] (Table 4.69).

Table 4.69 Calculation of the sum of damage

Designation	Formula	No.
Sum of damage	$D = \sum_I \left(\frac{n_I}{N_I}\right) = \frac{n_1}{N_1} + \frac{n_2}{N_2} + \frac{n_3}{N_3} + \ldots..$ n_I = number of load cycles for Class I N_I = number of load cycles at which a failure is to be expected under a Class I stress	(4.279)

In the case of bevel gears, it should be noted that the contact pattern spreads in the profile and lengthwise directions as load increases; in addition, the increased load causes the centre of the contact pattern to shift on the tooth flank (see Sect. 3.4.3). Unlike cylindrical gears, bevel gears thus exhibit no "cumulative damage" in the proper sense since, depending on the applied load, the damaged zone will not always be located at the same point on the tooth flank. This means that an increased sum of damage can actually be attained as often observed on bevel gears. This is particularly true when load components are increasingly located above the endurance limit since, due to the displacements of the axes, different operating positions for the most heavily stressed tooth flank regions are to be expected [THOM98].

4.3 Efficiency

4.3.1 Total Power Loss of a Gear Unit

In general, the efficiency of a system is the ratio of benefit to expenditure. Regarding a gear unit, the total efficiency η is the ratio of the power output P_{Ab} to the power input P_{An}. The total power loss P_V is the difference between power input and output, and can be divided into load-dependent losses P_{VP} and load-independent losses P_{V0}. These two losses may be separated between different components of the unit: load-dependent losses occur on the gears (P_{VZP}) and bearings (P_{VLP}); likewise, load-independent losses also occur on the gears (P_{VZ0}) and bearings (P_{VL0}), but additionally in seals (P_{VD}) and other components (P_{VX}), e.g., clutches and oil pumps [NIEM86.2] (Table 4.70).

Table 4.70 Calculation of the gear efficiency and total power loss

Designation	Formula	No.
Gear efficiency	$\eta = \frac{output}{input} = \frac{P_{Ab}}{P_{An}} = \frac{P_{An} - P_V}{P_{An}} = 1 - \frac{P_V}{P_{An}}$	(4.280)
Total power loss	$P_V = P_{VP} + P_{V0} = P_{VZP} + P_{VLP} + P_{VZ0} + P_{VL0} + P_{VD} + P_{VX}$	(4.281)

Load-independent losses arise by the displacement and squeezing of oil in gears, bearings and clutches, friction between seals and shafts, and by components of the transmission system like oil pumps or fans which may require additional power. The losses can either be quantified experimentally or, in the case of bearings and seals, determined according to manufacturers' specifications; in the case of gears, they can be estimated using Niemann/Winter [NIEM86.2]. More accurate approaches are suggested in Research Project No. 44 [FVA44] of the German Research Association for Power Transmission Engineering (FVA). However, practical experience indicates that exact calculation of load independent losses is difficult [DOLE03].

Losses under load can likewise be determined experimentally or can be calculated. Manufacturers' catalogues suggest calculation methods for bearings; for gears, there are numerous methods, based on different approaches, to determine the coefficient of friction. The special conditions of bevel and hypoid gears are taken into account in the method of Wech [WECH87] (see Sect. 4.3.3).

4.3.2 Influences on Gear Efficiency

Power losses and efficiency of gears are usually calculated using the following method (Table 4.71):

Table 4.71 Calculation of load-dependent power losses and efficiency

Designation	Formula	No.
Power loss	$P_{VZP} = F_R v_g = \frac{1}{p_e}\int_A^E \mu F_n v_g dg = \mu F_t v_t \frac{1}{p_e}\int_x \frac{F_n(x)}{F_n}\frac{v_g(x)}{v_t} dx = \mu P_A H_V$	(4.282)
	where: $P_A = F_t \cdot v_t$ = driving power at the gear	(4.283)
	μ = mean coefficient of friction	
	H_V = tooth mesh loss factor	
Gear efficiency	$\eta_{VZ} = \frac{P_A - P_{VZ}}{P_A} = 1 - \frac{P_{VZ}}{P_A} = 1 - H_V \cdot \mu_{mZ}$	(4.284)

The tooth mesh loss factor H_V is -in first approximation- a purely geometric variable representing the variation of the tooth normal force and the progression of the sliding velocity along the path of contact. The mean coefficient of friction μ_{mZ} is a variable which is additionally dependent on operating and lubricating parameters. The influence variables to calculate efficiency may thus be divided into geometric, operational and lubricant variables.

One of the main geometric variables is the ratio between sliding velocity and circumferential speed which is determined chiefly by hypoid offset and profile shift.

Both variables should be minimized as far as possible. Tooth mesh losses increase progressively with offset, especially in the range of higher loads and lower speeds. On the other hand, with lower loads and higher speeds, the influence tends to be small and, at idle, is negligible [WECH87]. Considering the risk of undercut, profile shift should also be kept to a minimum in order to balance sliding conditions on the tooth flanks.

Another influence variable is surface roughness which plays a role mainly at low speeds accompanied by low lubricant film thicknesses. When operating in this range, the roughness should be kept as low as possible [MICH87]. This can be achieved by means of special surface finishing processes or a running-in procedure to smooth the tooth flanks.

Under operating conditions, the magnitudes of rotational speed and lubricant temperature have the greatest effect on gear efficiency, while the influence of pressure is of rather subordinate importance. As circumferential speed and hence lubricant film thickness increase, tooth mesh losses decrease. They reach a minimum at dip lubrication since, with a continuing increase in speed, load-independent losses rise more rapidly than the fall of load-dependent losses [NIEM86.2]. With injection lubrication, the splash and hence the load-independent losses are significantly smaller and efficiency is correspondingly higher. In the overall picture of total efficiency, it is however necessary to account for the power requirement for oil pumps etc. Heat dissipation from the tooth flanks is significantly lower with injection lubrication, which has a negative effect on efficiency owing to the higher bulk temperatures in the contact, especially with high loads and low circumferential speeds. The influence of lubricant temperature on the efficiency is primarily an influence in viscosity. At low speeds with small lubricant film thicknesses, a higher viscosity due, for example, to lower oil temperature is more favourable, since the flanks are better separated by the lubricant film and solid to solid contact can be avoided. At higher speeds and thicker lubricant film, by contrast a lower viscosity is more favourable as it reduces losses from internal friction in the lubricant during contact and from moving the lubricant when dipping in the oil sump.

The direction of driving and the tooth flank in contact likewise have an effect on gear efficiency. The usual positive profile shift on the pinion results in an asymmetrical distribution of sliding velocity as the contact line during the addendum phase is longer than that during the dedendum phase. There is "pull-sliding" at tooth tip when the pinion is driving the wheel, which is regarded as more favourable in terms of friction behavior. Therefore, the "pinion driving wheel" case is better with respect to gear efficiency. In the case of hypoid gears with different effective pressure angles on the drive and coast tooth flanks, driving the gear on the drive tooth flank is more favourable [WECH87].

Viscosity, base oil type and additives are the main influences of the lubricant on gear efficiency [DOLE03]. Viscosity decisively affects the load-independent losses generated by displacing and squeezing the lubricant during mesh and by dipping of the gear teeth into the oil sump. In this respect, a low viscosity is more favourable than a high one. The different chemical structures of the different lubricant types such as mineral oils, polyalphaolefins, esters, polyglycols, etc., also affect efficiency, especially in EHD lubrication where the tooth flanks are predominantly separated by a lubricant film under load. Frictional losses can be reduced by as much as 40 % by using synthetic lubricants when compared to using mineral oil based lubricants. At boundary lubrication, the frictional properties of the tribological protective layer, resulting from the compound of the additives with the tooth flank material, are of primary interest. The lubricant and hence the bulk temperatures also play a role here.

The properties of a lubricant in terms of frictional behavior at no-load, EHD and boundary lubrication can be examined in the FZG efficiency test [FVA345].

4.3.3 Calculation of Gear Efficiency

Gear efficiency η_{VZ} is calculated from the tooth mesh loss factor H_V and the mean coefficient of friction μ_{mZ} according to Table 4.71. The tooth mesh loss factor represents the load sharing and the progression of the sliding velocity along the path of contact (Table 4.72).

Table 4.72 Calculation of the tooth mesh loss factor according to [NIEM86.2]

Designation	Formula	No.
Tooth mesh loss factor	$H_V = \frac{1}{p_e \cos\alpha} \int\limits_x \frac{F_t(x)}{F_n} \frac{v_g(x)}{v_t} dx$	(4.285)

The distribution of the tangential force on the meshing teeth, i.e., the variation of the normal force along the path of contact from A to E, Fig. 4.28, depends on the deflection of the teeth. This in turn depends on the load. However, according to [WECH87], it is possible to obtain a good approximation by assuming a load-independent, and hence purely geometrical, tooth mesh loss factor. Figure 4.28 shows the curves for sliding velocity, tooth normal force and the product of these two variables.

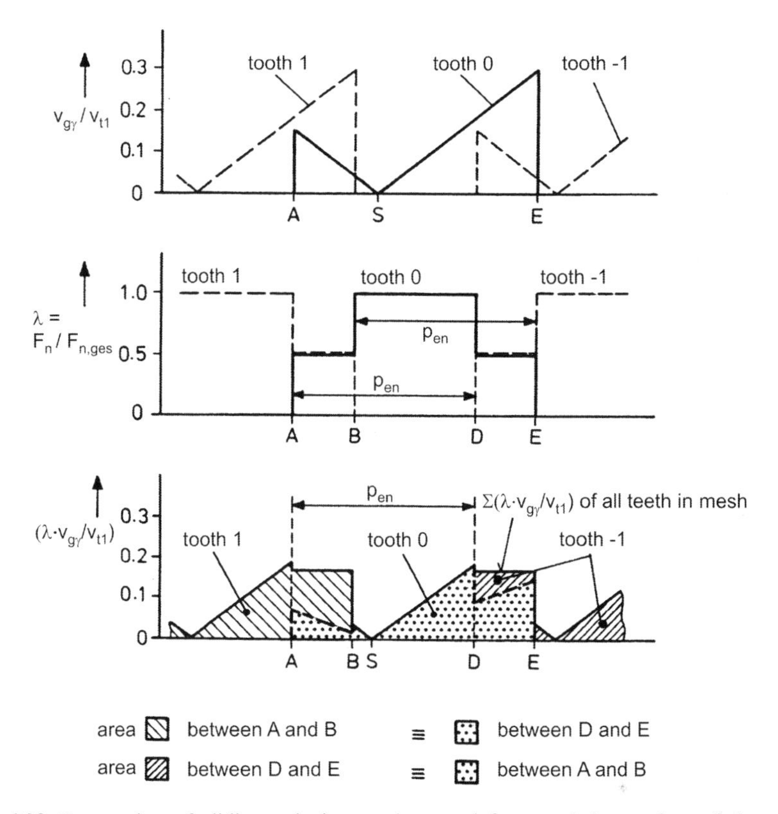

Fig. 4.28 Progression of sliding velocity, tooth normal force and the product of these two variables along the path of contact [WECH87]

Load distribution function λ The load distribution function λ assumed for the calculation of the tooth mesh loss factor is shown in Fig. 4.29. It is calculated as a function of sections $g_n(X)$ along the path of contact from A to E according to Table 4.73:

Table 4.73 Load distribution function according to [WECH87]

Designation	Formula	No.
Load distribution	$\lambda = \frac{F_n(x)}{F_n} = k_1 + k_2 \cdot x$	(4.286)
	where: $x = g_n\,(X) =$ section on the path of contact according to Table 4.62	

For the determination of the constants k_1 and k_2 see Table 4.74.

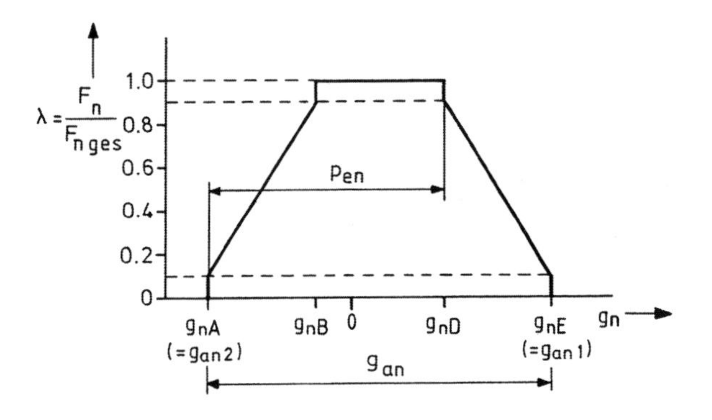

Fig. 4.29 Assumed load distribution for the determination of the tooth mesh loss factor [WECH87]

Table 4.74 Calculation of the constants k for the load distribution function

Range	$g_n(A) < g_n(x) < g_n(B)$	$g_n(B) < g_n(x) < g_n(D)$	$g_n(D) < g_n(x) < g_n(E)$
Constant k_1	$0.1 - \dfrac{0.8 \cdot g_n(A)}{g_n(B) - g_n(A)}$	1	$0.9 + \dfrac{0.8 \cdot g_n(D)}{g_n(E) - g_n(D)}$
Constant k_2	$\dfrac{0.8}{g_n(B) - g_n(A)}$	0	$-\dfrac{0.8}{g_n(E) - g_n(D)}$

Sliding velocity curve The progression of sliding velocity along the path of contact is determined according to Table 4.75 and is likewise calculated as a function of sections g_n *(X)* along the path of contact from A to E:

Table 4.75 Sliding velocity curve according to [WECH87]

Designation	Formula	No.
Sliding velocity	$\dfrac{v_g(x)}{v_{t1}} = \sqrt{D_1 \cdot g_n(X)^2 + D_2 \cdot g_n(X) + D_3}$	(4.287)
	where: $x = g_n\ (X)$ section of the contact point concerned on the path of contact according to Table 4.62	

D constants When calculating integration constants D_1 to D_3 according to Table 4.76, it is necessary to ensure that helix angles β_{s1} and β_{s2} are inserted with the correct sign. This means that β_{s1} is negative for a left-hand pinion and positive for a right-hand pinion.

Table 4.76 Calculation of the D constants for the sliding velocity curve

Designation	Formula	No.
Integration constants	$D_1 = \left[\sin^2\alpha_{sn} \cdot \left(\dfrac{\tan\beta_{s1}}{r_{s1}} - \dfrac{\tan\beta_{s2}}{r_{s2}}\right)^2 + \left(\dfrac{1}{r_{s1}} + \dfrac{1}{r_{s2}}\right)^2\right] \cdot \cos^2\beta_{s1}$	(4.288)
	$D_2 = 2\dfrac{\sin(\beta_{s1} + \beta_{s2})}{\cos\beta_{s2}} \cdot \sin\alpha_{sn}\cos\beta_{s1}\left(\dfrac{\tan\beta_{s1}}{r_{s1}} - \dfrac{\tan\beta_{s2}}{r_{s2}}\right)$	(4.289)
	$D_3 = \left(\dfrac{\sin(\beta_{s1} + \beta_{s2})}{\cos\beta_{s2}}\right)^2$	(4.290)

Integration of the sliding velocity curve over the path of contact yields the equations in Table 4.77:

Table 4.77 Integrals of the sliding velocity curve over the path of contact

Designation	Formula	No.
	$I_1(g_n(X)) = \int \frac{v_g(x)}{v_{t1}} dx = \int \sqrt{D_1 \cdot g_n(X)^2 + D_2 \cdot g_n(X) + D_3}\, dg_n(X) =$ $= \frac{2D_1 g_n(X) + D_2}{4D_1}\sqrt{G} + \frac{4D_1D_3 - D_2^2}{8D_1\sqrt{D_1}} \ln\left(2\sqrt{D_1 G} + 2D_1 g_n(X) + D_2\right)$	(4.291)
	$I_2(g_n(X)) = \int g_n \sqrt{D_1 \cdot g_n(X)^2 + D_2 \cdot g_n(X) + D_3}\, dg_n(X) =$ $= \frac{G\sqrt{G}}{3D_1} - \frac{D_2(2D_1 g_n(X) + D_2)}{8D_1^2}\sqrt{G} - \frac{D_2(4D_1D_3 - D_2^2)}{16D_1^2\sqrt{D_1}} \times$ $\times \ln\left(2\sqrt{D_1 G} + 2D_1 g_n(X) + D_2\right)$	(4.292)
	where: $G = \sqrt{D_1 \cdot g_n(X)^2 + D_2 \cdot g_n(X) + D_3}$	(4.293)

Tooth mesh loss factor H_V Given the equations for load distribution and the sliding velocity curve, the following formula may be used to calculate the tooth mesh loss factor H_V (Table 4.78):

Table 4.78 Solution of the tooth mesh loss factor equation

Designation	Formula	No.
Tooth mesh loss factor	$H_V = \{k_{1I} \cdot [I_1(g_n(B)) - I_1(g_n(A))] + k_{2I} \cdot [I_2(g_n(B)) - I_2(g_n(A))] +$ $+ k_{1II} \cdot [I_1(g_n(D)) - I_1(g_n(B))] + k_{1III} \cdot [I_1(g_n(E)) - I_1(g_n(D))] +$ $+ k_{2III} \cdot [I_2(g_n(E)) - I_2(g_n(D))]\}/p_{en} \cos\alpha_{sn} \cos\beta_{m1}$	(4.294)

Mean coefficient of gear friction μ_{mZ} The mean coefficient of gear friction to determine power losses or gear efficiency is calculated according to Table 4.79:

Table 4.79 Coefficient of friction according to [WECH87]

Designation	Formula	No.
Mean coefficient of friction	$\mu_{mZ} = 0.054 \cdot V_R V_S V_Z V_L \cdot \frac{(F_n \cos\beta_{b2}/b_2)^{0.05}}{K_{gm}{}^{0.6} \rho_n{}^{0.2} v_{\Sigma m}{}^{0.35}}$	(4.295)
	$V_R = X_R$ = roughness factor according to Table 4.42	
	V_S = lubricant system factor = 1.0 for injection or dip lubrication up to $t = 1/6 \cdot d_{ae2}$ = 0.92 for dip lubrication with an oil level up to $0 \ldots 1 \cdot b_2$ below the axis of the wheel	
	V_Z = viscosity factor $= (0.1\ \eta_{oil})^{-1/n}$ if 4 mPa-s < η_{oil} < 10 mPa-s $= (0.1\ \eta_{oil})^{-0,05}$ if η_{oil} > 10 mPa-s	
	$V_L = X_L$ = lubricant factor according to Table 4.42	
	$F_n = F_{bt2} = \frac{2000 T_2}{d_{b2}}$ = normal force of the virtual cylindrical gear	(4.296)
	K_{gm} = sliding factor	
	$\rho_n = \rho_{Cn}$ = equivalent radius of curvature according to Table 4.4	
	$v_{\Sigma m}$ = mean sum of velocities	

Since the calculation method for the coefficient of friction according to [WECH87] is based on his experimental results, the following limits must be taken into account regarding the validity range of this method (Table 4.80):

Table 4.80 Limits of the equation for the coefficient of friction according to [WECH87]

Designation	Formula	No.
Ranges of validity	$150\frac{N}{mm} < \frac{F_n}{\cos\beta_{b2} b_2} \leq 1200\frac{N}{mm}$ $0.2 < K_{gm} \leq 0.6$ $7\ \text{mm} < \rho_n \leq 16\ \text{mm}$ $1\ \text{m/s} < v_{\Sigma m} \leq 25\ \text{m/s}$	(4.297)

Sliding factor K_{gm} The sliding factor K_{gm} represents the ratio of the mean sliding velocity to the circumferential speed of the pinion. In order to determine K_{gm}, the integral of the relative sliding velocity along the path of contact is calculated and is divided by the length of the path of contact (Table 4.81):

Table 4.81 Calculation of sliding factor K_{gm}

Designation	Formula	No.
Sliding factor	$K_{gm} = 1/g_\alpha \int_A^E \frac{v_g(x)}{v_{t1}} dx = \frac{\lvert I_1(g_n(A)) + I_1(g_n(E))\rvert}{\lvert g_n(A)\rvert + \lvert g_n(E)\rvert}$ I_1 according to Table 4.77, $x = g_n(X)$ according to Table 4.62	(4.298)

Mean sum of velocities $v_{\Sigma m}$ Since the sum of velocities changes only slightly along the path of contact, it is possible to obtain a good approximation by using the mean sum of velocities $v_{\Sigma m}$ in the calculation. This is found by combining the respective sums of velocities in the profile and lengthwise directions, $v_{\Sigma\alpha}$ and $v_{\Sigma\beta}$ (Table 4.82):

Table 4.82 Mean sum of velocities

Designation	Formula	No.
Sum of velocities	$v_{\Sigma m} = v_{\Sigma\gamma}(g_n(0)) = \sqrt{(v_{\Sigma\alpha})^2 + (v_{\Sigma\beta})^2}$	(4.299)
	where: $v_{\Sigma\alpha} = v_{t1} \cdot 2\cos\alpha_{sn}\cos\beta_{s1}$	(4.300)
	$v_{\Sigma\beta} = v_{t1} \cdot (\sin\beta_{s1} - \sin\beta_{s2}\cos\beta_{s1}/\cos\beta_{s2})$	(4.301)
	$\beta_{s1,2} < 0$ for left hand helix angles $\beta_{s1,2} > 0$ for right hand helix angles	

4.4 Stress Analysis

4.4.1 *Preliminary Considerations*

Different rating standards are available for bevel gears (see Sect. 4.2.1). These calculation methods are suitable to quickly obtain an initial conclusion on the load capacity of a gear pair when certain basic variables are specified. Nevertheless, methods of this kind, which are based exclusively on analytical approaches, are only capable of an approximation to the complex relationships between load, deflection and stress on randomly curved tooth surfaces being in contact.

A more exact method, working without a virtual gear pair, calculates the equations of motion of a virtual gear cutting machine using some basic data (see Sect. 3.2.3) and derives from these equations the contact parameters of a bevel gear set [KREN07]. On the basis of these contact parameters, it is possible to calculate more accurately the load capacity parameters for the rating standards.

For a detailed description of the load capacity of a gear system, including the evolution of the contact pattern, a complex stress analysis is required. The following sections describe a calculation method of this kind for spiral bevel, helical bevel and hypoid gears. Calculation program [BAUM01], reflecting state-of-the-art technology in this field, is based on that method.

4.4.2 *Methods for the Determination of Tooth Meshing Stresses*

Apart from numerical techniques, where developments have been permitted to a large extent by the ever increasing power of computer technology, a series of experimental methods are known. These are used primarily for quality control in gear manufacturing and gearbox assembly, but can also be regarded as a criterion for performance parameters like load capacity, efficiency and quiet running of a gear reduction stage.

4.4.2.1 Experimental Methods

Contact pattern test with thin paper In the case of tooth flanks which are slightly curved or twisted along the face width, the load or pressure distribution on stationary flanks can be measured using thin paper where contact pressure has the effect of altering the transparency of a roughly 10 μm thick paper strip. The resulting differences in optical translucency serve as a measure of the unit load per unit face width. Measurement of load distribution between meshing teeth in different contact positions allows conclusions to be drawn on the load and stress parameters of the gears in mesh. Practical limitations arise mainly from the available space around the gears or from application of the load [FVA319].

Using strain gauges to measure tooth root stress Stresses in the tooth root zone of gear teeth can be derived from the measurement of deformations in the fillet. Depending on the stress condition, the actual stress can be calculated using the modulus of elasticity. Deformation measurements are made using strain gauges. More exact results are obtained with a chain consisting of a number of individual strain gauges placed consecutively along the tooth face width. The measurement principle lies in evaluating the change in electrical resistance of the strain gauges under deformation. The strain gauges are glued at the bottom of a tooth in the fillet area where the highest stress, and thus deformation, is expected (the middle of each strain gauge is around the 30° tangent in the fillet). A series of measurement results is obtained using the chain of strain gauges which results in a stress distribution curve compensating for any inaccuracy in strain gauge positioning from affecting the maximum stress value which is being sought. The advantage of this method is that the direct measurement of tooth root stress detects load differences along the face width even when a conventional contact pattern test shows a pattern extending almost evenly across the entire tooth face width. A disadvantage in terms of widespread use lies in the substantial preparation for the test usually required at the design stage. The method is therefore generally reserved for large gear sets, or for specialized tests [BECK91, BAUM95, KUNE95].

Thermography The power loss which occurs due to pressure and sliding when two tooth flanks mesh leads to local heating of the flanks. Recording the resulting thermograph provides a representation of the tooth flank contact stresses. Because of the good thermal conductivity of the gears, the accuracy of the information obtained in this way depends on the brevity of the interval between tooth meshing and the recording of the thermograph. As far as possible, the thermograph should be made while the gear set is running rather than after it has stopped. A thermal or infra-red camera is suitable for this purpose. Special analytical algorithms make it possible to associate contact stresses to the meshing contact which has just taken place. Contact pattern displacements due to manufacturing deviations such as pitch errors and wobble, which are covered and disguised when a contact pattern marking compound is used, are visible in the thermograph. Within the contact pattern, pressure distributions caused by the different tooth flank temperatures are revealed by differences in coloration [BECK91, GRAB90].

4.4.2.2 Calculation Methods

FEM calculations According to science of mechanics tooth meshing is a contact problem. A static external load presses the tooth flanks one against the other, causing them to deform locally thus creating a flat contact area. The shape, size and position of the contact area depend on tooth stiffness, tooth geometry and applied load, and determine the gear stresses which are of interest. As long as the tooth flanks are not contacting over the full face width, the correlation between load and deflection is non-linear (in addition to Hertzian deformation). An iterative process is therefore necessary to solve problems of this kind. Simply described, an initial state (length of contact) is assumed which leads to a first solution in terms of tooth deflection. This solution then determines a modified state, with new contact length and stiffness, which is used to generate a second solution. The iteration is repeated until the change between two steps does not exceed a very small prescribed value and a static equilibrium has been reached. This iteration is also necessary if only root stresses are calculated. Commercial finite element programs applicable to a wide range of tasks or calculation programs conditioned specifically for gears are used to solve contact problems of this type [SCHI00, BAUM01, SCHA03].

For a complete root and contact stress analysis using FEM, the FE model must be extended to include the entire surroundings of the gear, and a very fine mesh must be generated in the tooth zone, at least at points where high notch stresses are expected (tooth root zone). With regard to the applied load, all meshing positions must be taken into account. Despite high computation speeds and the ability to process large data volumes, calculations of this sort are currently performed only for specialized individual tasks and for comparative tests, owing to the high processing effort involved. As an example, Fig. 4.30 shows the results of a 3D FE simulation for a hypoid gear set in a rear axle gear unit.

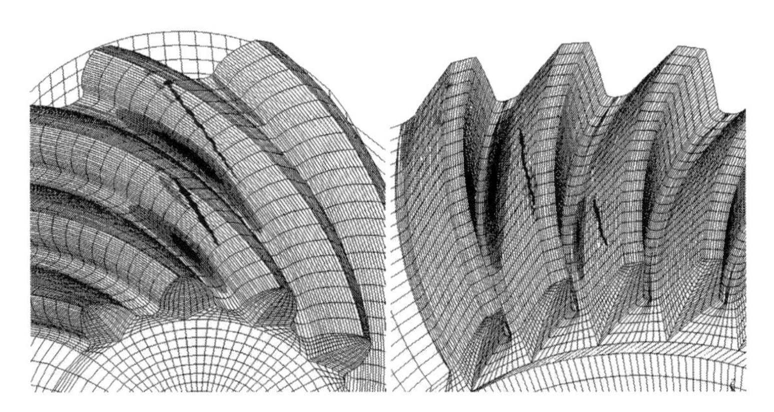

Fig. 4.30 Tooth contact analysis using FEM on a hypoid gear set (*left*: pinion; *right*: wheel) in a rear axle gear unit [ZFAG]

The aim of this simulation is to analyze the stress distribution in the tooth root during the contact. In Fig. 4.30, the resulting lines of contact are clearly visible. The available mesh fineness does not allow evaluation in terms of contact stress on the flank. The calculation was performed for 10 meshing positions (increments of 0.1 p_e); results are shown for one contact position on the pinion and the wheel.

Calculation method based on influence numbers A stress analysis based on influence numbers consists of two main parts: calculation of the load distribution (solving the contact problem) and calculation of the stresses resulting from this load distribution.

In a real gear set, load distribution is generally influenced by the ease-off, i.e., the combination of transmission error, crowning and operating positions, the elastic deflections of the gear teeth and of the neighboring components (such as shafts, gear bodies, bearings, housing) and by bearing clearances. Since the contact lines run diagonally on the tooth flank, tooth segments of varying stiffness are in simultaneous contact: and contact more or less runs from tooth tip to tooth root. Even in conjugate gears, this condition produces an uneven load distribution along the contact line. At the beginning or end of the contact, or if the meshing gears have a large overlap ratio, there are times when there is no load on certain segments, causing a significant stiffening effect in the immediate vicinity of the neighboring loaded tooth segments. Since several tooth pairs are simultaneously meshing in a gear set, they contribute to the transmission of load in the form of parallel-connected springs (Fig. 4.31). Each individual tooth pair takes over a part of the total load in proportion to its spring stiffness.

The load distribution may thus be calculated with the aid of deflections; there is no explicit exact solution.

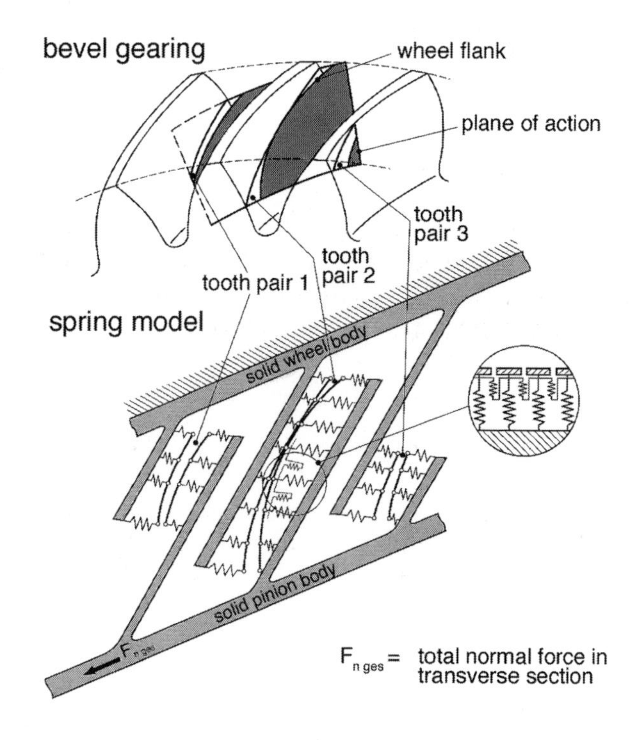

Fig. 4.31 Spring model of a bevel gear pair according to [NEUP83]

General approach to calculate load distribution Approaches for the determination of load distribution must reflect the complex stress–deflection conditions in tooth contact.

A model which has proved useful stems from cylindrical gear analysis, its basic concept being known from a series of publications [ZIEG71, SCHM73, OEHM75, SABL77, HOSE78, PLAC88]. In essence, it is assumed that all influences affecting load distribution may be superimposed: the total magnitude of the deflection is obtained by superimposing the individual influences. When a tooth pair is loaded, deflection takes place which is expressed by a change in the angle of rotation of the two gears. The total deflection $f_z(s)$ in any transverse section s of the gears in the direction of the tooth normal force thus results from the sum of the components of the pinion tooth (I) and the wheel tooth (II). For each transverse section the component of deflection is always expressed by an angle of rotation and an effective radius.

$$f_z(s) = f_{zI}(s) + f_{zII}(s) = \varphi_{zI}(s) \cdot r_{wI}(s) + \varphi_{zII}(s) \cdot r_{wII}(s) \tag{4.302}$$

The effective radius $r_w(s)$ is the lever arm which multiplies the tooth normal force $F_n(s)$ in contributing its share of the total applied torque.

The sum of the rotation angles from the single deflections in each transverse section has to remain constant if teeth *I* and *II* make contact along one contact line, i.e., theoretically they neither penetrate nor separate.

$$\begin{aligned} &\varphi_z(s) = \varphi_{zI}(s) + \varphi_{zII}(s) \cdot (r_{wII}/r_{wI}) = \varphi_z = \mathit{constant} \\ &\varphi_z = \text{total rotation angle in relation to the pinion} \end{aligned} \tag{4.303}$$

In the next step of evaluation, an arbitrary distance from the flank contact $f_k(s)$ is superimposed to the elastic deflections, in order to take into account those parts of the flanks which are not always in contact. Therefore, Formula (4.302) needs to be extended. Given an angle φ_z constant over the full face width, it is true to say that

$$f_z(s) = \varphi_z \cdot r_{wI}(s) + f_k(s) - g(s) \tag{4.304}$$

Component $g(s)$ is the residual distance from flank contact of the tooth pair which remains after elastic deflection.

Once all forces, deflections, tooth stiffness values and contact line deviations in the direction of the plane of action in the normal section have been accounted for, it is possible to describe the stress–deflection relationships of the meshing gear teeth by means of a system of linear equations.

The coefficient matrix is a system of mutual *influence numbers* for deflection. The resulting solution vector consists of a number of forces reflecting the load distribution and the total rotation angle of the gear pair. Also known are representations of the load-deflection relationships through a matrix of paired support points of a polynomial approximation [MITS83], a system of non-linear integral equations [KUBO81, BONG90], or a tri-diagonal matrix consisting of sectional or coupling stiffnesses.

The load distribution determined along the contact lines forms the basis for subsequent calculation of local stresses. Numerical, analytical and partly analytical methods may be used for the calculations.

4.4.3 Special Methods of Stress Analysis

4.4.3.1 Load, Stress, Safety Factor

Loads act on a structural assembly or a component thereof in the form of torques, bending moments, external forces, centrifugal and frictional forces, speeds, velocities and temperatures.

As a result of these loads, strains occur in the component, manifesting themselves as deflections, stresses or heating (Fig. 4.32).

According to the hypothesis of cumulative damage, each stress leads to a progressive damage state (see Sect. 4.2.7). If a certain limit is exceeded, this causes damage such as wear, pitting, scuffing etc. (see Sect. 4.1). The limit is usually a strength value which, when related to the stress, will produce a safety factor.

The main external loads acting on a gear system are torques and speeds of rotation. Over a particular period of time, these may be regarded either as a constant nominal load or as a collection of discrete loads.

The loads act to produce tooth forces, frictional forces and speeds at specific points on the gears, reactions and frictional forces on the bearings, and bending moments and transverse and longitudinal forces on the shafts. The crucial factors in judging the load capacity and running behavior of a set of gears are contact stresses on the tooth flanks, bending stresses in the tooth fillet and root, and scuffing and micro-pitting stresses combined to tooth deflection. Deformations under load of the gears, housing, shafts and possibly gear bodies cause displacements of the gear teeth, and hence a change in the load on the tooth flanks, consequently influencing these stresses.

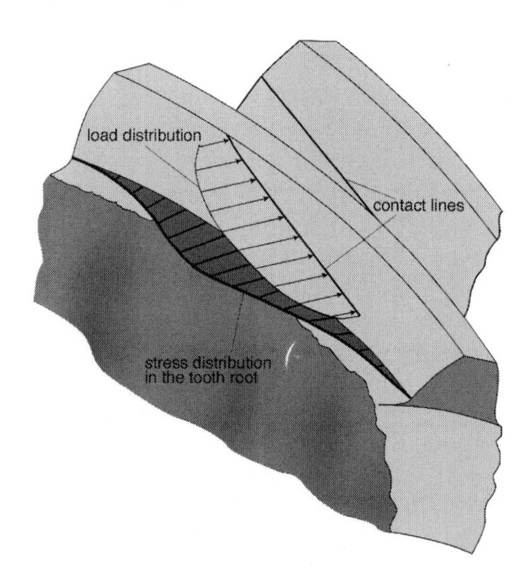

Fig. 4.32 Load and stress distribution on the bevel gear

4.4.3.2 Calculation of Load Distribution by the Generalized Influence Numbers Method

Basic relationships The influence numbers method is useful because it allows the simple integration of all elastic deflections in the gearbox, and handles arbitrary contact line deviations, including flank modifications, in an equally simple manner [HOSE78, BAUM91, BOSE95, BSL03]. The principle is based on the fact that deflections and displacements of a single gear or gear pair are composed of the influences of single loads occurring successively along the contact line. The contact lines are sub-divided into sections of equal size, to which uniform forces are assigned. A transition is made from an arbitrary transverse section *s* in Formulae (4.303) and (4.304) to the respective mid-point of a discrete section *j* or *i* of the contact line. The deflections at a point *i* of the contact line caused by a tooth normal force F_n (j) acting at point *j* may be calculated as:

$$f_{zij} = a_{ij} F_n(j) \quad \text{with i} = 1 \text{ to n and j} = 1 \text{ to n} \tag{4.305}$$

The system of equations for a tooth under load is:

$$\begin{aligned} a_{11} \cdot F_n(1) + \ldots + a_{1j} \cdot F_n(j) + \ldots + a_{1n} \cdot F_n(n) &= f_{z1} \\ &\cdot \\ a_{i1} \cdot F_n(1) + \ldots + a_{ij} \cdot F_n(j) + \ldots + a_{in} \cdot F_n(n) &= f_{zi} \\ &\cdot \\ a_{n1} \cdot F_n(1) + \ldots + a_{nj} \cdot F_n(j) + \ldots + a_{nn} \cdot F_n(n) &= f_{zn} \\ r_{w1\,\mathrm{I}} \cdot F_n(1) + \ldots + r_{wj\,\mathrm{I}} \cdot F_n(j) + \ldots + r_{wn\,\mathrm{I}} \cdot F_n(n) &= M_{t\,\mathrm{I}} \end{aligned} \tag{4.306}$$

For a tooth pair, it may be stated that:

$$f_{zij} = f_{zij\,I} + f_{zij\,II} \tag{4.307}$$

In the case of a single tooth pair, a system of equations equivalent to that for the single tooth is obtained. After introducing conditions according to Formulae (4.303) or (4.304), a line of the system of linear equations for section *i* of a contact line with *n* sections possesses the following form:

$$\sum_{j=1}^{n} \left(e_{ij} F_n(j) \right) + f_{ki} - g_i = \varphi_z \cdot r_{wj\,\mathrm{I}} \tag{4.308}$$

e_{ij} is the sum of the influence numbers for the deflection of the pinion tooth and wheel tooth:

$$e_{ij} = a_{ij\,I} + a_{ij\,II} \tag{4.309}$$

At the same time, the sum of all sub-moments on the pinion $M_t(j) = F_n(j) \cdot r_{wj\,\mathrm{I}}$ for one contact line (p-th tooth pair) must be equal to the total moment $M_{tI}(p)$.

$$\sum_{j=1}^{n} F_n(j) \cdot r_{wj\ \mathrm{I}} = M_{t\mathrm{I}}(p) \tag{4.310}$$

If several tooth pairs are simultaneously involved in the transmission of forces ($\varepsilon_\gamma > 1$), the system of equations (4.306), (4.308) may be extended accordingly.

For meshing tooth pair p with n_p individual loads, each point i is governed by the following equation:

$$\sum_{j=1}^{n_p} e(p)_{ij} F_n(p)_j + f_k(p)_i - g(p)_i = \varphi_z \cdot r_{wI}(p)_i \tag{4.311}$$

where:

$e(p)_{ij}$	influence number (p-th tooth pair, point i, load at j)
$F_n(p,j)$	load (p-th tooth pair, load at j)
φ_z	total rotation angle in relation to the pinion
$r_w(p)_i$	effective radius at the pinion (p-th tooth pair, point i)
$f_k(p)_i$	contact line deviation/distance from flank contact (p-th tooth pair, point i)
$g(p)_i$	residual distance from flank contact (p-th tooth pair, point i)

The cumulative sub-moments on the pinion $M_t(p,j) = F_n(p,j) \cdot r_{wIj}$ being equal to the total moment M_{tI}, the following is written:

$$\sum_{p=1}^{z_k} \sum_{j=1}^{n_p} F_n(p)_j \cdot r_w(p)_j = M_{tI} \tag{4.312}$$

The accuracy of the load curve is dependent to a great degree on segment width Δb (see Fig. 4.37). This should be in the order of $0.5 \cdot m_{mn}$, less for specialized studies.

It is generally possible to solve this problem using an iterative procedure, first assuming a single flank positioning over the whole contact line of ($g(p)_i = 0$), and, after solving the system of equations, setting negative forces to zero ($F_n(p)_j = 0$) for a recalculation in the negative force range. As a rule, the method converges after 1 to 2 iterations.

Extensive experience is available for a practicable solution of the system of equations. Computing times are short if effective algorithms are selected [BOSE95]

Determination of the deflection influence numbers, method and influences When a gear tooth is loaded, each tooth segment is subjected to a corresponding displacement. Assuming that Hooke's law applies, there will be a linear relationship between loads and displacements. If a number of loads are applied along a contact line, each of them will cause a displacement at each considered point along the contact line. The resultant displacement of a considered point under the action of all loads is equal to the vector sum of all individual displacements.

The linear relationship between the external load at a point and the displacement at this or another point may be represented by constant coefficients which are the deflection influence numbers (e_{ij}). The first index characterizes the point at which the deflection occurs; the second characterizes the external load by means of which the relevant component of the deflection is caused.

The symmetry of the influence numbers $e_{ij} = e_{ji}$ is given by the independence in the sequence in load application on the final state of the energy of deflection. The displacement at point i in the direction $F_n(i)$ by a force $F_n(j)$ is equal to the displacement at point j in the direction $F_n(j)$ by a force of equal magnitude $F_n(i)$.

The number of deflection influence numbers required to solve the system of equations (4.306), (4.308) is accordingly reduced from k^2 to $(k+1)\,k/2$.

The deflection influence numbers theoretically include the influences of all elastic deflections within the concerned gear system. The most important components are:

- tooth deflections
- gear body deflections
- shaft deflections
- bearing deflections
- housing deflections.

Owing to a presumed linear behavior in load-deflection, these components can be determined individually and added. This linearity does not exist for some components of tooth deflection such as Hertzian contact flattening, nor for the deflections of the shaft bearings. In these cases an approximate linearization in the employed range can be adopted, followed by iterative calculation. In principle, it is possible to represent all components of the total deflection within a gear system by means of deflection influence numbers.

After an estimation of the cost-benefit ratio, the model has proved useful for practical calculations, and perfectly adequate for the cases considered here, which are tooth and gear body deflections. Shaft, bearing and housing deflections can be determined at the respective constant loads over the face width and are allowed for in subsequent calculations by the resulting contact line deviations. Recalculation with an altered load distribution improves the result. Whereas simple methods are sufficient for shaft bending and shaft torsion, housing deflections can generally be determined only by means of FEM calculations.

Bearing deflection in roller bearings can also be determined according to [WICH67] and estimated according to [INFA06], while deflection in plain bearings can be determined according to [LAST78] [DIN31652].

Actual tooth deflection is again understood as the sum of various components. Deflections due to bending and shear stresses and to Hertzian contact flattening at contact points are regarded as crucial. On solid gear bodies, the magnitudes of compressive deflections are usually negligible in comparison to other deflections.

Determining tooth deflection Solving the contact problem by means of the influence numbers method supposes that pinion and wheel tooth are sub-divided into several segments along the face width at which different deflections may occur. The load along the contact line is dissected into individual forces corresponding to these sub-divisions, with the load acting in the middle of each considered segment. These segments are interlinked by means of imaginary springs, which is true for all form alterations with the exception of Hertzian contact flattening (deformation). Besides, the segments are created by means of spherical sections of radius r through the bevel gear tooth (Fig. 4.33).

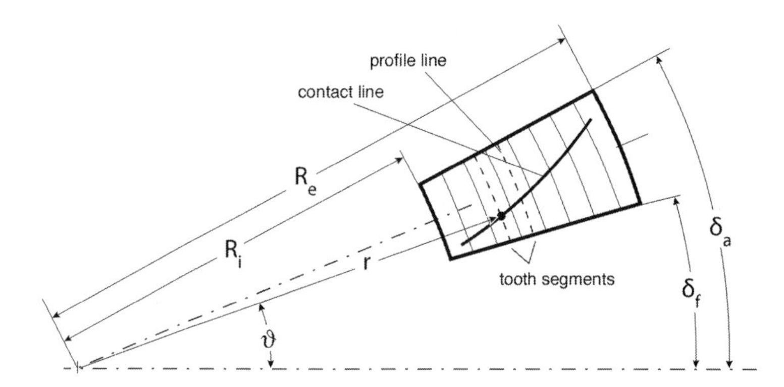

Fig. 4.33 Profile lines and contact line on the wheel tooth

An arbitrary line of intersection between the sphere and the tooth flank/fillet yields a profile line. A tooth segment is accordingly formed by two spheres and the profile lines. A point of intersection between a profile line in the middle of a segment and a contact line (calculated with the load-free tooth contact analysis) is the working point of an individual force and is described by polar coordinates r and ϑ. The grid thus created on the wheel tooth is transformed to the pinion tooth in correspondence with the contact line. The position of any desired point on a profile line can be calculated by using fitting surfaces and tooth thickness, such that the exact dimensions of the models for stress calculation on the tooth segments are available. Since models of this kind are not adaptable to analytical calculation, the spherical surfaces of the segment sides are disregarded, and are approximated by plane surfaces.

The method, which is detailed subsequently, includes the deflection of tooth segments lying at a distance from the working point, at which the force is acting with the aid of an applied generalized fading function E_∞, and infers the real deflection through compliances calculated for tooth segments $q\ (y)$.

The fading behavior of the deflection produced by the action of an individual force on various different tooth form types has been studied and systematically assessed with the aid of numerous FEM calculations. These have formed the basis for the determination of a universal relative fading function E [GAJE86, KUNE99], which allows an approximate definition of the influence function for a particular gear member. Calculations were first made for cylindrical gears and then transferred to bevel gears in experimental and numerical studies [BAUM91, BAUM95].

The fading function for a tooth of infinite width is expressed as:

$$E_{\infty}(x^*) = 0.146 \cdot \left[\cos\left(0.027 \cdot x^{*3} - 0.333 \cdot x^{*2} + 1.545 \cdot x^*\right) + 1\right] \quad (4.313)$$

$$\text{where: } x^* = |x| / m_{mn} \quad \text{with} \quad x^* > 6 : \; E_{\infty}(x^*) = 0$$
$$x^* = \text{relative distance to the point where the force is applied} \quad (4.314)$$

The fading function $E_{\infty}(x^*)$ covers the general course of specific tooth deflection in the vicinity of the point at which the load is applied, irrespective of gear parameters, for a tooth of infinite width (Fig. 4.34). The reference variable is the elastic compliance q at a point $y = y_F$ on a tooth segment having uniform gear geometry and a width $b = 1 \cdot m_{mn}$. The difference in size of the absolute deflection or compliance, due to the special geometry given by, e.g., the number of teeth and profile shift, accounts for the actual compliance of the particular tooth q $(y = y_F)$.

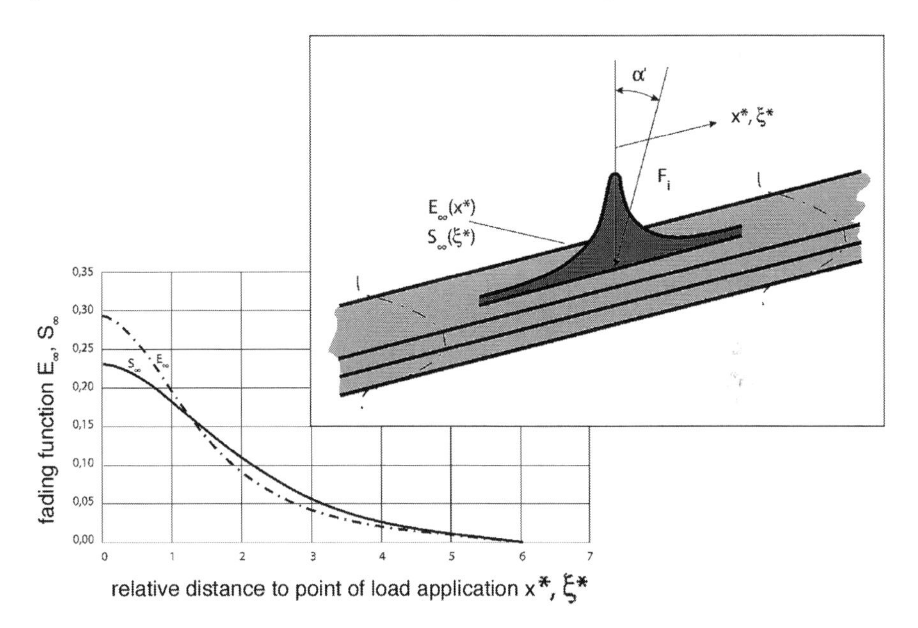

Fig. 4.34 General fading function for tooth deflection E_{∞} (Formula 4.313) and tooth root stress S_{∞} (Formula 4.329) [KUNE99]

Different meshing positions, with different bending lever arms, are accordingly accounted for through a corresponding compliance function $q(y = y_F)$. On skewed and spiral bevel gears, the contact lines run diagonally across the tooth face, such that contacting tooth segments show different compliances caused by different bending lever arms and tooth thicknesses. The influence function, which is therefore also determined diagonally across the tooth, is established for each of the tooth segments not lying below the load application point using their compliance (q (y)) in response to loading at point y_F (see also Fig. 4.36).

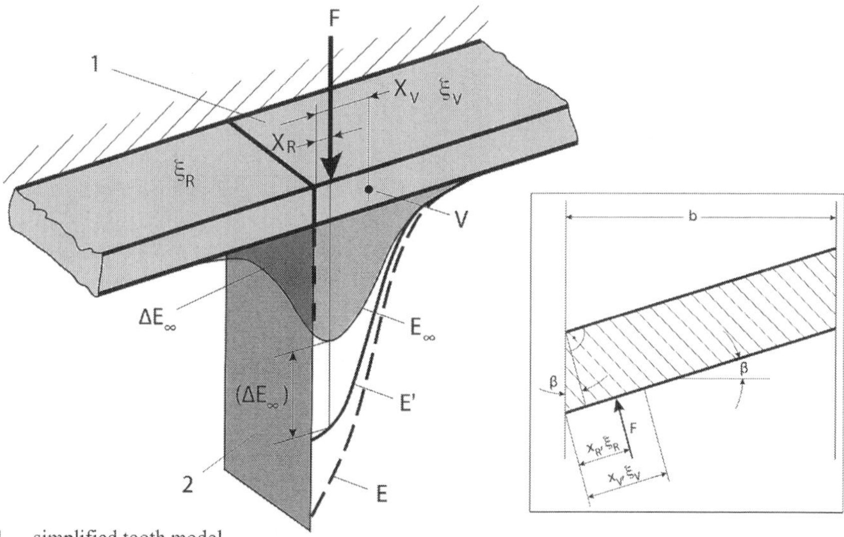

1 simplified tooth model
2 mirror plane (boundary)
V point concerned
E_∞ deflection without boundary influence
ΔE_∞ deflection to be mirrored
E' total deflection with boundary influence at ß=0°
E total deflection with boundary influence and gear face influence ß≠0° for acute or obtuse angle

Fig. 4.35 Tooth deflection calculation model

The influence of the finite face width b and the tooth boundary on the fading function E_∞ of the infinitely long tooth is accounted for by mirroring the deflection curve on the imaginary tooth face of the infinite tooth strip. Deflections lying outside the face width b are superimposed to the deflections within it (Fig. 4.35) [JARA50].

$$E = E_\infty + \Delta E_\infty \tag{4.315}$$

The result can be further improved by introducing a corrective function W_f.

$$E = \left[E_\infty\left(x^* = \left|x_V^* - x_R^*\right|\right) + E_\infty\left(x^* = x_V^* + x_R^*\right)\ \right]\ W_f\left(x_V^*, x_R^*\right) \tag{4.316}$$

where

$$x_V^* = \frac{|x_V|}{m_{mn}}; \quad x_{V\max}^* = 6; \quad x_R^* = \frac{|x_R|}{m_{mn}}, \tag{4.317}$$

x_V, x_R according to Fig.4.35

The influence of face canting on skewed and spiral gear teeth is allowed for by means of approximation W_f.

$$W_f = 1 + f_G\, f_L \tag{4.318}$$

An additional function, f_L, represents the theoretical fading behavior of the boundary influence on spiral bevel gears, while size function f_G realizes the magnification or reduction of the deflection in the vicinity of the tooth boundary at a particular spiral angle when compared to a straight bevel gear.

The boundary influence of the deflection fades at distance L_R from the boundary. This region may be defined as a function of the helix angle and in relation to the normal module, and results (empirically) from:

$$L_R^* = \frac{L_R}{m_{mn}} = 2.1 \cdot \pi \cdot \tan|\beta| \tag{4.319}$$

At $0 < x^*_V \leq L^*_R$ and $0 < x^*_R \leq L^*_R$ (in the relevant fading region L^*_R), variable x_{rel} for fading function f_L (x_{rel}), on which the deflection influence number is determined, may be calculated according to Formula (4.320):

$$x_{rel} = \frac{L_R^* - (x_V^* + x_R^*)/2}{L_R^*}$$
$$x_{rel\,min} = 0 \tag{4.320}$$
$$x_V^*,\ x_R^* \text{ see Formula (4.317);}$$

The fading function f_L is likewise determined empirically, and looking from the tooth tip a distinction is drawn between the acute angled side of the tooth face and the obtuse angled side:

$$\text{for } \beta_{St} > 0:\ f_L(x_{rel}) = 8.26\,x_{rel}^5 - 15.6\,x_{rel}^4 + 10\,x_{rel}^3 - 2\,x_{rel}^2 + 0.333\,x_{rel} \tag{4.321}$$

$$\text{for } \beta_{St} > 0:\ f_L(x_{rel}) = 1.027 \cdot x_{rel} \tag{4.322}$$

The size function f_G is determined as:

$$f_G(\beta_{St}) = 0.645\,\overset{\frown}{\beta}{}_{St}^3 + 1.454\,\overset{\frown}{\beta}{}_{St}^2 + 1.176\,\overset{\frown}{\beta}_{St} + 0.0435 \tag{4.323}$$

where:

$\overset{\frown}{\beta}_{St} = +|\beta|$ for loading on the acute angled side

$\overset{\frown}{\beta}_{St} = -|\beta|$ for loading on the obtuse angled side

$\overset{\frown}{\beta}_{St}$ in radians

The influence numbers may therefore be calculated as:

$$a_{ij} = E \cdot q_{ij} \tag{4.324}$$

E bending influence function, Formula (4.316)

a_{ij} influence number (describes the influence of force F_{nj} acting at point j on the deflection at point i)

q_{ij} compliance of a tooth segment of width $\Delta b = 1$ m_{mn} according to Formula (4.325) at point i as a result of loading on the lever arm at point j

Compliance The elastic compliance $q(y)$ is understood as the sum of the effects of a normal force on the tooth and the adjacent part of the gear body. During calculation, it is first differentiated as bending deflection (B) of the tooth, shear deflection (S) of the tooth and deflection of that part of the gear body (R) adjacent to the tooth. Deformation due to Hertzian contact flattening is treated separately.

$$q = q_B + q_S + q_R \tag{4.325}$$

Bending compliance q_B is determined on the basis of the equation for the deflection curve η of a trapezoidal-shaped beam of variable cross-section (cf. [WEBE53]). In the case of hypoid gears, the tooth profile becomes asymmetrical and an imaginary mid-line of the tooth is tilted by the limit pressure angle α_{lim} (see Sect. 2.2.5.3). In calculating the torque at a section given by distance y, it is necessary to add the bending component arising from the non-centric compressive force component of the normal force on the tooth (Fig. 4.36).

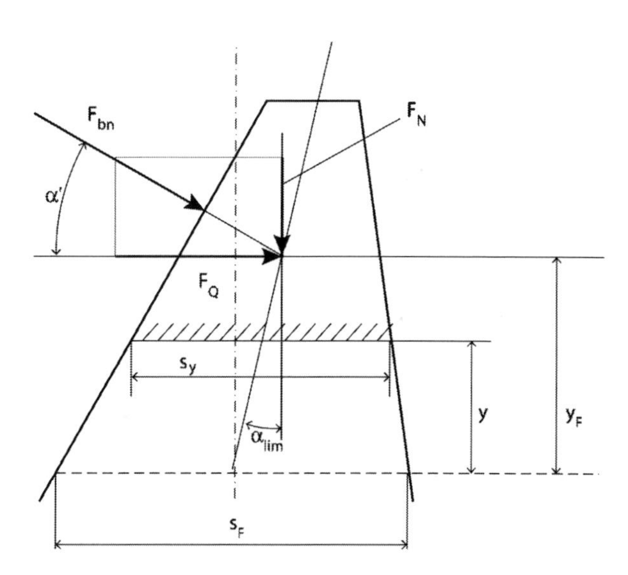

Fig. 4.36 Determination of boundary deflection using the mirroring method (valid, by analogy, for tooth root stresses)

Determining Hertzian deformation With bending deflection, shear deflection and deflection of the part of the gear body adjacent to the tooth, a strong mutual influence exists between the individual tooth segments. In the case of Hertzian contact, it is assumed that only the particular segment of the tooth subjected to the load, will experience a deformation. Consequently, there is no need to allow for mutual influences and in terms of Hertzian contact the tooth is composed of decoupled segments of finite width. The tooth profiles are approximated by their radii of curvature and replaced by short cylinders (Fig. 4.37). Because of the bevel gear geometry, the radii of curvature for each point on a contact line differ. The Weber and Banaschek approach [WEBE53] may be used as a method of calculation.

The influence numbers are calculated for the pinion and wheel respectively, by analogy with Formula (4.324), the influence being set to $E = 1$ (no mutual influence).

$$a_{ij}(H) = q_{Hj} \tag{4.326}$$

H Hertzian deformation
q_H elastic compliance of a tooth segment of width Δb at point j due to loading at point j

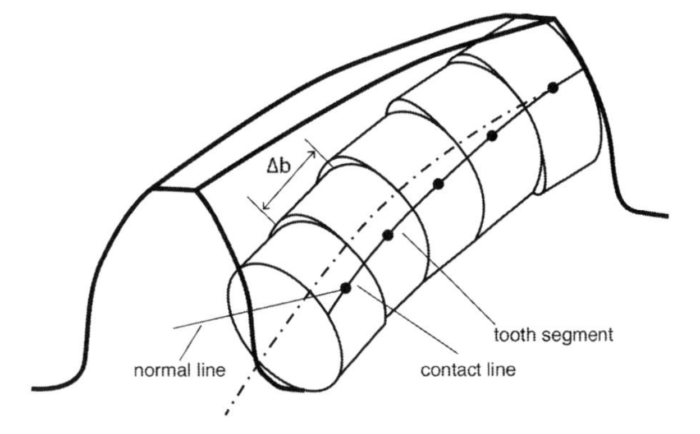

Fig. 4.37 Calculation model for flank deformation and contact stress on a bevel gear tooth

Hertzian contact deformations do not vary linearly with load, but on a diminishing scale. In a first approximation, an initial value corresponding to a mean gear load of $\sigma_{Hmax} \approx 800$ N/mm^2 for all tooth segments can be used to linearize the relationship between load and deformation. An improvement of the load distribution obtained in this way is achieved by appropriate iteration.

4.4.3.3 Including the Gear Environment

The gear environment is understood as comprising those gear system elements which surround the gear. These include the gear bodies, shafts, bearings and housing. These components or sub-assemblies have deviations in size due to manufacturing which cause deviations in position. They are additionally deflected as a result of gear loading. Since the gear teeth of the pinion and wheel are "connected" to these elements, the relative positions of the pinion and gear teeth and, hence the meshing conditions, the contact pattern and consequently the stresses on the gear members are changed. It is therefore essential to include the environment of the gear in a stress analysis.

The factors include displacements and tilting of the pitch cones due to:

- deflection of the shafts, bearings, gear bodies and housing
- bearing clearance
- positional deviations of the bearing bores caused in manufacturing
- errors in gear assembly

Load-dependent deflections are caused by applied loads including torque, tooth forces of the bevel gear stage, and additional loads on the shafts such as forces of a cylindrical gear stage, shear forces and bending moments at the ends of the shafts.

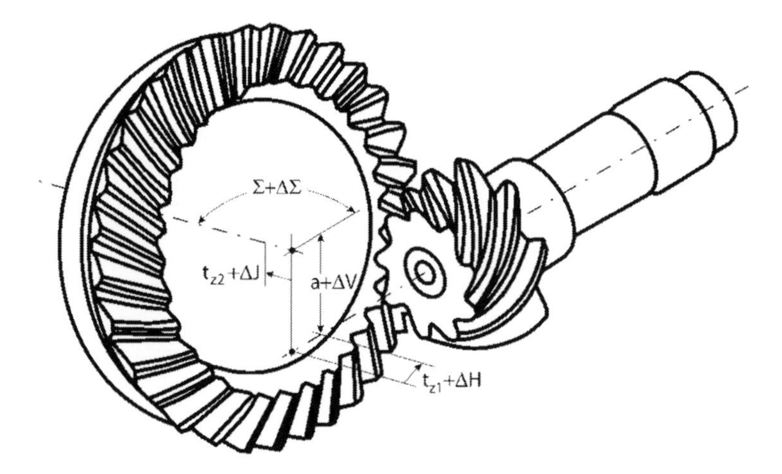

Fig. 4.38 Displacements of the gear axes

The relative positions of the pinion and wheel are, for example, described by the four possible forms in axis displacement and their deviations in a hypoid bevel gear set (Fig. 4.38):

- shaft angle Σ — shaft angle deviation $\Delta\Sigma$
- hypoid offset a — hypoid offset deviation ΔV
- pinion apex distance t_{z1} — pinion axial displacement ΔH
- wheel apex distance t_{z2} — wheel axial displacement ΔJ

The non-linear relationship between load and deflection can be taken into account by section-wise linearization or by an iterative procedure.

4.4.3.4 Calculating Tooth Flank Stress

The maximum normal stress acting in the contact area between meshing tooth flanks, also known as Hertzian stress, is employed as an equivalent variable in characterizing tooth flank stress, as in [DIN3990, DIN3991, ISO6336, ISO10300] (Fig. 4.39).

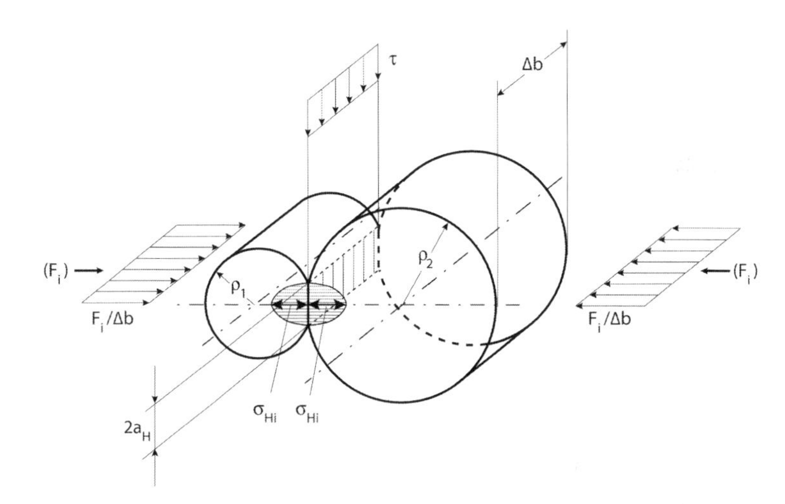

Fig. 4.39 Hertzian stress between two cylinders

In a manner similar to the calculation of local tooth deflection, the original Hertzian model of two contacting cylinders is modified in a model of two contacting 'inclined' cones pointing away from one another, and subdivided into sections (cylindrical segments). The local flank stress resulting from the load determined for the ith flank region F_{ni} (normal force) may be calculated as:

$$\sigma_{Hi} = \sqrt{\frac{E_i \cdot F_{ni}}{2\pi \Delta b (1 - \nu^2) \rho_i}} \tag{4.327}$$

Consideration of all discrete points i of a contact line provides the pressure distribution for a tooth in one meshing position. Repeating the procedure through the zone of action provides the total flank stress for a tooth during its engagement (Fig. 4.40).

For more advanced calculations, it is also possible to draw in all the components of the spatial stress state and calculate a comparative stress state depth wise. Shear stress τ caused by friction due to the normal load may also be superimposed to the stress components.

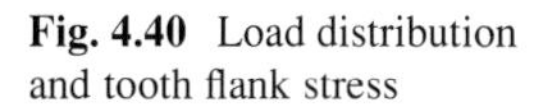

Fig. 4.40 Load distribution and tooth flank stress

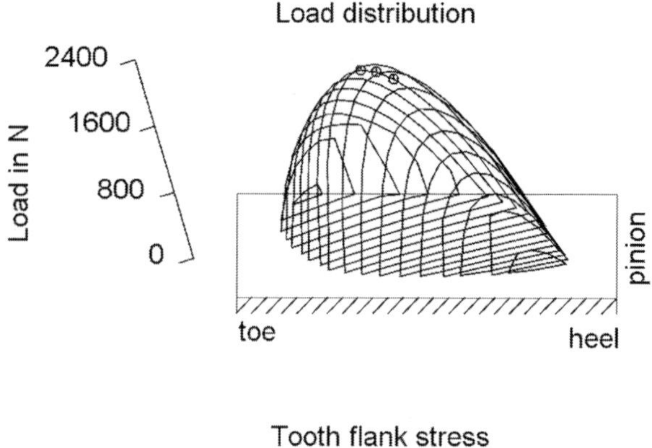

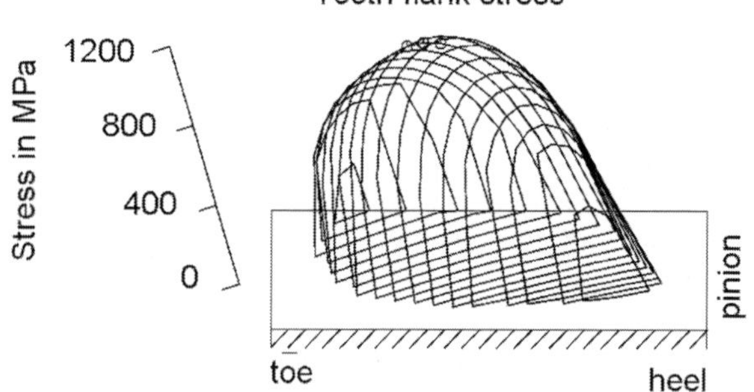

4.4.3.5 Calculating Tooth Root Stress

The required tooth root stress may be determined from the previously calculated load distribution.

Three-dimensional methods of numerical simulation (FEM, BEM, Finite Strips) can be used to calculate the local tooth root stress (stress due to the notch effect) in the face width direction using the known load distribution along the contact line. Since the stress distribution differs from the load distribution, the calculation must be performed not only for the meshing position with the highest local load, but for several positions, preferably for the complete zone of action.

In the second solution, the influence numbers method is used spatially to determine the nominal tooth root stress curve from the known load distribution, as in the deflection calculations. The local tooth root stress is then calculated for the tooth segments formed via spherical sections, using a plane method (DIN 3990, BEM, FEM).

The second solution is described below, making use of stress concentration factors according to DIN 3991 or employing a plane BEM method [HUEN01] (singularities method [LINK78]).

Calculation of tooth root nominal stress distribution The method is based on the same principles and conditions as the method for the determination of load distribution along a contact line. The main factor under consideration is the tooth root bending stress. The much smaller shear and compressive stresses may be obtained in a similar manner.

Each of the individual loads along the contact line (results of the load distribution calculation) generates a characteristic stress curve (influence function) in the tooth root (for example at the 30° tangent) across the face width, and so induces a stress component at an arbitrary point up to a distance of roughly $6{\cdot}m_{mn}$ from the application point at which the force is exerted. Superposition of all components of the individual forces provides the stress value for this point. If influence function S is then discretized and expressed in the form of mutual influence numbers s_{ij}, the nominal tooth root stress for a tooth segment i is obtained as:

$$\sigma_{\mathrm{F\ nom,\ i}} = \sum_{\mathrm{j=1}}^{\mathrm{n}} s_{\mathrm{ij}} \cdot F_{\mathrm{nj}} \tag{4.328}$$

Consideration of all discrete points in the tooth root provides the nominal stress distribution in a tooth for one meshing position.

Determination of the stress influence numbers The stress influence numbers may be calculated with the aid of a three-dimensional FE method [SCHA03] or on the basis of approximate solutions [SCHI00, BAUM01]. The method described here stems from investigations by Baumann [BAUM91, BAUM95] and Kunert [KUNE95, KUNE99], and relies on the generalization of results from a large number of FEM calculations, which have also to a large extent been confirmed experimentally.

As in the case of tooth deflection, the general fading function (see Fig. 4.34) covers the general tooth root stress curve resulting from a single force, irrespective of gear parameters, for a tooth of infinite length:

$$\text{for } |\xi^*| < 6\text{: } S_\infty(\xi^*) = 0.115\left[\cos\left(0.01\cdot|\xi^*|^3 - 0.16\xi^{*2} + 1.092\cdot|\xi^*|\right) + 1\right] \tag{4.329}$$

$$\text{for } |\xi^*| \geq 6\text{: } S_\infty(\xi^*) = 0 \quad \text{with } \xi^* = \xi/m_{mn} \tag{4.330}$$

The influence function for a tooth of finite width b and spiral angle β is in turn obtained by mirroring fading function S_∞, taking into account the obtuse and acute angled sides of the tooth face at the toe or heel by means of factor W_σ (Formula

4.324), and additionally determining the influence of the change in stress state on the tooth face when compared to the center zone (plane stress state/plane deformation state), expressed by the boundary stress reduction R.

Mirroring $$S = S_\infty + \Delta S_\infty \tag{4.331}$$

Multiple mirroring is necessary for gear widths $b < 6 \cdot m_{mn}$. If this limit goes below significantly, larger errors in the results must be anticipated.

Form of the tooth faces $$W_\sigma(\beta_{St}) = 1 \pm 0.72 \cdot |\beta_{St}| \tag{4.332}$$

+ for the acute angled side of the tooth face
− for the obtuse angled side of the tooth face
β in radians

Boundary stress reduction $$R(\xi^*) = \tanh(\xi_R^* + 1) \tag{4.333}$$

Influence function $$S = \left[S_\infty\left(\xi^* = \left|\xi_V^* - \xi_R^*\right|\right) + S_\infty\left(\xi^* = \xi_V^* + \xi_R^*\right)\right] \cdot W_\sigma \cdot R \tag{4.334}$$

where: $\xi_V{}^* = \xi_V / m_{mn}$ and $\xi_R{}^* = \xi_R / m_{mn}$ (see Fig. 4.35)

Influence numbers s_{ij} may be determined using the stress reference value N and function value S:

Influence numbers $$s_{ij} = S \cdot N_{ij} \tag{4.335}$$

s_{ij} describes the influence of the force F_{nj} exerted at point i on the stress at point i

Stress reference value $$N_{ij} = \frac{6 y_{Fj} \cdot \cos \alpha_j'}{s_{Fni}^2 \cdot \Delta b} \tag{4.336}$$

N_{ij} stress reference value for a tooth segment of width $\Delta_b$at point i as a result of loading at point j

The stress reference value N includes the working point, lever arm h, force application angle α' (see Fig. 4.36) of the single force applied at point j, and the tooth thickness s_{Fn} in the normal section of the tooth root at each considered point i. This value is calculated for a tooth segment modeled as a cantilevered beam.

Once the nominal stress curve has been calculated in three dimensions, the local stress (notch stress) on a plane model is determined in a second step. The stress concentration due to the notch effect in the tooth root of a gear is dependent on the fillet notch parameter $2\rho_{Fn}/s_{Fn}$ and on the lever arm of the load y_F/s_{Fn}, and is thus different for each tooth segment of the numerical model. The stress-increasing effect of the notch (in the fillet) is expressed for each case (now beeing plane) by means of a specific stress concentration factor Y_S. This is defined as the ratio of local stress to nominal stress:

Stress concentration factor $$Y_S = \frac{\sigma_{local}}{\sigma_{nominal}} \tag{4.337}$$

These factors can be calculated exactly for the fillet geometry of the considered tooth segment, with relatively little numerical computation effort, using 2D or 3D FE or BE methods (e.g., including grinding shoulders), or can be determined according to empirical relationships (e.g., in ISO 10300) as a function of notch parameter and lever arm. Since the stress concentration factor is likewise dependent on the lever arm y_F due to the influence of the shear stress, this must be established as an imaginary value in such a way that the ratio of bending to shear stress in the root of the tooth, determined by three-dimensional methods (calculated in the same way as bending stress), is in the same proportion [LINK96]. In a justifiable approximation [BAUM95, HUEN01], y_F can be identified with the real lever arm of the load at this point (individual force from the load distribution calculation). The stress correction factor according to Formula (4.72) for a section i of the concerned bevel gear tooth is then:

Stress correction factor

$$Y_{Si} = \left(1.2 + 0.13\frac{s_{Fni}}{y_{Fi}}\right) \cdot \left(\frac{s_{Fni}}{2 \cdot \rho_{Fi}}\right)^{\frac{1}{1.21 + 2.3/\left(\frac{s_{Fni}}{y_{Fi}}\right)}} \tag{4.338}$$

The local tooth root stress may now be calculated for each point i from the nominal stress and the stress concentration factor:

Local tooth root stress $$\sigma_{Fi} = \sigma_{F\ \mathrm{nom,\ i}} \cdot Y_{Si} \tag{4.339}$$

By repeating this procedure for the passage of a tooth through the entire zone of action, the complete tooth root stress is obtained as illustrated in Fig. 4.41.

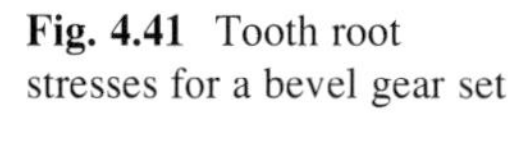
Fig. 4.41 Tooth root stresses for a bevel gear set

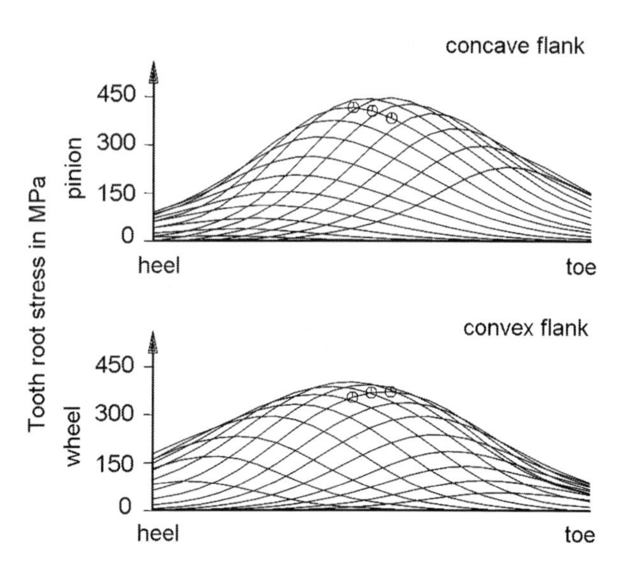

4.4.3.6 Calculation of Scuffing Strain, Frictional Loss and Efficiency

Scuffing strain Strain leading to scuffing damage on the tooth flanks is regarded as a critical instantaneous local contact temperature, consisting of the bulk and the flash temperature (method according to Blok [BLOK37]) or a mean surface temperature, formed from the sum of the bulk temperature and integrated (mean) flash temperature (integral temperature method according to Michaelis [MICH87]).

Blok's approach for the determination of the contact temperature uses the local parameters of load, radii of curvature, tangential speeds and coefficient of friction at each point on the tooth flank. For bevel gears, the cylindrical gear model used for the original approach can be transformed by converting the bevel gear tooth into finite tooth segments. The load variables are derived from the load distribution calculations in Sect. 4.4.3.2, the geometric and kinematic variables stem from the mathematical description of the tooth flanks and the speed of rotation. The local coefficient of friction is determined according to the current state of the art. Influences like, for example pull or push sliding, are not taken into account.

The instantaneous local contact temperature $\vartheta_{\text{con i}}$ is formed as the sum of the temperature of the bevel gear teeth prior to engagement (bulk temperature ϑ_{M}) and the maximum value of the temperature during the contact (flash temperature ϑ_{fla}). Unlike the methods described in Sect. 4.2.6, this method determines the temperatures in discrete calculations for finite tooth segments of the respective contact lines. The required values, such as tangential speed and tooth surface curvatures, are determined for the individual calculation points, not on the basis of an approximate virtual cylindrical gear but exactly from the tooth flank geometry and kinematics. The contact temperature at the discrete point i is thus calculated on the basis of the previously calculated loads per unit face width and of locally determined influence variables for each point, such as the coefficient of friction.

The analysis of discrete points i on the currently engaged contact lines yields the flash temperature distribution for one meshing position; repeating the process through the zone of action yields the complete scuffing stress for a tooth during one tooth engagement (Fig. 4.42).

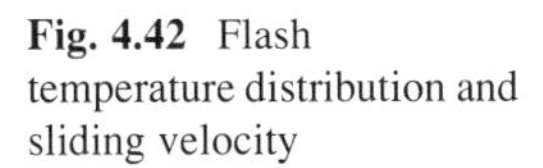
Fig. 4.42 Flash temperature distribution and sliding velocity

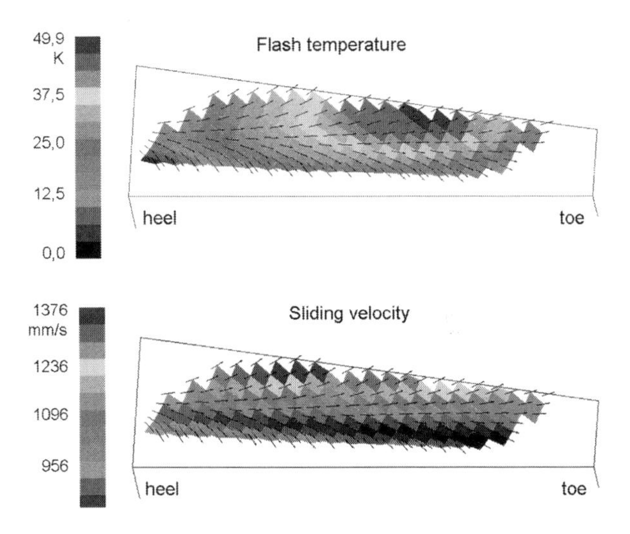

If one rather assumes that a mean value of the flash temperatures along the path of contact is decisive for scuffing, i.e., a lower temperature acting for a longer time, it is possible to integrate the temperature curve $\vartheta_{fla}(x)$ from the locally calculated flash temperatures ϑ_{flai} for a finite flank section (tooth segment) along the path of contact g_α, and then derive the integral temperature ϑ_{int}:

$$\vartheta_{int} = \vartheta_M + C_2 \cdot 1/g_\alpha \frac{\int_{x=0}^{x=g_\alpha} \vartheta_{fla}(x) \cdot dx}{g_\alpha} \approx \vartheta_M + C_2 \cdot \frac{1}{k} \sum_{i=1}^{k} \vartheta_{fla\,i} \quad (4.340)$$

where:
$C = C_2 = 1.5$ for bevel gears (see Formula (4.206))
$C = C_{2H} = 1.8$ for hypoid gears (see Formula (4.259))

The integral temperature is calculated for each tooth segment. The maximum value $\vartheta_{int,max}$ of the tooth segment temperatures ϑ_{int} represents the determinant value for the meshing tooth pair.

Frictional losses and efficiency The local values for load F_{ni}, tangential speeds v_{t1i}, v_{t2i} and local coefficient of friction μ_i are known from the previous stress calculations. It is therefore possible to calculate an instantaneous local frictional power loss $P_{fric,i}$ for the relevant point i of the teeth:

Frictional power loss $P_{\text{fric},i} = \mu_i \cdot F_{ni} \cdot \left| \vec{v}_{t1i} - \vec{v}_{t2i} \right|$ (4.341)

The instantaneous power loss $P_{\text{fric,j}}$ due to rolling/sliding of all engaged tooth pairs in a meshing position j is then the sum of all local losses resulting from this meshing position:

Total power loss $P_{\text{fric},j} = \sum_i P_{\text{fric},ij}$ (4.342)

The work transformed into frictional heat during the 'passage' through one meshing position j over time increment Δt_j is $W_{\text{fric,j}}$. Time increment Δt_j may be calculated from the discretization of the meshing points on the profile lines and the speed of rotation of the pinion as:

Friction work $W_{\text{fric},j} = P_{\text{fric},j} \cdot \Delta t_j$ (4.343)

Total friction work $W_{\text{fric}} = \sum_j W_{\text{fric},j}$ (4.344)

$j =$ all meshing positions within p_e

Driving power $W_{an} = M_{t1} \cdot \omega_1 \cdot \sum_j \Delta t_j$ (4.345)

Gear efficiency $\eta = 1 - \frac{W_{\text{fric}}}{W_{an}}$ (4.346)

4.4.3.7 Transmission Error and Tooth Stiffness Calculation

Tooth flank modifications, manufacturing deviations, and positional deviations from the gear environment and applied load lead to modifications of the path of contact and transmission errors, which affect the way bevel gears mesh under load. This may in turn produce significant effects on the dynamics of the transmission chain (see Sect. 5.4). Since tooth segments of varying stiffnesses are engaged in each meshing position, the transmission error, specified as an arc/distance (at radius r_{wm}) or as an angle, fluctuates as a function of the rotation position.

The transmission error in the normal tooth section represents angle φ_z described in Sect. 4.4.3.2 and related to the pinion, and may be taken directly from the solution vector of the system of equations used to calculate load distribution (Formula (4.312)). The calculation is performed for each mesh position i and is projected in the transverse section such that the transmission error for the entire tooth engagement is known. The total transmission error φ_{ges} is formed from the sum of the load-free transmission error φ_{korr} and the rotation angle error under load $\varphi_z \cdot \cos\beta_m$.

Total transmission error $\varphi_{ges} = \varphi_{korr} + \varphi_z \cdot \cos\beta_m$ (4.347)

Tooth stiffness governs the excitation of additional internal dynamic forces and gear noise as well as the resonance frequency of the gear pair. Tooth stiffness represents the ratio between an applied force and the resulting deflection. Theoretically, under load, all points on the contact lines are in contact; this state is approximately reached at the rated load of the gear. The tooth stiffness $c_{\gamma theo}(\varphi_1)$ acting in the transverse section at meshing position φ_1 is the sum of all tooth section stiffnesses c_i in the normal section.

$$\text{Tooth stiffness} \quad c_{\gamma theo}(\varphi_1) = \cos\beta_m \cdot \sum_{i=1}^{n} c_i(\varphi_1) \tag{4.348}$$

$$\text{with:} \quad c_i(\varphi_1) = \frac{F_{ni}}{\varphi_z(\varphi_1) \cdot r_{wm} - f_{ki}} \tag{4.349}$$

F_{ni} tooth force acting at segment i
f_{ki} distance from flank contact at segment i

By calculating the above tooth stiffness for each segment along the face width b and introducing corrective factor C_M (by analogy with DIN 3990) for contact elasticity components not covered in the calculation, such as surface roughness and lubricant, it is possible to obtain the specific contact stiffness c'_γ at meshing position φ_1 (Fig. 4.43):

$$\text{Specific contact stiffness} \quad c'_\gamma(\varphi_1) = C_M \cdot \frac{c_{\gamma theo}(\varphi_1)}{b} \tag{4.350}$$

$$\text{with} \quad C_M \approx 0.8$$

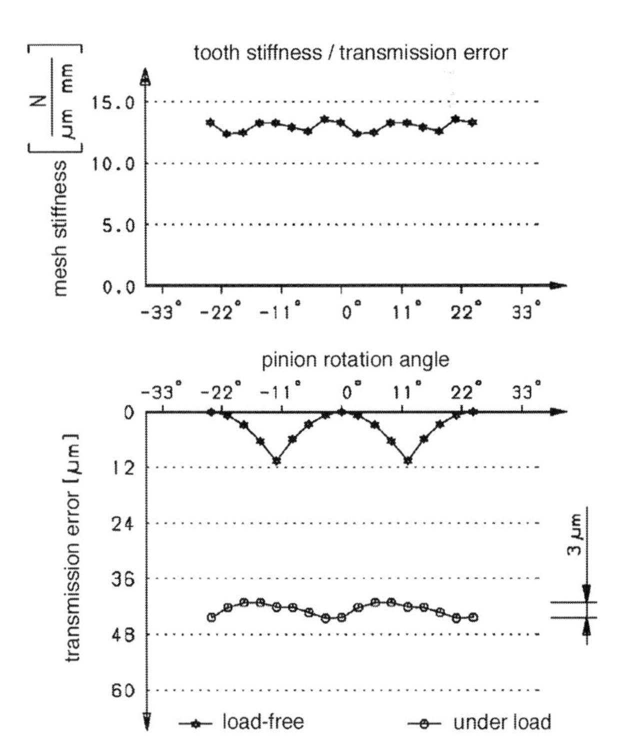

Fig. 4.43 Tooth stiffness and transmission error curves

4.4.3.8 Load Capacity and Safety Factor

Judging the load carrying capacity of the gears in a gear system is analogous to what is done in the standardized calculation methods (see Sect. 4.2), namely one compares the calculated stresses to an acceptable strength value and specifies a safety factor (Formulae (4.137), (4.171) and (4.205)):

$$\text{Safety factor} \quad S = \frac{strength}{strain} \tag{4.351}$$

A safety factor is calculated by comparing the calculated stress to a maximum value from which damage results such as permanent deformation, crack initiation, brittle fracture of the surface zone, forced fracture, material fatigue (tooth fracture, pitting, micro-pitting) and scuffing.

The complex stress analysis relies on a local calculation method (unlike the standard methods based on analytical/empirical approaches), in which stresses are determined for discrete locations on the tooth flank and tooth root. At a constant external load, the maximum value of the stress is usually decisive and is compared to a strength value. If a load spectrum is applied, damage due to the different load stages, which affect the size and position of the contact pattern, can be determined for each applied load.

The strength values and S-N curves from [ISO6336] were determined experimentally and are linked to the calculation methods by converting the maximum permissible torque determined in the test. The tooth flank and tooth root zone are considered at the same time and special influences are accounted for by means of influence factors. The strength value itself also contains non-quantifiable influences. These must be accounted for in the local analysis and, if necessary, allowed for in a local strength value. To ensure the local character of both the stress and strength values, the location with the maximum load need not necessarily be the location with the minimum safety factor.

This procedure is based on experimentally-backed studies on the load capacity calculation for hypoid gears according to [WIRT06] (see Sect. 4.2).

Tooth root load capacity Safety factor against fatigue fracture at point *i*

$$\text{Safety factor (tooth root breakage)} \quad S_{\mathrm{Fi}} = \frac{\sigma_{\mathrm{FPi}}}{\sigma_{\mathrm{Fi}}} \tag{4.352}$$

where:

σ_{Fi} local tooth root stress at point *i* (maximum value for the entire contact) according to Eq. (4.339)

σ_{FPi} local strength value at point *i* in the tooth root (strength value for point *i*)

$$\text{Local strength value} \quad \sigma_{\mathrm{FPi}} = \sigma_{\mathrm{FE}} \cdot Y_{\delta\mathrm{relTi}} \cdot Y_{\mathrm{RrelT}} \cdot Y_{\mathrm{X}} \cdot Y_{\mathrm{NT}} \tag{4.353}$$

where:

σ_{FE} allowable stress number (bending) according to Table 4.19
$Y_{\delta relTi}$ relative sensitivity factor according to Table 4.20 for point i
Y_{RrelT} relative surface factor according to Table 4.21
Y_X size factor according to 4.22
Y_{NT} life factor according to Fig. 4.20

Because the stress correction factor for bevel gears varies over the face width, the supporting effect (stress reduction due to the supporting effect of the neighboring points) is also variable and must be determined for each point i. On the other hand, the influence of the module, which varies along the face width, is neglected in determining the size factor.

The safety factor against permanent deformation, crack initiation or brittle fracture as a result of maximum loads may be determined by reference to [DIN3990] or [LINK96] and may be calculated using the specified local tooth root stress value .

Tooth flank load capacity Safety factor against pitting damage at point i

$$\text{Safety factor (pitting)}\ S_{Hi} = \frac{\sigma_{HPi}}{\sigma_{Hi}} \tag{4.354}$$

where:

σ_{Hi} local contact stress at point i according to Formula (4.327)
σ_{HPi} local strength value at point i on the tooth flank

Strength value for point i

$$\sigma_{HPi} = \sigma_{Hlim} \cdot Z_X \cdot Z_L \cdot Z_R \cdot Z_v \cdot Z_W \cdot Z_{NT} \cdot Z_{Si} \cdot Z_{Hypi} \tag{4.355}$$

where:

σ_{Hlim} allowable stress number according to Table 4.30
Z_X size factor according to Table 4.30
Z_L lubricant factor according to Table 4.31
Z_R roughness factor according to Table 4.32
Z_v speed factor according to Table 4.33
Z_W work hardening factor according to Table 4.34
Z_{NT} life factor according to Fig. 4.22
Z_{Si} slip factor for point i according to [FVA411]
Z_{Hypi} hypoid factor according to Table 4.36 for point i

Factors Z_{Si} and Z_{Hypi} depend on local speed conditions (tangential speed and slip) and therefore change at each point i on bevel gear tooth flanks.

The safety factor regarding permanent deformation, crack initiation or brittle fracture of the surface as a result of maximum loads may again be determined by reference to [DIN3990] or [LINK96] and calculated using the maximum specified local tooth flank contact stress value.

Scuffing load capacity Safety factor against scuffing according to [BLOK37]:

$$\text{Safety factor (scuffing) } S_B = \frac{\vartheta_S - \vartheta_{\text{oil}}}{\vartheta_{B\,i} - \vartheta_{\text{oil}}} \tag{4.356}$$

where:

ϑ_{Bi} local contact temperature according to Eq. (4.172) at point i
ϑ_{oil} lubricating oil temperature before contact entry
ϑ_{S} approximate scuffing temperature according to Table 4.47

Safety factor against scuffing by the integral temperature method is calculated according to Table 4.60

Local treatment of micro-pitting resistance is theoretically possible, and is also useful, due to the dependence of load capacity with speed conditions and local tooth flank contact pressure. At present, however, no studies backed by experimental test results are available to verify local load capacity.

References

[AGMA2003] ANSI/AGMA 2003-A86: Rating pitting resistance and bending strength of generated straight bevel, Zerol bevel, and spiral bevel gear teeth (1986)

[ANNA03] Annast, R.: Kegelrad-Flankenbruch, Diss. TU München (2003)

[BAUM91] Baumann, V.: Untersuchungen zur Last- und Spannungsverteilung an bogenverzahnten Kegelrädern. Diss. TU Dresden (1991)

[BAUM95] Baumann, V., Thomas, J.: Grundlagen zur Ermittlung der Zahnflanken- und Zahnfußbeanspruchung bogenverzahnter Kegelräder auf Basis experimentell gestützter Näherungsbeziehungen. FVA-Heft **429** (1995)

[BAUM01] Baumann, V., Bär, G., Haase, A., Hutschreiter, B., Hünecke, C.: Programm zur Berechnung der Zahnfußbeanspruchung an Kegelrad- und Hypoidgetrieben bei Berücksichtigung der Verformungen und Abweichungen der Getriebeelemente. FVA-Heft **548** (2001)

[BAYE96] Bayerdörfer, I., Michaelis, K., Höhn, B.-R., Winter, H.: Method to assess the wear characteristics of lubricants – FZG test method C/0,05/90:120/12, DGMK Information Sheet, Project No. 377, November 1996

[BECK91] Becker, E., Dorn, G., Stammberger, K.: Messung und Korrektur des Breitentragens von Zahneingriffen im Groß- und Sondergetriebebau. Antriebstechnik **30**(8) (1991)

[BLOK37] Blok, H.: Theoretical study of temperature rise at surface of actual contact under oiliness lubricating. In: Proceedings of General Discussion Lubrication. IME, London 2 (1937)

[BLOK67] Theyse, F.H.: Die Blitztemperaturhypothese nach Blok und ihre praktische Anwendung bei Zahnrädern. In: Schmiertechnik, Ausgabe Jan/Feb 1967

[BONG90] Bong, H.-B.: Erweiterte Verfahren zur Berechnung von Stirnradgetrieben auf der Basis numerischer Simulation und der Methode finiter Elemente. Diss. RWTH Aachen (1990)

[BOSE95] Börner, J., Senf, M.: Verzahnungsbeanspruchung im Eingriffsfeld – effektiv berechnen. Antriebstechnik **34** (1995)

[BSL03] Börner, J., Senf, M., Linke, H.: Beanspruchungsanalyse bei Stirnradgetrieben – Nutzung der Berechnungssoftware LVR. DMK 2003, Dresden (2003)

[DIN3990] Tragfähigkeitsberechnung von Stirnrädern (1994)

[DIN3991] Tragfähigkeitsberechnung von Kegelrädern ohne Achsversetzung (1988)

[DIN31652] Hydrodynamische Radial-Gleitlager im stationären Betrieb (1983)

[DIN51354] FZG-Zahnrad-Verspannungs-Prüfmaschine (1990)

[DNV41.2] Det Norske Veritas: Calculation of Gear Rating of Marine Transmissions, Classification Notes No. 41.2 (1993)

[DOLE03] Doleschel, A.: Wirkungsgradberechnung von Zahnradgetrieben in Abhängigkeit vom Schmierstoff, Diss. TU München (2003)

[FRES81] Fresen, G.: Untersuchungen über die Tragfähigkeit von Hypoid- und Kegelradgetrieben (Grübchen, Ridging, Rippling, Graufleckigkeit und Zahnbruch), Diss. TU München (1981)

[FVA2] FVA-Informationsblatt zum Forschungsvorhaben No. 2/IV Pittingtest (1997)

[FVA44] FVA-Forschungsheft No. 237 zum Forschungsvorhaben No.44/IV (1986)

[FVA54] FVA-Informationsblatt zum Forschungsvorhaben No. 54/7. Testverfahren zur Untersuchung des Schmierstoffeinflusses auf die Entstehung von Grauflecken bei Zahnrädern (1993)

[FVA125] Abschlussbericht zum FVA-Forschungsvorhaben No. 125/III. Zahnflankenlebensdauer (1995)

[FVA188] Abschlussbericht zum FVA-Forschungsvorhaben No. 188/II. Zahnfuß-Betriebsfestigkeit (1996)

[FVA243] FVA-Informationsblatt zum Forschungsvorhaben No. 243. EP-Fressen (1996)

[FVA319] FVA-Informationsblatt zum Forschungsvorhaben No. 319. Lastverteilungsmessung mit Feinpapier (2000)

[FVA345] FVA-Informationsblatt zum Forschungsvorhaben No. 345. Wirkungsgradtest (2003)

[FVA411] Abschlussbericht zum FVA-Forschungsvorhaben No. 411. Hypoid-Tragfähigkeit (2008)

[GAJE86] Gajewski, G.: Ermittlung der allgemeinen Einflussfunktion für die Berechnung der Lastverteilung bei Stirnrädern. Forschungsbericht TU Dresden (1986)

[GRAB90] Grabscheidt, J., Hirschmann, K.H., Kleinbach, K., Lechner, G.: Qualitätsbeurteilung von Getrieben durch Thermografie. Antriebstechnik **29**(5) (1990)

[HOSE78] Hohrein, A., Senf, M.: Untersuchungen zur Last- und Spannungsverteilung an schrägverzahnten Stirnrädern. Diss. TU Dresden (1978)

[HUEN01] Hünecke, C.: Untersuchungen zur Zahnfußbeanspruchung bogenverzahnter Kegelräder ohne und mit Achsversatz auf Basis der genauen Zahngeometrie. Diss. TU-Dresden (2001)

[INFA06] Wälzlagerkatalog. Schaeffler KG (2006)

[ISO6336] Calculation of load capacity of spur and helical gears, Part 1–3, 2nd edn (2006); Part 6, 2nd edn (2011)

[ISO10300] Calculation of load capacity of bevel gears, Part 1–3, 2nd edn (2014)

[ISO10825] Gears – Wear and damage to gear teeth (1995)

[ISO23509] Bevel and hypoid gear geometry (2006)

[ISO/TR13989] Calculation of scuffing load capacity of cylindrical, bevel and hypoid gears, Part 1–2, (2000)

[JARA50] Jaramillo, T.J.: Deflections and moments due to a concentrated load on a cantilever plate of infinite length. J. Appl. Mech. **17**; Trans. ASME **72** (1950)

[KAES77] Käser, W.: Beitrag zur Grübchenbildung an gehärteten Zahnrädern, Einfluss von Härtetiefe und Schmierstoff auf die Flankentragfähigkeit, Diss. TU München (1977)

[KNAU88] Knauer, G.: Zur Grübchentragfähigkeit einsatzgehärteter Zahnräder, Einfluss von Werkstoff, Schmierstoff und Betriebstemperatur, Diss. TU München (1988)

[KREN07] Krenzer, T.: The bevel gear. Rush NY **14543** (2007)

[KUBO81] Kubo, A.: Estimation of gear performance. In: International Symposium of Gearing & Power Transmission, Tokyo (1981)

[KUNE95] Kunert, J., Trempler, U., Wikidahl, F.: Weiterentwicklung der Grundlagen zur Ermittlung der Lastaufteilung und Lastverteilung bei außenverzahnten Gerad- und Schrägstirnrädern durch Verformungs- und Spannungsmessung. FVA-Heft **458** (1995)

[KUNE99] Kunert, J.: Experimentell gestützte Untersuchungen zum Verformungs- und Spannungsverhalten an außenverzahnten Stirnrädern für eine verbesserte Beanspruchungsanalyse. Diss. TU Dresden (1999)

[LAST78] Lang, O.R., Steinhilper, W.: Berechnung und Konstruktion von Gleitlagern mit konstanter und veränderlicher Belastung. Springer, Berlin (1978)

[LINK78] Linke, H.: Ergebnisse und Erfahrungen bei der Anwendung des Singularitätenverfahrens zur Ermittlung der Spannungskonzentration an Verzahnungen. Wiss. Zeitschrift der TU Dresden, Heft 3/4 (1978)

[LINK96] Linke, H.: Stirnradverzahnung. Hanser-Verlag, München (1996)

[MICH87] Michaelis, K.: Die Integraltemperatur zur Beurteilung der Fresstragfähigkeit von Stirnradgetrieben. Diss. TU München (1987)

[MITS83] Mitschke, W.: Beitrag zur Untersuchung des Einflusses der Radkörpergestaltung auf die Lastverteilung der Verzahnung. Diss. TU Dresden (1983)

[NEUP83] Neupert, B.: Berechnung der Zahnkräfte, Pressungen und Spannungen von Stirn- und Kegelradgetrieben. Diss. TH Aachen (1983)

[NIEM86.2] Niemann, G., Winter, H.: Maschinenelemente Band II. Springer, Berlin (1986)

[OEHM75] Oehme, J.: Beitrag zur Lastverteilung schrägverzahnter Stirnräder auf der Grundlage experimenteller Zahnverformungsuntersuchungen. Diss TU Dresden (1975)

[PLAC88] Placzek, Th.: Lastverteilung und Flankenkorrektur in gerad- und schrägverzahnten Stirnradstufen. Diss. TU München (1988)

[SABL77] Sablonski, K.I.: Das Zahnradgetriebe – Lastverteilung. Verlag Technik, Kiew (1977)

[SCHA03] Schäfer, J.: Kontaktmodell mehrfacher Zahneingriff – Berechnung von Stirnradpaarungen mit Mehrfacheingriffen mit Hilfe der Finiten-Elemente-Methode. FVA-Heft **723** (2003)

[SCHI00] Schinagl, S.: Ritzelkorrektur, Programmbeschreibung (RIKOR G). FVA-Heft **481** (2000)

[SCHM73] Schmidt, G.: Berechnung der Walzenpressung schrägverzahnter Stirnräder unter Berücksichtigung der Lastverteilung. Diss. TU München (1973)

[SCHR00] Schrade, U.: Einfluss von Verzahnungsgeometrie und Betriebsbedingungen auf die Graufleckentragfähigkeit von Zahnradgetrieben, Diss. TU München (2000)

[THOM98] Thomas, J.: Flankentragfähigkeit und Laufverhalten von hart-feinbearbeiteten Kegelrädern, Diss. TU München (1998)

[WEBE53] Weber, C., Banaschek, K.: Formänderung und Profilrücknahme bei gerad- und schrägverzahnten Rädern. Schriftreihe Antriebstechnik, Heft 11, Vieweg, Braunschweig (1953)

[WECH87] Wech, L.: Untersuchungen zum Wirkungsgrad von Kegelrad- und Hypoidgetrieben, Diss. TU München (1987)

[WICH67] Wiche, E.: Radiale Federung von Wälzlagern bei beliebiger Lagerluft. Konstruktion **19** (1967)

[WIRT06] Höhn, B.-R., Michaelis, K., Wirth, Chr.: Entwicklung eines Berechnungsverfahrens zur Grübchen- und Zahnfußtragfähigkeit von Hypoidrädern. FVA-Forschungsreport 2006, Würzburg (2006)

[ZFAG] Zahnradfabrik Friedrichshafen AG

[ZIEG71] Ziegler, H.: Verzahnungssteifigkeit und Lastverteilung schrägverzahnter Stirnräder. Diss. RWTH Aachen (1971)

Chapter 5
Noise Behavior

5.1 Causes of Noise Generation

The main causes of noise generation in a gearbox are the gears themselves. Gear teeth meshing periodically excite vibrations in a gear system, on the one hand causing the emission of direct air-borne noise and on the other hand conveying structure-borne noise to the surface of the housing, where it is radiated as indirect air-borne noise (Fig. 5.1).

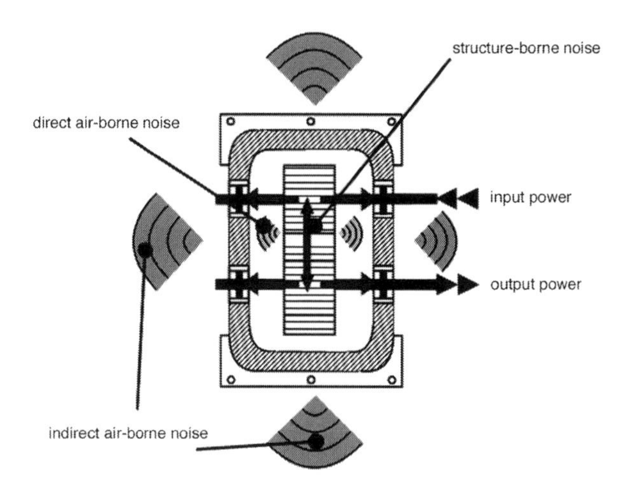

Fig. 5.1 Noise generation in a gearbox

To clarify the mechanisms which effectively determine the vibration behavior of gears, Fig. 5.2 shows the torsional vibration model of a single stage cylindrical gear set. The observations which follow apply to both cylindrical and bevel gears. The

J. Klingelnberg (ed.), *Bevel Gear*, DOI 10.1007/978-3-662-43893-0_5

model is essentially composed of the two gear bodies of the gear set connected by means of a coupling element. This torsional vibration model is linked to the gear housing via spring and damper elements representing the bearing and housing stiffnesses and their damping properties.

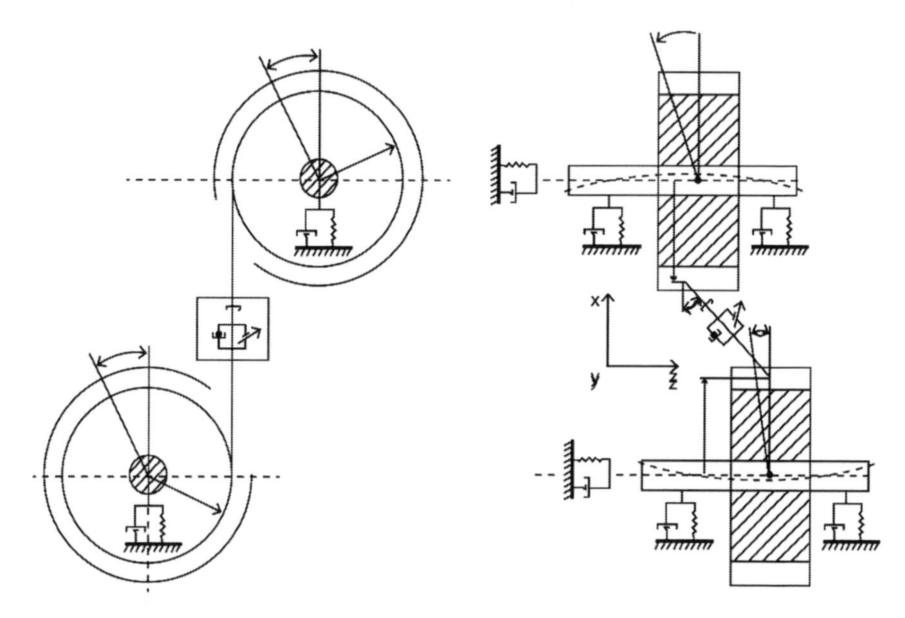

Fig. 5.2 Vibration model of a single stage cylindrical gear set

The three most important variables influencing internal excitation and the behavior of the vibration system are path-induced excitation, meshing impact excitation and parameter induced excitation like variation in the stiffness of meshing gear teeth [VDI90]. It should be noted that excitation by meshing impact at contact entry, although listed separately here, is also in fact a form of path-induced excitation [TESC69], [TOSH61].

Path-induced excitation on cylindrical gears is attributable either to geometrical deviations in the involute profile caused in manufacturing or to deliberately induced modifications of the flank form [TOPP66]. In the case of bevel gears, this means figuratively that flank topography deviates from the theoretically exact conjugate gear. This may be due either to deliberate design of the flank topography or the ease-off, or to manufacturing-induced deviations in the flank topography. Tooth flank topography deviations induced by gear cutting also occur on bevel gears, as very few cutting processes are available to produce conjugate gears. Such deviations are usually accepted in order to benefit from other advantages of a gear cutting method, for example high productivity.

Apart from deviations in tooth flank topography, manufacturing- and load-induced pitch deviations also lead to path-induced excitation. Figure 5.3 exemplifies the relationships between tooth flank form deviations and pitch deviations for the quasi-static transmission error of a cylindrical gear pair with involute profiles. It is apparent from the shape of the profile deviation that the two exact profiles make contact only around the pitch circle. At this point the transmission error becomes zero. Profile deviations in areas above and below the pitch circle cause a parabolic curve as transmission error per each tooth mesh (Item 1 of Fig. 5.3).

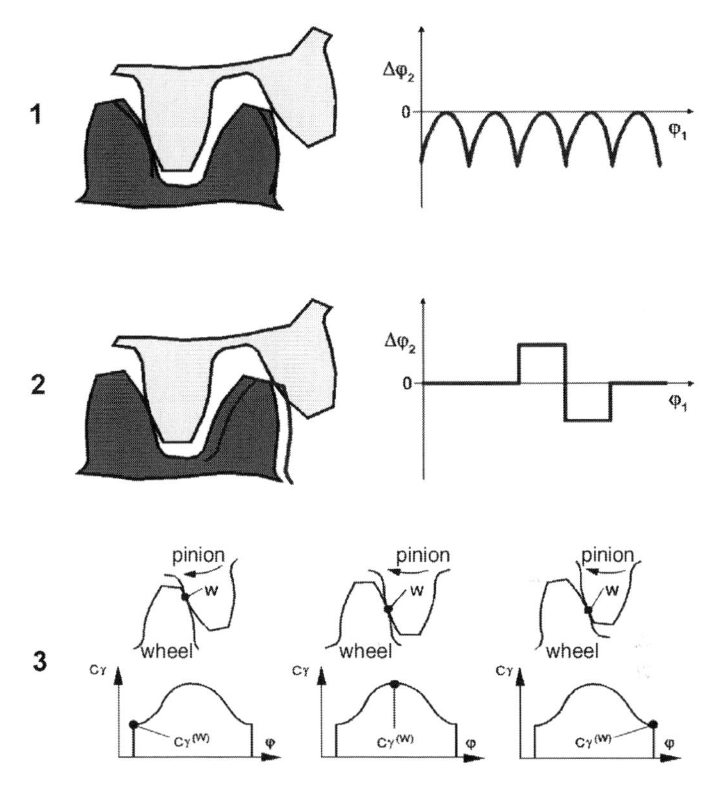

Fig. 5.3 Excitation mechanisms for gear noise.

1 Path-induced excitation, caused by flank modification, load- or assembly-induced topography deviations.

2 Impact excitation: pulsatile path-induced excitation, caused by manufacturing- or load-induced pitch deviations.

3 Parameter excitation: stiffness variation, caused by variable meshing tooth pair stiffness along the path of contact

Item 2 in Fig. 5.3 depicts the influence of a pitch deviation on the load-free transmission error. Since the tooth profile is not affected by the pitch deviation, the transmission error changes in the form of a step-shaped leap, whose amplitude corresponds to the pitch deviation converted to an angle. The resulting discontinuity in transmission error at the transfer point causes enormous accelerations, even in the load-free state. In operation, this leads to pulsatile loads on the tooth flanks. The term "pulse" is therefore preferred to the term "path-induced excitation". The importance of the pulse effect may be affected by form and pitch deviations. More thorough investigations in the relationships between variations in gear tooth geometry and load-free transmission error have been carried out, particularly by [FAUL68], [OPIT69] and [NAUM69]. Apart from pitch and tooth flank topography deviations which may excite vibrations in a gear set, stiffness variation of the kind shown in Item 3 of Fig. 5.3 constitute a third cause in tooth mesh noise generation. The depth wise variation in tooth thickness causes variations in mesh stiffness of a tooth pair, which in turn leads to the cyclical excitation of the gear set [BOSC65].

Whereas numerous studies on cylindrical gears have demonstrated a clear correlation between contact geometry, stiffness and noise behavior [BOSC65], [MÖLL82], [WECK92], stiffness-related evaluation of the running behavior of bevel gear sets has hitherto been confined to the macro-geometry [BECK99]. One reason for this may be seen in the inadequate accuracy of the calculation methods used to determine the tooth pair stiffness curves while taking the exact contact parameters into account. Parametric vibration generation being of special importance for all further dynamic considerations, new approaches to calculate dynamic tooth forces, discussed in greater detail in Sect. 5.4, are currently being explored. Thus, greater importance has been devoted to the transmission error curve than to the tooth pair stiffness curve in the noise optimization of bevel gear sets. A comparison of the transmission errors of cylindrical and bevel gears shows that transmission error amplitude in bevel gears frequently reach values more than twice as large as those in cylindrical gears [WECK99], [WECK00], [WECK02].

Vibrations may also be generated in gear teeth by the surface structure of the tooth flanks where waviness, essentially a series of form deviations, is generally regarded as a separate case, being normally smaller than crowning, pressure and spiral angle and tooth flank bias deviations.

Various measures may be employed to influence the noise behavior of a gear system. Currently, the most effective measures are primary active measures, i.e. reducing meshing vibration [DIET99]. The next chapter describes possible methods and tools which can be used to systematically minimize noise excitation when designing bevel and hypoid gears.

5.2 Noise Excitation by Means of Gear Tooth Design

5.2.1 Optimizing the Macro Geometry

Demands made on noise emission levels, especially for automotive gearing, are increasing as a result of generally improved sound comfort in vehicles. Vehicle inspectors and media representatives judge engine, wind and driving noise very critically in tests. Vehicles are already fitted with noise-damping materials at many points, but these involve increases in weight and cost. The most important goal of the bevel gear designer is therefore to minimize the excitation arising from the gear teeth themselves. Apart from the macro geometry (gear body), the micro geometry (ease-off), surface structure and gear quality are all of considerable importance for noise behavior. Their influences are dealt with in Sects. 5.2.2 and 5.3.

There are a number of ways to minimize running noise. The first step is to optimize the macro geometry. Here it is important to note that measures taken to increase the contact ratio may also lead to a reduction in load capacity (see Table 5.1). Several basic measures to improve noise properties are listed below. The outside diameter of the wheel has been kept constant in the different possible variations considered. The quantitative statements remain valid even if changing a parameter improves some properties of a gear but makes other properties disproportionately worse. The design of a bevel gear set must always take many criteria into account, and is hence a compromise between various requirements. In order to achieve the maximum possible use of the material, it is desirable to balance flank and root load according to safety factors. As will be evident from Table 5.1, it will be beneficial in terms of noise emission to compare the results on the basis of the total contact ratio and to modify the compromise design in favour of lower noise excitation. In a further step, the transmission error and hence the effective contact ratio are calculated by means of a micro geometry optimization. The examples used in the diagrams in this chapter have been selected from a large number of production gear sets and are representative of the complete hypoid gear range employed in rear axle drive passenger cars.

Table 5.1 Effects of macro geometry design parameters on gear evaluation

	Having effects on:		
Increasing of:	**Contact ratio**	**Tooth root load capacity**	**Manufacturing costs**
Hypoid offset, a	+	+	–
Number of teeth, z	+	–	–
Whole depth, h_m	+	–	–
Module, m_{mn}	–	+	–
Spiral angle, β_m	+	–	–
Pressure angle, α_e	–	+	–
Tool radius, r_{co}	–	–	+
Face width, b	+	+	–

Contact Ratio The transmission error and hence the effective total contact ratio can be determined with sufficient accuracy by means of a tooth contact analysis (Fig. 5.4).

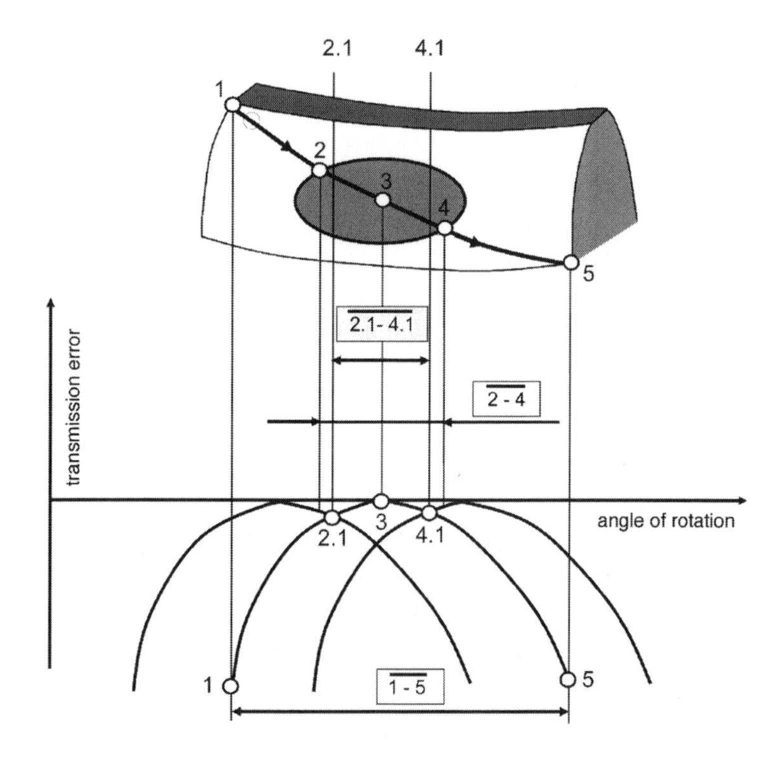

Location along path of contact	Designation
Point 1 to Point 5	Maximum contact of a single tooth
Point 2.1 to Point 4.1	Pitch
Point 2 to Point 4	Effective tooth contact, dependent on load
$\frac{\text{single tooth contact}}{\text{pitch}}$	Maximum total contact ratio
$\frac{\text{effective tooth contact}}{\text{pitch}}$	Effective total contact ratio

Fig. 5.4 Definition of the effective total contact ratio

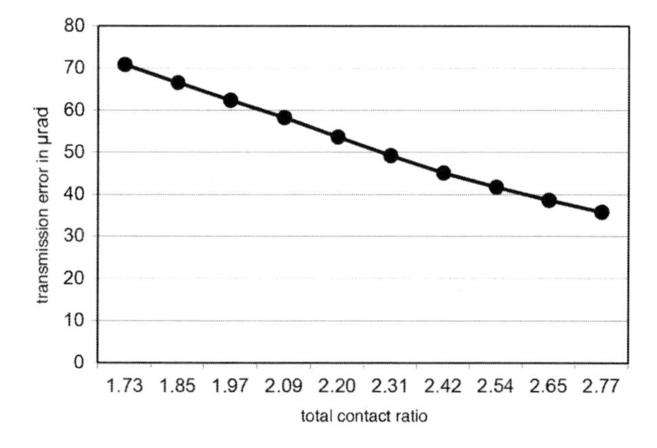

Fig. 5.5 Example of the dependence between transmission error and contact ratio

Gears with the same transmission ratio but differing tooth numbers were chosen for the curve shown in Fig. 5.5. The macro- and micro geometry were kept mainly constant for the calculation. There is good correlation between the total contact ratio and the transmission error.

Hypoid Offset It is helpful to use the relative offset a_{rel} (see 2.2.5.1) to calculate the offset in relation to noise excitation, since the offset must always be considered in proportion to the wheel diameter. As is evident from Fig. 5.6, the offset contributes only negligibly to the increase in contact ratio if the sum of the face widths and the sum of the spiral angles of pinion and wheel are kept constant. If the face width and the spiral angle solely of the wheel are kept constant, there is much more influence. In this case, a larger relative offset increases the face width and the spiral angle on the pinion, and hence the total contact ratio (Fig. 5.7).

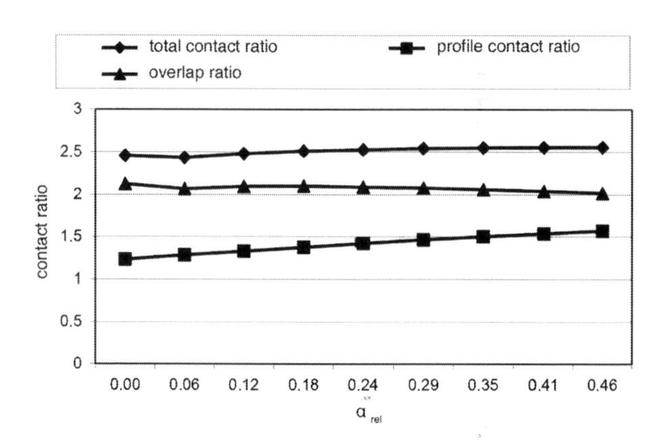

Fig. 5.6 Contact ratio as a function of the relative offset a_{rel}, given a constant sum of pinion and wheel faces widths and spiral angles

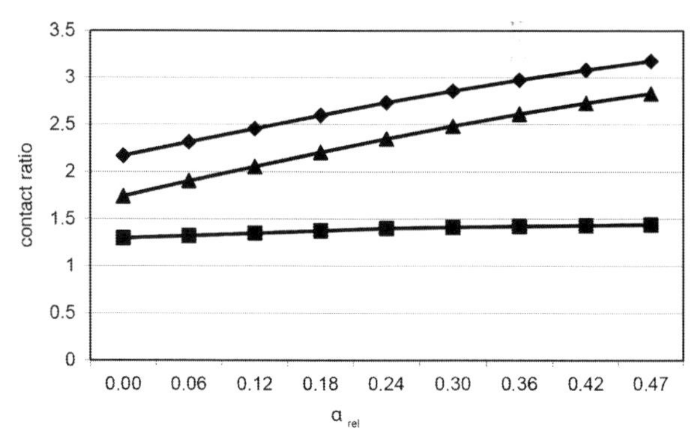

Fig. 5.7 Contact ratio as a function of the relative offset a_{rel}, given a constant face width and spiral angle on the wheel

Unlike gears without offset, longitudinal sliding motion over the tooth flank is present in a hypoid gear set (see Sect. 2.4.3). Sliding increases with an increasing relative offset a_{rel}, and causes friction which usually exercises a damping effect on noise excitation. Drawbacks are poorer efficiency and the resulting power loss. A limit is reached when the heat generated by sliding can no longer be dissipated by convection through the gear housing or through a radiator, which both leads to overheating of the gear set (Figs. 5.8 and 5.9).

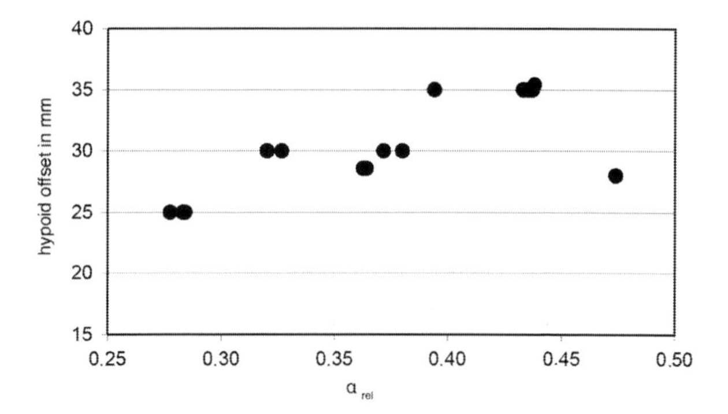

Fig. 5.8 Hypoid offset and relative offset a_{rel} in selected gear examples

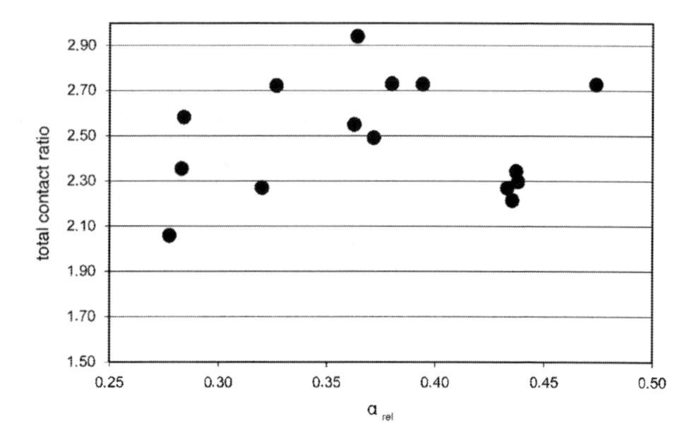

Fig. 5.9 Total contact ratio as a function of the relative offset a_{rel} in selected rear axle drives of passenger cars

Numbers of Teeth The number of teeth used for the driving pinion in rear axle gear sets generally range between 7 and 19, and those for the wheel between 30 and 50. Accordingly, transmission ratios vary from $i=2.5$ to $i=6$. The exceptions are axle gears which drive the transmission shafts of four-wheel drive vehicles with transversely front-mounted engines (Fig. 1.3). These have transmission ratios from $i=1$ to $i=2$ with tooth numbers from 20 to 40. The sums of the tooth numbers for the selected examples vary from 41 to 65. Figure 5.10 shows the chosen sum of tooth numbers and the resulting total contact ratios for 15 production ground rear axle gears. In all cases, the designers have aimed to find a compromise between quiet running, load capacity and manufacturing costs. Various design strategies have been employed, depending on the vehicle. Figure 5.11 shows the relationship between the sum of tooth numbers and the contact ratio, given a constant transmission ratio and identical wheel diameter. The increase in the contact ratio with an increasing number of teeth is clearly apparent.

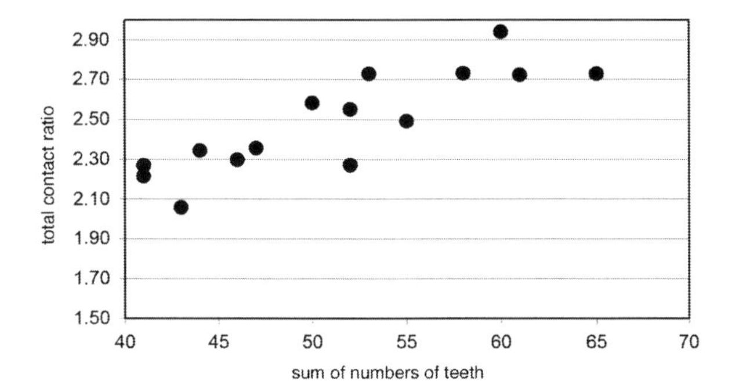

Fig. 5.10 Total contact ratio of sample gears as a function of the sum of pinion and wheel tooth numbers

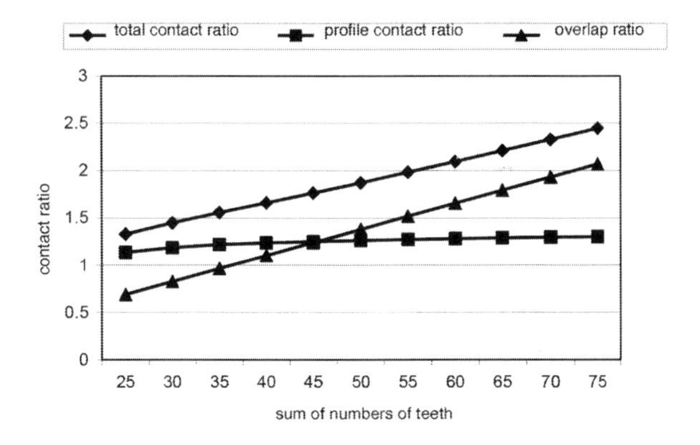

Fig. 5.11 Total contact ratio as a function of the sum of pinion and wheel tooth numbers

Whole Depth The whole depth is calculated as the sum of the addendum and dedendum. The difference between the pinion addendum and wheel dedendum, rectified to allow for profile shift, is the clearance. Owing to the desired magnitude of the tooth root radius or to the curved tooth root shape induced by tilting of the tool, the distance between the tip and root must be at least 0.25–0.30 m_{mn} in order to prevent interference. The whole depth is therefore usually 2.25–2.30 m_{mn} (cf. Sect. 3.1). Larger whole depth factors increase the lever arm of the point at which the force is applied and, hence, the bending moment on the tooth; smaller whole depth factors increase contact pressure because the tooth flank area diminishes. To increase the contact ratio when the space for installing the gears is limited, the whole depth can be made larger than recommended using standard factors. Gears whose whole depth is enlarged in this way are referred to collectively as "high profile gears". Different whole depth factors are chosen to allow reworking of the flanks. Work pieces can be reworked if the clearance in the basic design is sufficiently large. Reworking of the pinion or wheel reduces the thickness of the tooth. The resulting larger backlash is compensated by reducing the installed distances, and a functioning gear set is still obtained.

Different addendum and dedendum factors are used for different vehicle gear systems. Figure 5.12 shows the extent to which these design strategies differ, using the 15 rear axle gear sets from Fig. 5.10 as examples.

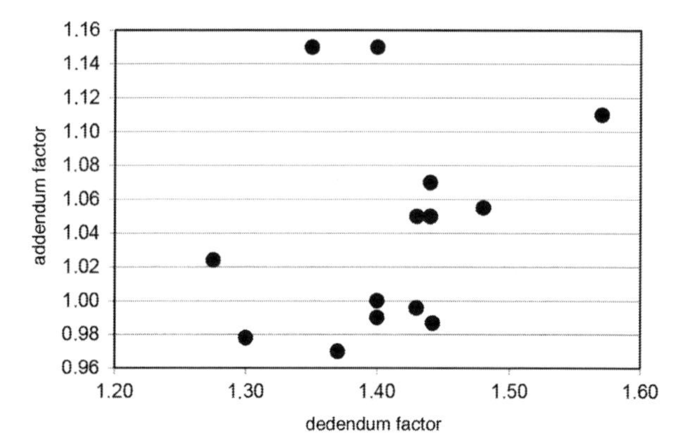

Fig. 5.12 Addendum and dedendum factors of ground gears for rear axle drives of passenger cars

Spiral Angle The spiral angle is frequently used to alter the ratio between the axial and radial tooth forces. The spiral angle is also a variable which can be used to modify the stress on the tooth flank and tooth root in a given space without changing the transmission ratio. Enlarging the spiral angle increases the developed face width and therefore improves the overlap ratio. A disadvantage is the smaller normal module with consequentially reduced safety factors against tooth root breakage. Balanced safety factors for tooth flank and tooth root load capacity are generally achieved for ordinary case hardening steels at a spiral angle of about 35–38° which allows optimum use of the material [SCHW94]. Spiral angles of 25–45° are therefore in general use. In hypoid gear sets, the pinion spiral angle is larger than that of the wheel. The pinion spiral angle is usually around 50° and varies from 25° to 35° for the wheel (cf. Sect. 3.1). In Fig. 5.13, the spiral angle varies in order to reveal the increase in contact ratio with an increase in spiral angle.

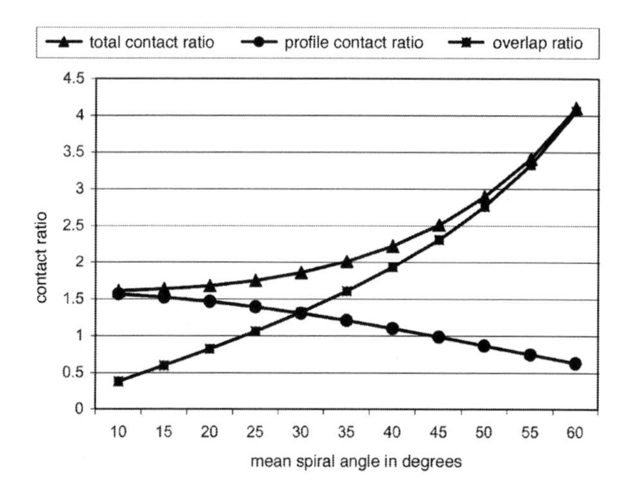

Fig. 5.13 Total contact ratio, profile contact ratio and overlap ratio on modified flanks as a function of spiral angle

Pressure Angle In Fig. 5.14 a parametric study is performed, for a transmission ratio of i = 1:3, with the pressure angle as the design criterion. The macro geometry and ease-off are kept identical. When the pressure angle increases from 15° to 25° the total contact ratio falls from 2.379 to 2.24. The overall influence of the pressure angle on the total contact ratio is rather small, and the effect on anticipated running noise appears negligible. The influence on transmission error may be even smaller, as small pressure angles help so-called diamond contact patterns to form—diamond contact patterns exhibit asymmetric contact pattern lengths in the profile direction—, slightly altering the path of contact and possibly nullifying the advantage of a higher total contact ratio. In gear sets where noise under load is closely related to changes in tooth stiffness, the pressure angle may exert an appreciable influence on noise emission. Pressure angles larger than 20° have proved positive in noise reduction in such gear sets.

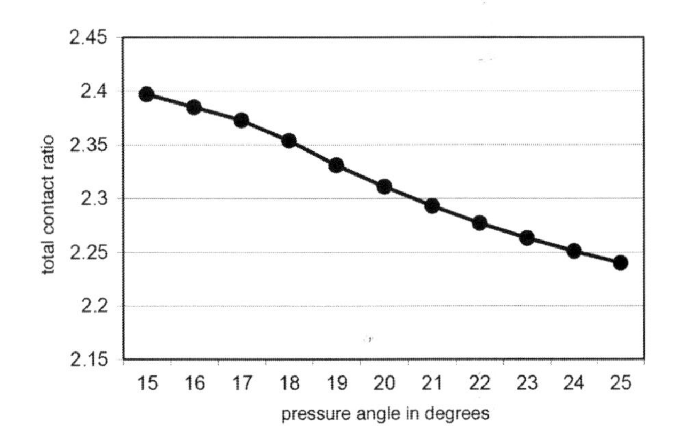

Fig. 5.14 Total contact ratio as a function of the pressure angle

Tool Radius The chosen tool radius influences load carrying capacity, contact pattern sensitivity to relative displacement, noise behavior and, last but not least, the manufacturing costs of a gear set. A number of studies have already demonstrated the influence of tool radius on stresses in the tooth root and flank [SCHW94], [BAGH73]. Effects on the position of the contact pattern were also taken into account in these studies. The gear tooth properties as a function of the tool radius are summarized under the terms small-cutter and large-cutter design, as described in Sect. 3.1. The ratio of tool radius to mean cone distance is the distinguishing criterion. This ratio, r_{c0}/R_{m2}, varies from 0.71 to 1.2 for for the sample gears. For a tool radius less than 75 % of the cone distance, the load capacity of the gear set increases but designing the ease-off becomes more difficult.

In single-indexing cutting processes, tool tilt needed to generate lengthwise crowning must be increased for smaller r_{c0}/R_{m2} ratios such as to ensure that the effective difference in radius between the inside and outside cutting edges is kept constant. In turn, increased tilt requires a larger blade angle on the inside cutting edge and a smaller angle on the outside cutting edge. Practical limits are approximately a maximum of 35° for the inside blade and a minimum of 8° for the outside blade (cf. Sect. 2.2.5.3). Tool radius does not vary continuously since cutters are manufactured in fixed sizes and offer only limited blade radial adjustment (Fig. 5.15).

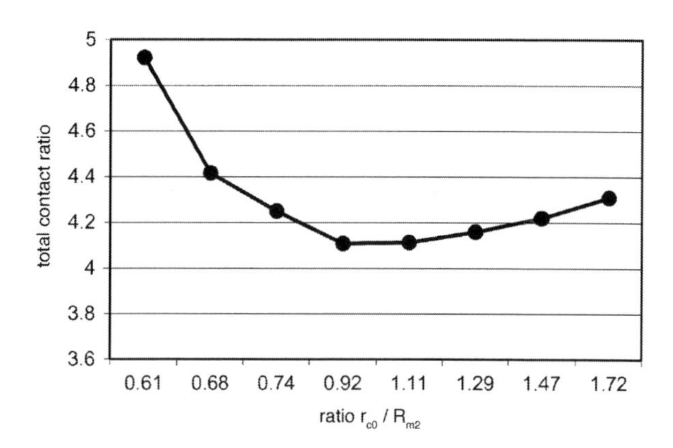

Fig. 5.15 Effects of tool radius on total contact ratio, given identical macro geometry

5.2.2 Optimizing the Micro Geometry

Software packages have been produced to determine the influence of micro geometry on running behavior. The programs include tooth contact simulation and analysis, and calculate the theoretical transmission error curve. Transmission error sensitivity can be determined by varying individual machine settings and tool parameters in the tooth contact analysis. Initially, both transmission error and the effective total contact ratio are determined load-free. Although gears are never truly load-free in operation, practice has shown that load-free simulation correlates well with operation at low loads. For example, loads in a passenger car axle gear set will be low when a vehicle is moving at a constant speed of less than 50 km/h. In this low load operating state, vehicles emit relatively little wind noise and tire noise such that other noise sources are not masked. At speeds between 80 and 100 km/h, tooth mesh frequencies of the axle unit are between 300 and 600 Hz; they are easily heard by the human ear and often perceived as a nuisance. Combining the greatest possible effective total contact ratio in low load operation with adequate displacement capability in full load operation is an important goal for the development of bevel gear micro geometry.

5.2.3 *Influence of Gear Crowning*

When calculating the amount of crowning required, it is necessary to account for the "environment" of the gear set (see Sect. 4.4.3.3), where displacements are analyzed in four directions, i.e. variations along the pinion axis, along the wheel axis, along the pinion offset and of the shaft angle (Fig. 4.38) [HAGE71]. Crowning must be chosen in such a way that tooth flank damage due to unfavorable load concentration does not occur in a representative load spectrum for the gear set (Fig. 5.16).

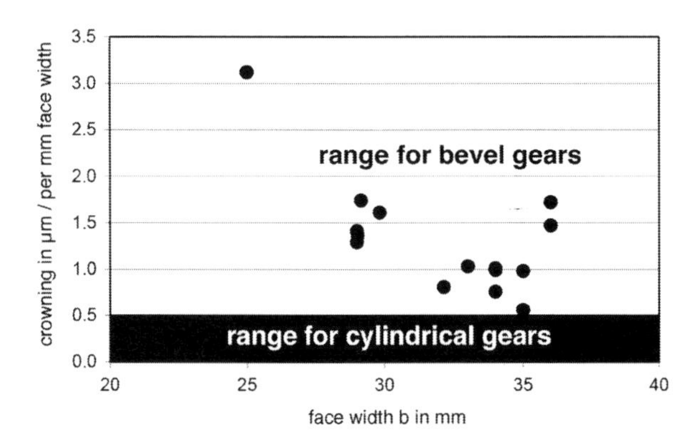

Fig. 5.16 Applied crowning for ground bevel gears compared to similarly loaded cylindrical gears

The way in which crowning is designed has a dominant influence on the running noise of a gear set. A number of important relationships are accordingly shown in Figs. 5.17, 5.18 and 5.19. The gear set being modified is a randomly selected bevel gear with 35 mm face width and 10 mm whole depth, of the kind generally used in passenger car rear axle units. The plotted transmission error values are intended as a qualitative indication of the curves for the dependent variables. Other gears may exhibit different behaviors.

As the results show, transmission error decreases with diminishing crowning or if the gear set has a "bias-in" characteristic (see Sect. 3.4.3). The only changes made to generate the curves in Figs. 5.17 and 5.18 were to the ease-off, all other parameters being kept constant. For the sake of simplicity, only circular crowning was used. Modern gear design and manufacturing offers a number of additional possibilities allowing crowning not only in circular form, but with different curvatures along the path of contact. The objective in crowning modifications is to keep crowning in the contact area for low-load operation as small as possible, while significantly increasing crowning in the outside areas to provide adequate displacement behavior. Modifications for this purpose are possible on both the pinion and the wheel (see Sect. 3.4.5 and Fig. 3.30).

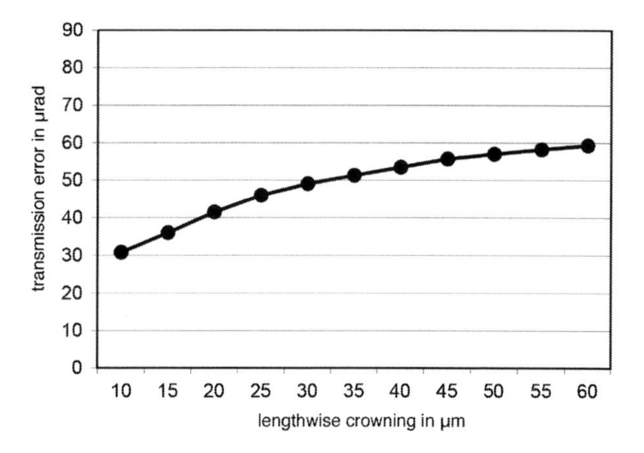

Fig. 5.17 Transmission error when lengthwise crowning is modified

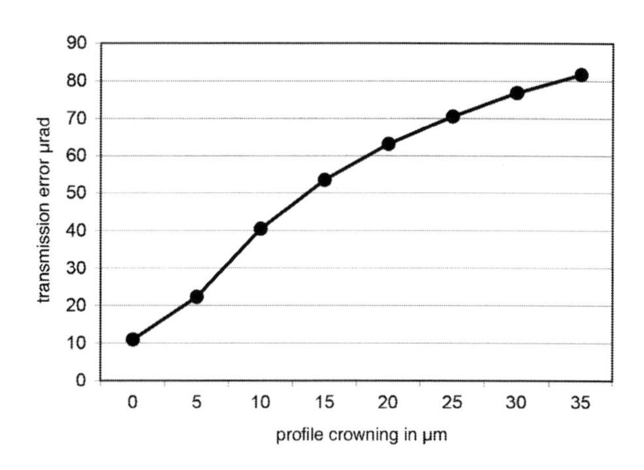

Fig. 5.18 Transmission error when profile crowning is modified

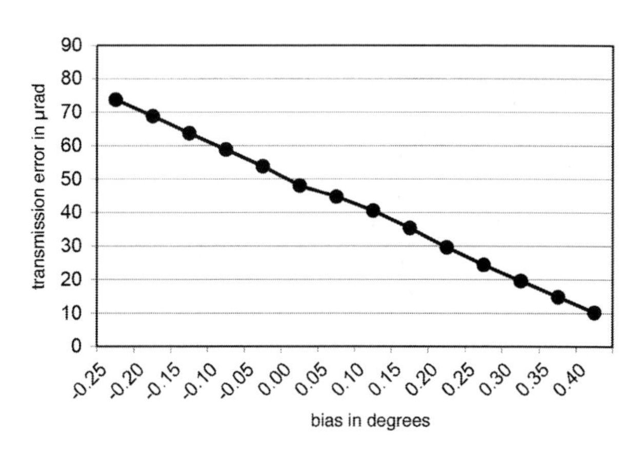

Fig. 5.19 Transmission error when bias varies

5.3 Noise Excitation Governed by Manufacturing

5.3.1 Influence of Gear Deviations on Transmission Error

Tooth flank topography of bevel gear teeth in the area of the contact pattern can be approximated by a second order surface. The transmission error in the tooth meshing region is therefore close to a parabola. If a gear set is manufactured exactly such that there are deviations neither in macro- nor in micro-geometry, the resulting transmission error curve will be as shown in Fig. 5.20 where the pinion has 16 teeth. The corresponding order spectrum shows the amplitudes of the mesh harmonics, which decrease exponentially from the first to the nth order. Side bands or background emission components are not found in the spectrum.

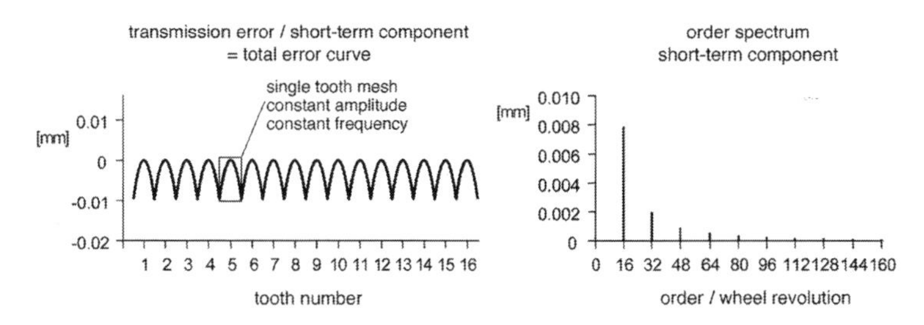

Fig. 5.20 Transmission error curve of an ideal bevel gear with no manufacturing or assembly deviations

Deviations during manufacturing may be classified as follows:

- radial run-out
- pitch deviations
- topography deviations
- surface structures and roughness
- damage

Radial run-out usually occurs on the teeth because of faulty clamping during the soft or hard machining process, or as a result of heat treatment.

The causes of pitch deviations are diverse. Periodic pitch deviations may, for example, result from inaccuracies in the axial motions in the spindles of the gear cutting machine. Periodic pitch deviations may likewise occur during the heat treatment of wheels with bolt holes in the mounting surface, the period corresponding to the number of bolts. Periodic pitch deviations may also be attributable to strain in the gear during hardening of the teeth. Material removal in the cutting process modifies the residual stresses in the gear which were introduced, for example, during forging of the blank. Depending on the form of the forged blank, a so-called memory effect may occur. An originally quadrangular form of the blank may reappear in the gear after heat treatment.

When grinding bevel gears, periodic and non-periodic pitch deviations may be introduced by the choice in pitch strategy to compensate for tool wear. Grinding wheel wear leads to a gradual narrowing of the tooth slot. In a grinding operation without wear compensation, the space width in the last slot is then ground smaller than that in the first one. To reduce this effect, it is possible either to use wear compensation or to apply the following strategy: After the first tooth slot is ground the workpiece is not rotated by one pitch but several pitches. Depending on the chosen grinding/pitch period, corresponding periodic pitch deviations may then occur on the work piece. Modern gear making machines include a device to compensate for manufacturing deviations of this kind (see Sect. 6.5.5.2).

Apart from radial run-out and pitch deviations, bevel gears also exhibit topography deviations. These are caused by incorrect settings of the gear cutting machine or geometry deviations of the machine axes. Such errors can, however, be compensated using corrective programs (see Sect. 7.1.5).

In addition to systematic topography deviations, there may also be stochastic deviations. These include, for example, deviations caused by wear-induced changes in the form of the tool. Topographic deviations appear mainly in the form of angular and crowning deviations of the tooth flank and in bias. Micro geometry deviations in the flank form also occur, for instance in the form of excessive surface structures and roughness. Among other causes, surface structures or waviness of the tooth flanks may be caused by excessive generating feed during gear tooth cutting, also referred to as cutter flats. Machine vibrations may produce chatter marks, which are likewise categorized as surface structures. Gear grinding may generate grooves in the tooth flank surface because of the presence of grinding wheel structures or wrong selections in the dressing parameters; this is no longer classified as roughness but as waviness.

The last category in gear tooth deviations is damage, usually caused around the edges by incorrect transportation of the gear sets after soft machining.

Radial Run-Out and Pitch Deviations Radial run-out and cumulative pitch errors cause long term transmission errors, and may therefore be considered together. The cumulative pitch error of a gear usually appears in a sinusoidal form. A pitch deviation leads to a shift in the relative position of following tooth flank topographies. If the tooth is otherwise free of deviations, a transmission error curve of the kind shown in the top-left half of Fig. 5.21 results. It is evident that individual tooth mesh transmission errors are shifted vertically by their respective component in cumulative pitch error, while retaining their basic form. By filtering the entire signal, it is possible to obtain a curve like the one shown in the bottom-left half of Fig. 5.21. It is apparent from a more detailed analysis of the individual tooth mesh transmission errors that tooth engagement is either shortened or prolonged.

Apart from cyclical changes in the meshing period, a further effect of cumulative pitch error is a cyclical change in the min-max values of the individual tooth mesh transmission errors [FAUL68], [NAUM69], [LAND03].

Translating these observations into signal processing terminology, cumulative pitch error may be said to cause amplitude and frequency modulations of tooth mesh transmission error. Such modulations are expressed spectrally by sidebands in the meshing order of the type shown in the bottom-right half of Fig. 5.21. The distance between the lower and upper sidebands and the meshing order are directly correlated with the period of the cumulative pitch error. In a single sinusoidal cumulative pitch error curve, the amplitude and the duration of an engagement respectively exhibit one maximum and one minimum during one revolution. In consequence, the sideband distance is exactly one mesh order (lower sideband = 15th wheel order, mesh order = 16th wheel order, upper sideband = 17th wheel order).

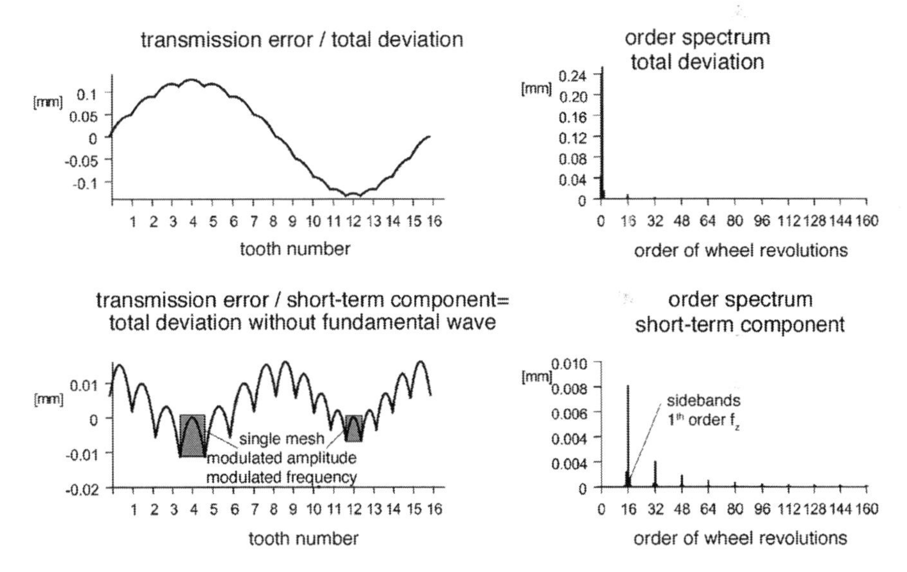

Fig. 5.21 Transmission error with long- and short-term components

A cumulative pitch error with a triple sinusoidal form produces three minima and three maxima per revolution for mesh amplitude and period (Fig. 5.22). The distances of the top and bottom sidebands then rise from one to three wheel orders, as can be seen from the bottom-right half of Fig. 5.22. Detailed discussions of the basic principles in amplitude and frequency modulation may be found in [DALE87], [REIC60] and [THOM80].

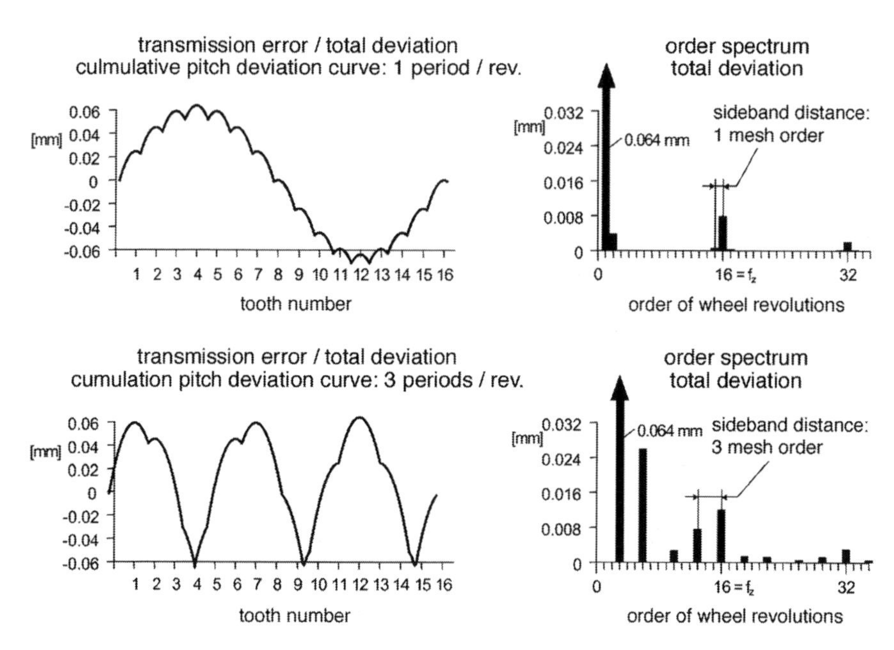

Fig. 5.22 Transmission error with single and triple long-term components

Topography Deviations Unlike radial run-out and pitch deviations, deviations in gear tooth topography from the desired form affect only the amplitudes and harmonics of the mesh order. Figure 5.23 shows an ease-off, the contact pattern with the path of contact, the transmission error and the corresponding order spectrum of a bevel gear set with conventional topography design. The ease-off shows simple lengthwise and profile crowning, usual for non-optimized designs. The transmission error is a parabolic curve; the amplitudes of the first to the fourth mesh order correspond to 80, 20, 9 and 6 % of the transmission error amplitude.

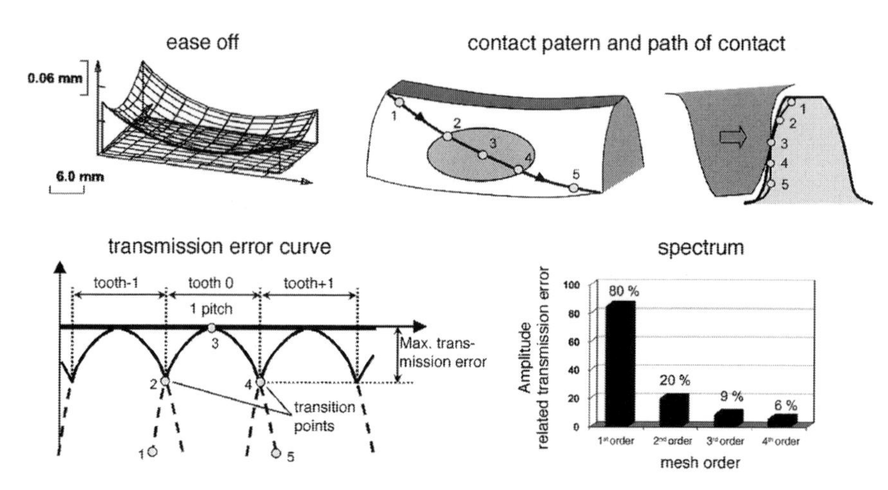

Fig. 5.23 Ease-off, contact pattern and transmission error for a bevel gear set with lengthwise and profile crowning

Incorrect machine settings or a defective tool cause deviations from the desired topography of the kind shown in Fig. 5.24. In this case, the manufacturing deviation tilts the ease-off, shifting the contact pattern into the heel-tip zone on the tooth of the wheel. The form of the transmission error changes from a parabolic to a saw-tooth curve. Although in this example the absolute value of the maximum mesh transmission error remains unchanged when compared to the original value, the relative amplitude levels for the first to the fourth mesh orders are altered. The first order amplitude has only 65 % of the maximum mesh transmission error as compared to the former 80 %. Conversely, the values for the other three orders increase. Studies have shown that topography deviations which can be described by second order surfaces mainly affect changes in the first four mesh orders. It is evident from the example shown that an analysis of the transmission error in one mesh cycle allows conclusions to be drawn on the type of topography deviations involved. Since bevel gears are usually tested in sets, rather than being compared individually to a master gear, it is not possible to assign the topography deviation explicitly to the pinion and wheel by analyzing the transmission error.

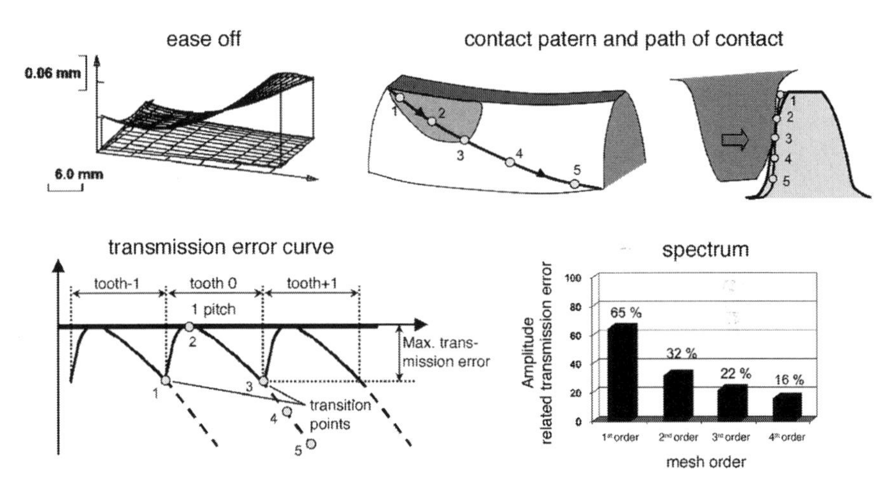

Fig. 5.24 Ease-off, contact pattern and transmission error of a bevel gear set with topography deviations

Surface Structure and Roughness Apart from angular and crowning deviations and flank bias, micro geometry deviations may also occur in the flank topography. As an example, Fig. 5.25 shows the transmission error of a bevel gear set without and with micro geometry tooth flank waviness. In accordance with the graphs in Figs. 5.20 and 5.23, the original transmission error curve is parabolic (Fig. 5.25 top-left) with amplitude distributions falling exponentially as the order rises (Fig. 5.25 top-right). If waviness is superimposed to the original parabolic form, Fig. 5.25 bottom-left, the change in the amplitudes of the first to the fourth order is virtually negligible, Fig. 5.25 bottom-right. Contrary to the mesh without waviness, however, this mesh also shows amplitude peaks from the fifth to tenth order. Depending on the waviness period, such amplitude peaks may occur at even higher orders.

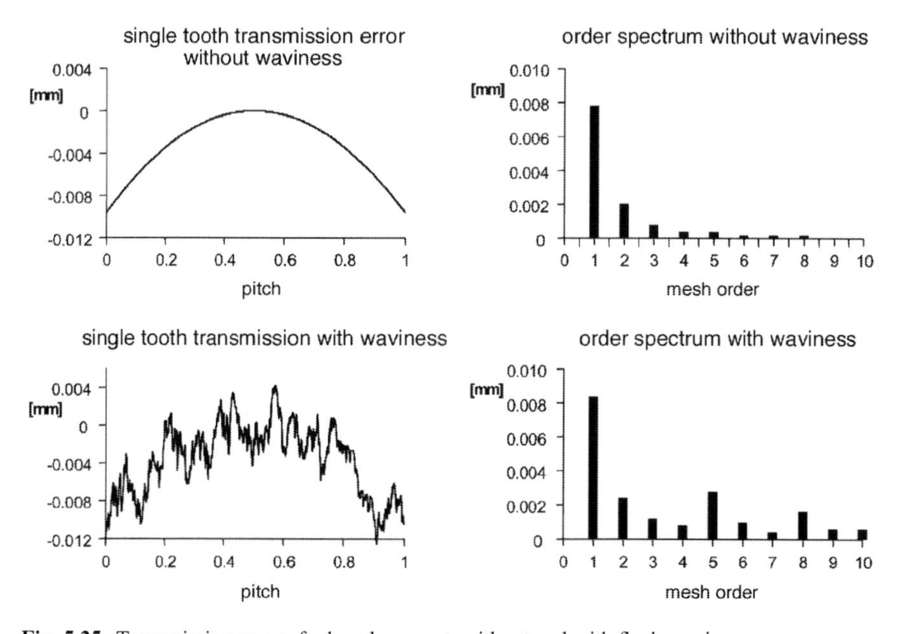

Fig. 5.25 Transmission error of a bevel gear set, without and with flank waviness

Tooth-to-Tooth Pitch Deviations and Damage Tooth-to-tooth pitch deviations and damage are disturbances leading to a temporary change in an otherwise regular transmission error curve. Figure 5.26 shows the transmission error curve of a 16-tooth gear for which one tooth engagement displays a clear amplitude peak. This disturbance causes a pulsatile change in the total transmission error and is expressed in the order spectrum by a significant increase in spectral lines throughout the range of orders. The amplitudes of these additional, pulse-dependent spectral lines are most pronounced below the first mesh order, decreasing in level with ascending order.

Knowledge of the influence of different geometry deviations on the transmission error curve and its spectral composition allows the identification of geometry deviations when analyzing the results of a single-flank test.

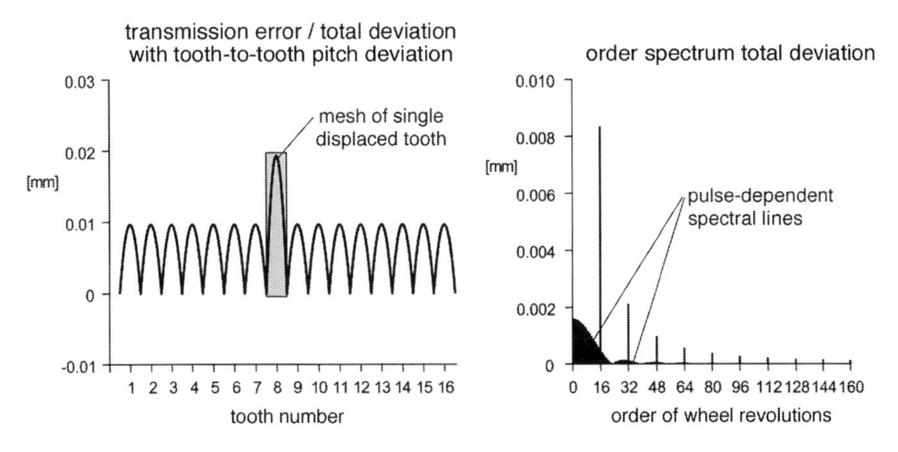

Fig. 5.26 Transmission error curve with one tooth-to-tooth pitch deviation

5.3.2 *Manufacturing Process Influence on the Transmission Error*

Hard finishing of bevel gears relies mainly on lapping, grinding and hard skiving. Additional finishing processes may be used to improve certain gear properties [LAND03]. Among others, we find shot peening, Trowalizing (vibratory grinding) and structural lapping.

Lapping Lapped gear sets are characterized by relatively high values in tooth to tooth transmission error and a sinusoidal or double sinusoidal cumulative pitch error curve. These arise because lapping cannot substantially reduce radial run-out after heat treatment. The primary effect of lapping is to level out disturbances in the individual tooth contacts. Disturbances may be caused by crowning or undesired material accumulations. The transmission error curve is therefore usually a sinusoidal or parabolic curve, generally displaying only slight waviness [SMIT85].

As an example of a transmission error measurement on a lapped gear set with a transmission ratio of 11/39, Fig. 5.27 presents the mesh order spectrum and the average working variation along with the contour diagram of the tooth meshes on the wheel. The order spectrum essentially contains the amplitudes of the first to the fourth order, and correlates well with the representation in Fig. 5.23. The sidebands are distanced by one or two wheel mesh orders. This correlates with the analysis of the relevant pitch measurements with a roughly sinusoidal and double sinusoidal cumulative pitch error curve.

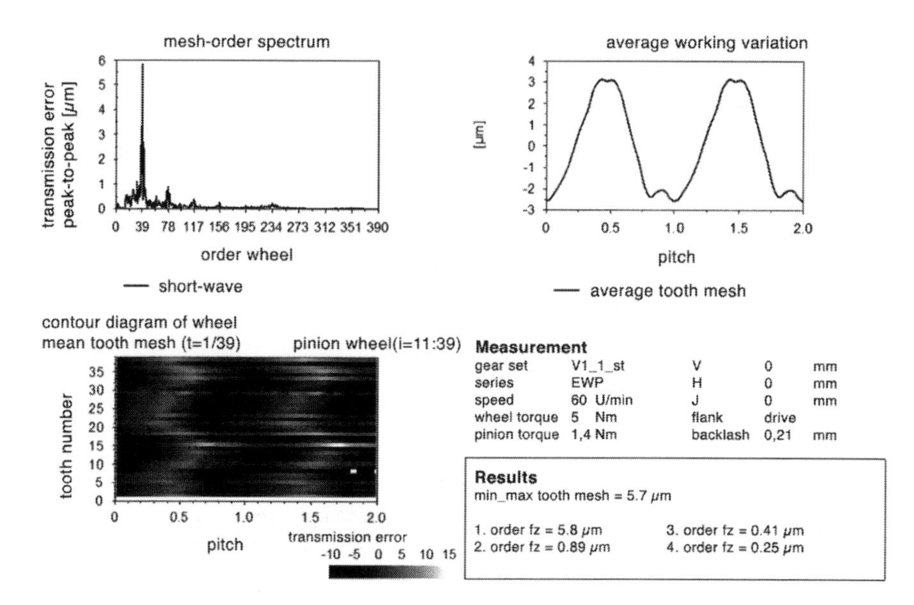

Fig. 5.27 Transmission error of a lapped gear set

Hard Skiving Skived bevel gears attain high pitch accuracies through a continuous indexing process. There are virtually no differences between the flank topographies of the individual teeth; flank waviness occurs only at extremely high generating feeds. Flank roughness typically attains very low Ra values in the range of 2 μm [LAND03].

Similarly to Fig. 5.27, Fig. 5.28 shows the average working variation results for a sample skived gear set with the gear ratio of 11/39. The analysis of the pitch deviations reveals a clear periodicity with a period of two pitches caused by an incorrectly adjusted cutter head. In addition to the amplitudes of the first mesh order (39th wheel order) and their harmonics, this periodicity causes strong sidebands in the order spectrum, at a distance of half the mesh order. Typically skived gear sets with correctly adjusted cutter head, show low amplitudes of the background components between the tooth mesh orders only.

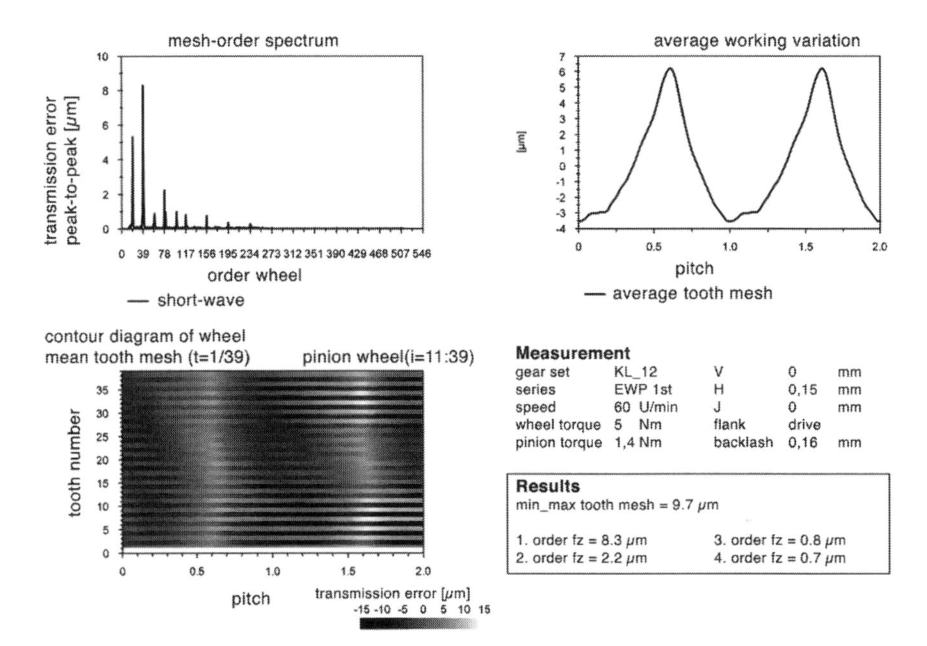

Fig. 5.28 Transmission error of a gear set skived with a modified cutter head

Grinding Ground gear sets are manufactured exclusively in a single indexing process. Modern bevel gear grinding machines achieve pitch accuracies comparable to the quality of skived gear sets. Deviations from the desired form in flank topography are usually in the order of a few micrometers.

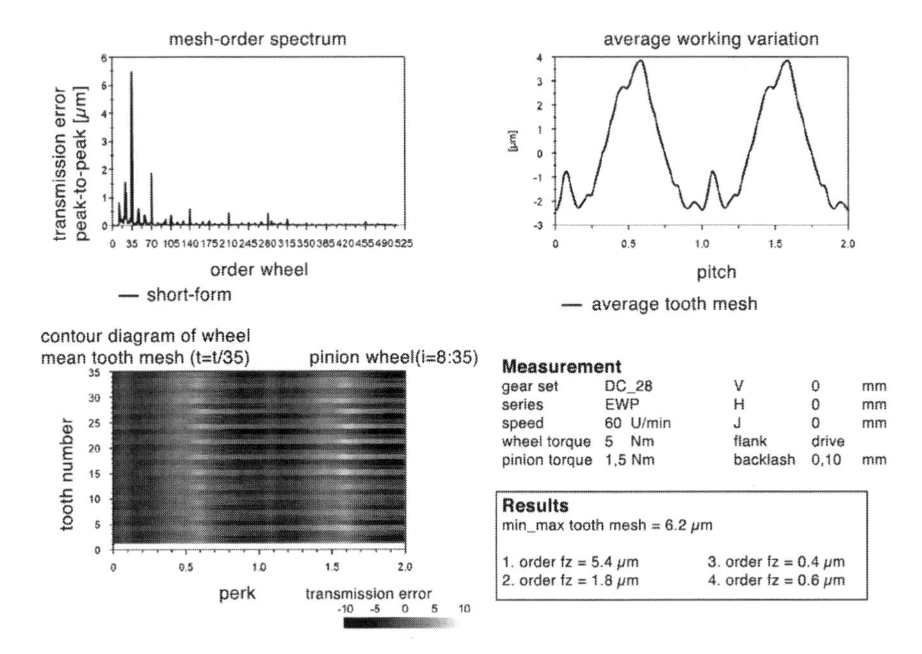

Fig. 5.29 Transmission error of a ground gear set

Despite the accuracy of the grinding process, periodicities may occur in the pitch deviation if, for example, the wrong pitch strategy is chosen during the grinding process. In the above case, an unfavourable ratio with the number of teeth was selected. Figure 5.29 shows the results in transmission error measurement for a ground gear set with a transmission ratio of 8/35. Pronounced sidebands to the mesh harmonic are apparent in the order spectrum; they are at a distance of about 12 wheel orders, corresponding to roughly one-third of a mesh order. During grinding of this sample gear, there was a division of three pitches between each slot grind, clearly corresponding with the order analysis.

An additional feature of the gear set in Fig. 5.29 lies in the clear amplitudes of the fourth, sixth and eighth mesh orders, which are attributable to waviness in the wheel tooth flank topography. In the example, flank waviness resulted from a choice of unfavourable parameters when dressing the grinding wheel. Due to the nature of the process, there are usually no generating flats on ground gears. However, generating flats may also occur if grinding wheels with large imbalances are used in combination with high generating feeds.

Comparison of Lapping, Hard Skiving and Grinding The final micro-geometry of lapped gear sets is more affected by the preceding heat treatment than that of gear sets finished with grinding or skiving. The closed-loop principle can be applied to grinding and skiving (see Sect. 7.2.5), generating the desired gear tooth geometry with residual deviations of only a few μm. In lapping, heat treatment distortions can be removed only to a limited extent. Radial run-out and cumulative pitch errors are therefore much higher by comparison with ground and skived gear sets. It is also impossible to eliminate the individual heat treatment distortions of particular teeth in the lapping process. The ease-off for each tooth pair of a lapped gear set will therefore be different, whereas in grinding and hard skiving one tooth is virtually identical to the next. This has an effect in particular on the meshing transmission error, as shown in Fig. 5.30 (upper diagrams).

If one compares the waterfall diagrams for individual tooth mesh transmission errors, it is evident that, unlike hard skived and ground gear sets, each individual tooth engagement in a lapped gear set is different. It is also clear from the contour diagrams (lower diagrams) that the maxima of the individual mesh transmission errors for the lapped gears are non-linear, as shown by the curve of the white line [LAND03].

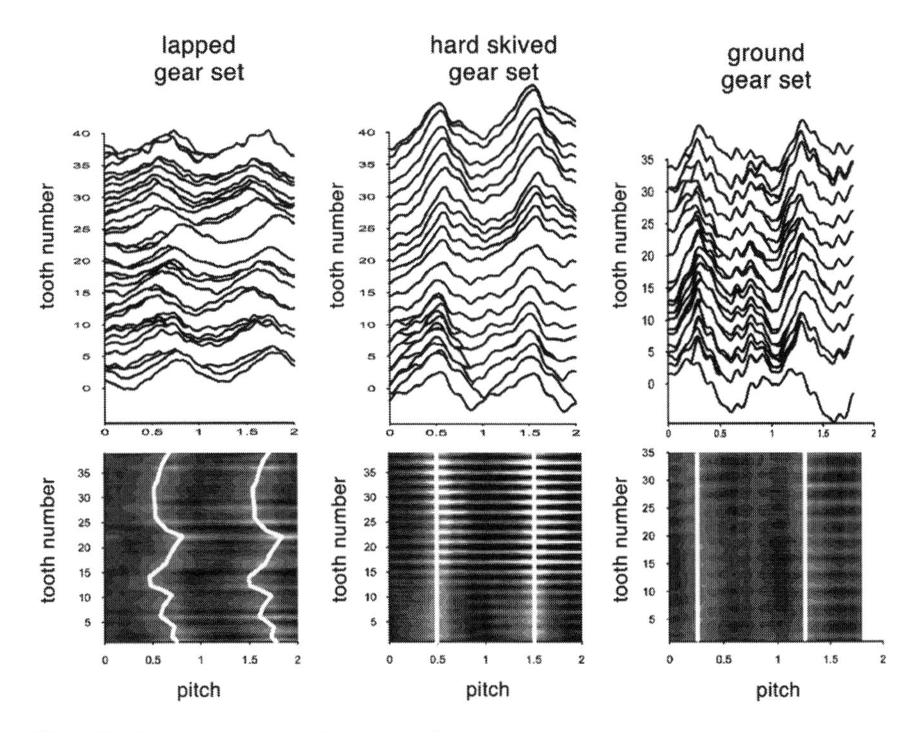

Fig. 5.30 Comparative waterfall diagrams of transmission errors

The better pitch quality and smaller deviations from the desired tooth flank topography in the hard skiving and grinding processes may, however, prove a drawback when noise behavior is analyzed. As opposed to a very exact period, broader band excitation may result from the differing periods with lapped gears. The narrow band excitation in certain frequencies, regarded as unfavorable, can be reduced and the acoustic impression consequently improved.

Structural Lapping The term structural lapping designates the subsequent lapping of skived or ground gear sets, with significantly shorter cycle times than in the normal lapping process. Structural lapping reduces micro geometry deviations in the tooth flanks. The negative influences of surface structures and waviness on noise quality can be partly or even completely eliminated. However, detailed investigations of structural lapping have revealed that the changes in surface structure are accompanied by changes in topography [LAND03]. The improvement in noise behavior obtained by structural lapping is therefore due to two different phenomena: on the one hand, flank waviness is removed; on the other hand, the originally circular crowning in the contact zone is reduced, creating differences in crowning along the face width and in the profile direction.

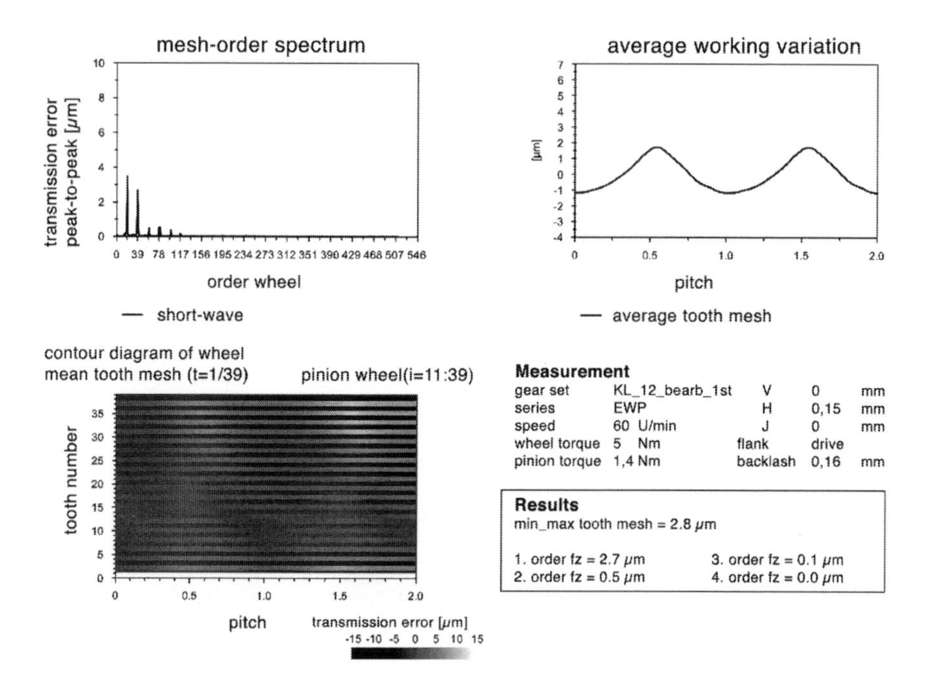

Fig. 5.31 Transmission error of a skived and structurally lapped gear set

Figure 5.31 shows the transmission error of a skived gear set (Fig. 5.28) after structural lapping. If one compares the results of Fig. 5.31 to those of Fig. 5.28, a reduction in the maximum transmission error, attributable to partial crowning removal, is apparent. The still detectable 39/2 order in the transmission error spectrum indicates that structural lapping cannot eliminate all the geometry deviations of a gear. Analysis of the pitch measurement after structural lapping of the skived gears revealed that the periodicity in pitch deviation, though reduced in amplitude, is still present.

Figure 5.32 shows the transmission error of a ground gear set (Fig. 5.29) after structural lapping.

The comparison of Figs. 5.29 and 5.32 shows that structural lapping can also significantly reduce the maximum transmission error on ground bevel gears. It is also evident that structural lapping has largely removed flank waviness. The average working variation curve assumes an approximately sinusoidal form after structural lapping.

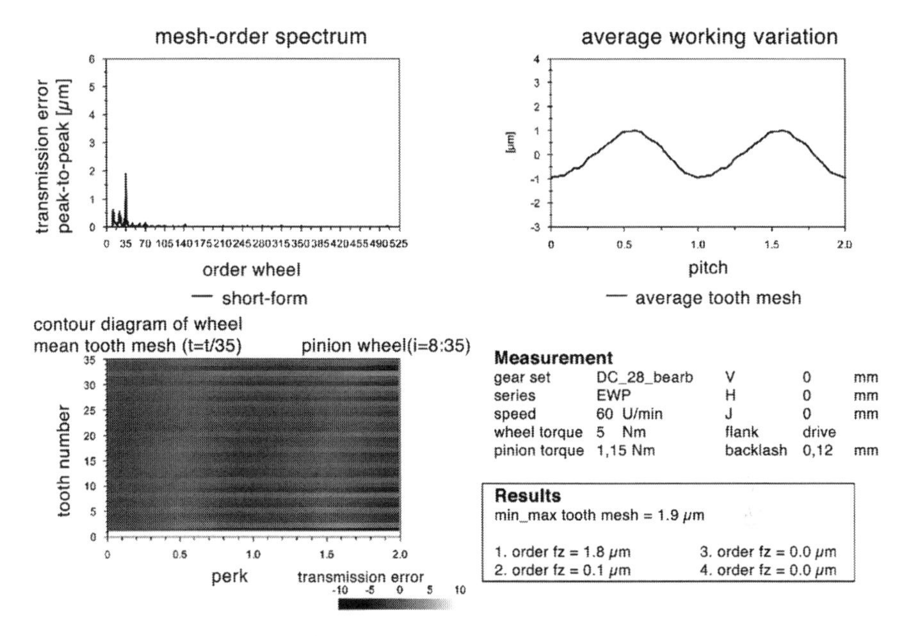

Fig. 5.32 Transmission error of a ground and structurally lapped gear set

5.4 Dynamic Noise Excitation

5.4.1 Dynamic of Bevel Gear Running Behavior

Section 5.1 describes a torsional vibration model approximating the complex mechanisms involved in tooth meshing with the aid of spring and damper elements. In the case of bevel and hypoid gears, the suitability of the reduced equivalent model with a single degree of freedom is limited since the displacements of a bevel gear set and the associated changes in the relative positions of the tooth flanks are not reproduced adequately.

The displacements result from the effects described in Sect. 3.4.2. Much effort must therefore be expended in developing a suitable equivalent mechanical model. For power trains, for example, one can develop the equivalent model as a multiple body system (MBS) [BREM92]. In the context of multiple body mechanics, the individual components of the power train are reproduced as either rigid or elastic bodies, which are subsequently linked together with the aid of force elements [BREC07a].

The basic equivalent model consists of the input and pinion shafts, the two output shafts, the ring gear and the differential housing. Initially, the input and output shafts are ideally mounted in order to reduce the degrees of freedom of the individual bodies in the MBS model (Fig. 5.33). At the joining points between the input and the pinion shaft and between the two output shafts and the differential

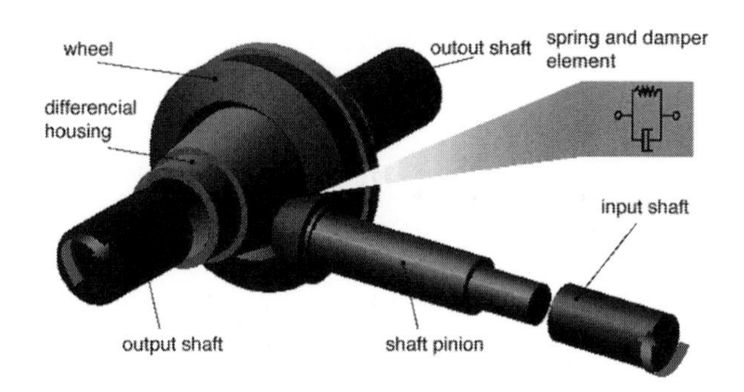

Fig. 5.33 Example of a MBS model of a bevel gear stage with spring and damper elements in the tooth contact

housing, the bodies are linked by force coupling elements in the form of torsional spring elements. The user can specify the speed of rotation of the input shaft and the braking torque on the output shafts. The pinion shaft and differential housing are modeled using force coupling elements. These allow for the effects of the tooth forces on the bending vibrations of the shafts, and transfer the resulting bearing reaction forces and moments to the housing structure.

The central element of the equivalent mechanical model is the tooth mesh, in which the excitation mechanisms described in Sect. 5.1 take effect. The wheel and pinion bodies are linked in tooth contact with the aid of two force elements realized as spring and damper elements. In order to enhance the accuracy with which the bevel gear stage is reproduced, the tooth meshing forces are determined with the help of a tooth contact analysis and utilized in the spring and damper element of the multiple body simulation. The response of the system to the input torque and speed are determined with the aid of the MBS model, resulting in new positions of rotation for the pinion and wheel, which in turn serve as input variables for the tooth engagement.

5.4.2 Calculating Load-Free and Load-Dependent Running Behavior

It is crucial for the quality of the numerical model that transmission errors should be determined as precisely as possible. The best possible way to reproduce three-dimensional contact conditions during tooth mesh is a tooth contact analysis (see Sect. 3.3) accompanied by calculation of the transmission error under load (see Sect. 4.4.3.7). Damping during tooth contact is extremely hard to quantify and may thus be simplified to a constant. This leads to sufficiently accurate results when compared to measurements [GEIS02], [GERB84].

Using tooth contact analysis, the load-free and load-dependent transmission errors may be determined individually [NEUP83]. The tooth contact analysis method is consequently suitable for the detailed analysis of the operational behavior in a discrete operating state.

It is possible to combine analyses under static load with a dynamic simulation. Repeated tooth contact analyses can be used to generate a graph showing the elastic characteristics of the tooth under various torques as a function of the position of roll and of the load-dependent transmission error, which can then be made available to the dynamic simulation. The varying tooth pair stiffness may be calculated from the difference between the load-free and loaded transmission error curve. Figure 5.34 gives examples of load-free and load-dependent transmission error curves. It is evident that compliance depends on the angle of rotation of the wheel φ_G. The resulting fluctuations in tooth pair stiffness are important in terms of dynamic excitation.

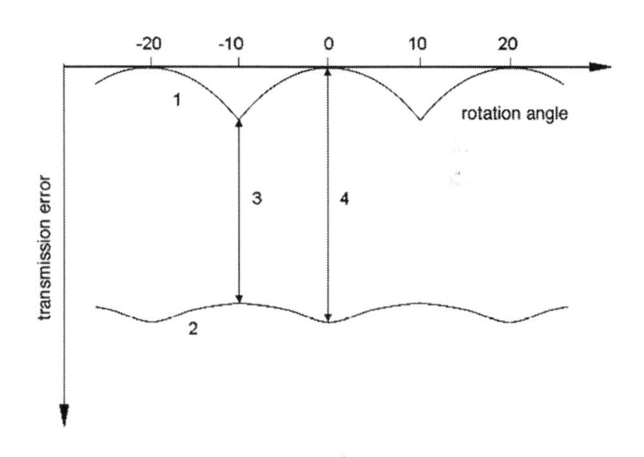

Fig. 5.34 Load-dependent and load-free transmission error curves.
1 load-free transmission error,
2 transmission error under load,
3 difference at $-10°$ rotation angle,
4 difference at $0°$ rotation angle

In this manner, it is possible to calculate, for all operating states, the system properties determined for discrete operating states by quasi-static analysis (see Fig. 4.43), and to interpolate them for the currently required state as part of the dynamic simulation. The resulting tooth pair stiffness characteristic for a complete tooth pitch p and moment M_z is shown in Fig. 5.35.

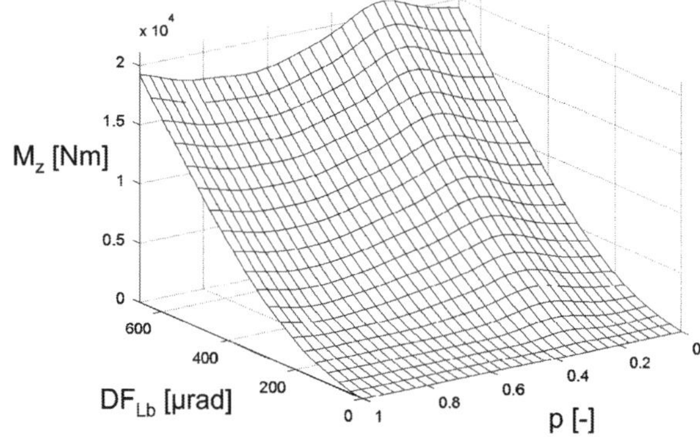

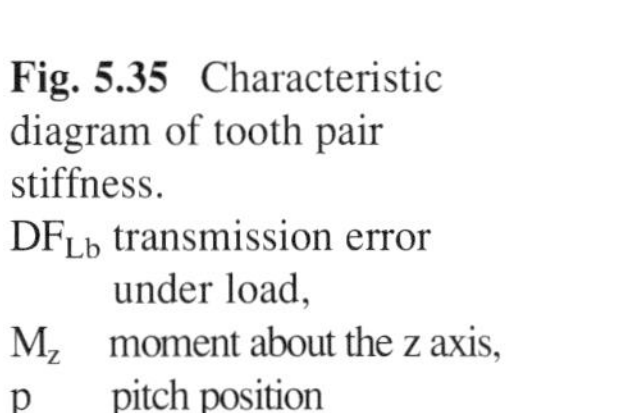
Fig. 5.35 Characteristic diagram of tooth pair stiffness.
DF_{Lb} transmission error under load,
M_z moment about the z axis,
p pitch position

If the effects of a load-dependent displacement of the pinion on tooth mesh are also to be accounted for, the tooth pair stiffness characteristics need to be extended to include further dimensions. The multidimensional characteristic diagrams are then functions of the pitch position, the load-dependent transmission error and the pinion position, which is explicitly determined by means of the vertical, horizontal and axial displacements and the shaft angle.

Using the angle of rotation of the wheel φ_G and pinion φ_P, and taking the transmission ratio into account, the total transmission error can then be determined for the mechanical model. The load-free transmission error, as a function of the rotation angle and of the vertical, horizontal and axial position of the pinion in relation to the wheel, is subtracted from the total transmission error, yielding the load-dependent transmission error. The tooth pair stiffness and the related tooth force is then analysed as a function of the current roll position, the load-dependent transmission error and the displacement of the pinion in relation to the wheel.

5.4.3 Test Rig for Rear Axle Gears

Acoustic measurements of gearboxes are usually performed in specialized noise test rigs. In many cases, these test rigs have special acoustic properties, allowing structure-borne noise as well as air-borne noise to be measured.

The goal in the acoustic measurement of gearboxes is to analyse noise behavior objectively. Measurements of air-borne and structure-borne noise serve to

- determine the initial state,
- clarify the causes of noise,
- plan noise reduction measures and
- demonstrate the improvements achieved

A wide variety of methods and instruments are available to measure gear noise. These range from simple manual sound level meters to processor-controlled measuring stations.

In a general purpose gear test rig, it is possible to perform tests on both the load capacity and the acoustics of gear assemblies. An acoustic measuring room surrounding the specimen is used for air-borne noise measurements [BREC07b].

Apart from the classic acoustic measurements of air-borne and structure-borne noise, measuring systems for such variables as transmission error and angular acceleration are employed. Such measuring systems can be used for both input to and output from gear assemblies.

The test array can additionally be subjected to excitation from drive and braking motors. Elements with low stiffness and a large damping effect therefore need to be interposed between the power output and the specimen, in order to mechanically filter the excitation.

5.4.4 Test Results

In order to determine the effective relationships between excitation from tooth mesh of bevel gear stages and the acoustic behavior of the entire system, gearbox and power train, two hypoid gear sets are mounted successively in a rear axle unit and their noise behavior is measured and compared. Gear set A is characterized by low vibrational excitation and gear set B by high excitation.

Experimental studies may be classified as measurements at static operating points and measurements while running up to speed. The measured variables are transmission error, structure-borne noise at the bearing points of the rear axle and sound pressure level.

Figure 5.36 indicates differences in the excitation behavior of gear set A for four static operating states on the basis of order spectrum. The operating speed and the torque load acting on the wheel were varied. The difference in speed between the left and right output sides was zero in all the tests. All the spectra presented show peaks at tooth mesh frequency and its higher harmonics. Here, tooth mesh frequency corresponds to the 41st order in relation to gear revolution. The first and second harmonics of tooth mesh frequency represent the 82nd and 123rd orders. No amplitude peak can be detected in higher spectra for all orders. The causes were identified as the compensating gears of the differential and the drive shaft. Because of the great distance between the exciting gear and the measuring systems, these harmonics were no longer detected. Instead, all spectra exhibit amplitudes at 12 times the rotary frequency of the drive and output motors. These peaks are in some cases higher than those of the meshing teeth. In the 123rd order in relation to gear rotation, the second harmonic of tooth meshing and the first harmonic of the 12 times the drive frequency coincide, explaining the conspicuous peak.

The excitation at 12 times the frequency of rotation is caused by the electric motors of the test rig. The asynchronous motors have two pole pairs, imparting 12 pulses per revolution to the test rig.

Despite the interference from motor excitation with the measured signal, clear trends in the transmission error may be noted when the speed of rotation or the torque is increased. At both torque loads, an increase in the speed of rotation of $n_{An} = 352\ min^{-1}$ to $n_{An} = 1{,}056\ min^{-1}$ leads to a significantly reduced amplitude at tooth mesh frequency (41st order). The greater inertia of the oscillatory system means that, as the speed of rotation increases, excitation is transferred less in the form of distance fluctuations and more in the form of force or torque pulses. Likewise, the transmission error falls with rising torque, thereby increasing the effective contact ratio (see Fig. 5.4).

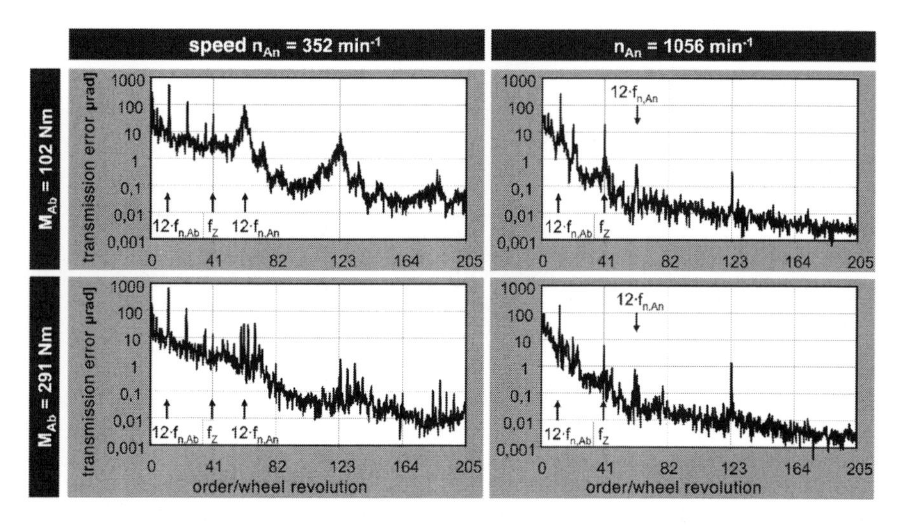

Fig. 5.36 Order spectrum of gear set A for measured transmission error under load

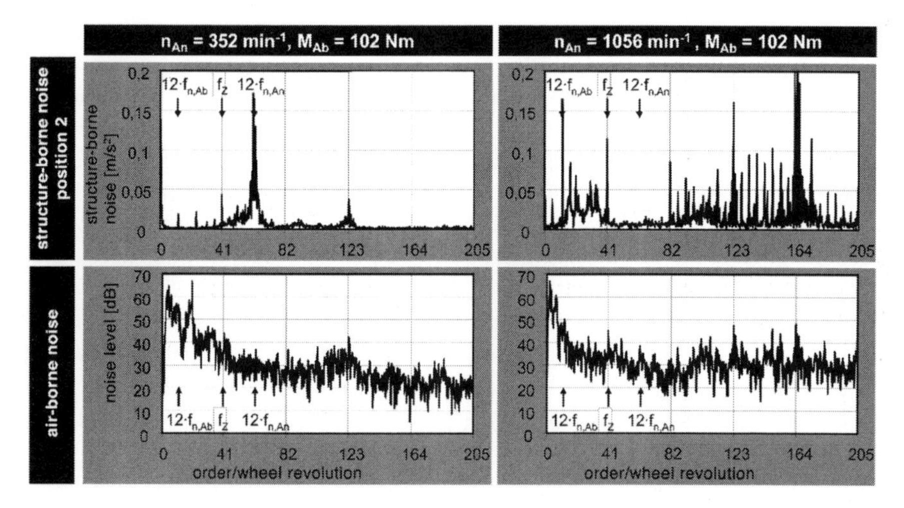

Fig. 5.37 Order spectrum of gear set A for measured structure-borne and air-borne noise under load

Whereas the amplitude of the measured transmission error decreases with an increase in the speed of rotation, higher speeds produce a rise in the amplitudes of structure-borne and air-borne noise (see Fig. 5.37). The higher speed of rotation introduces more energy in the rear axle vibration system, thereby increasing power loss through heat and vibrations.

Studies with static operating points reveal that the structure-borne noise signal depends largely on the chosen point. Results also indicate that acoustic behavior should not be judged by considering only one operating point. The operating point selected in the test should be that which has the greatest possible correlation with acoustic assessment in the vehicle. It is for this reason that acoustic behavior during speed run-ups is also examined.

The Campbell diagrams in Figs. 5.38 and 5.39 show the structure-borne noise signals on the output side and the air-borne noise signal at a distance of 1 m versus frequency and pinion speed. Resonant frequencies are represented as vertical lines; the ascending lines indicate excitation by the gear and the electric motors. These frequencies rise in proportion with the increase in rotation speed from 300 min^{-1} to 5,000 min^{-1}.

The Campbell diagrams for gear set A are characterized by excitations from the drive motor and the bevel gear stage, along with their harmonics. Predominant for gear set B, on the other hand, is the vibrational excitation of the gear. The Campbell diagrams demonstrate clearly that assessment of acoustic behavior on the basis of structure-borne and air-borne noise makes sense only at higher speeds. The meshing frequency harmonics are, likewise, strongly marked at higher speeds. By comparison, the spectra at the constant operating points, as shown in Fig. 5.36, exhibit virtually no harmonics of the meshing frequency.

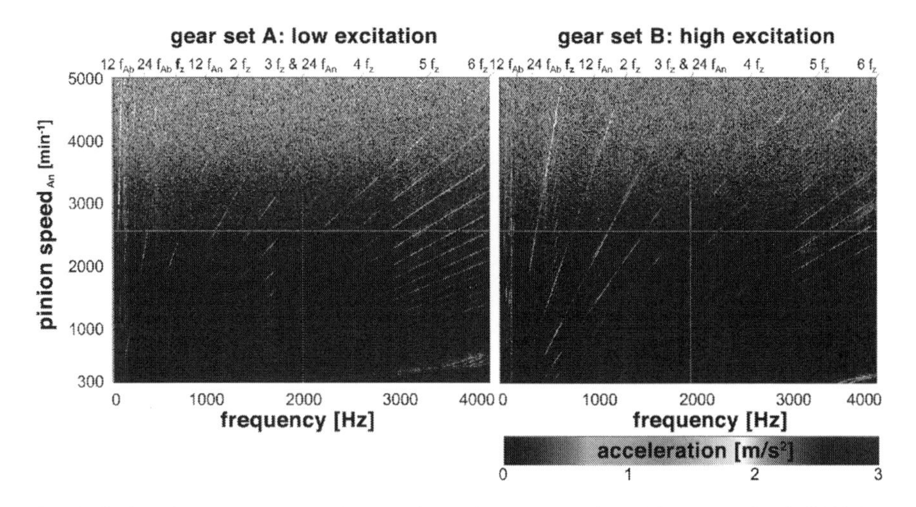

Fig. 5.38 Speed run-up: structure-borne noise at the output bearing with a torque load of 583 Nm

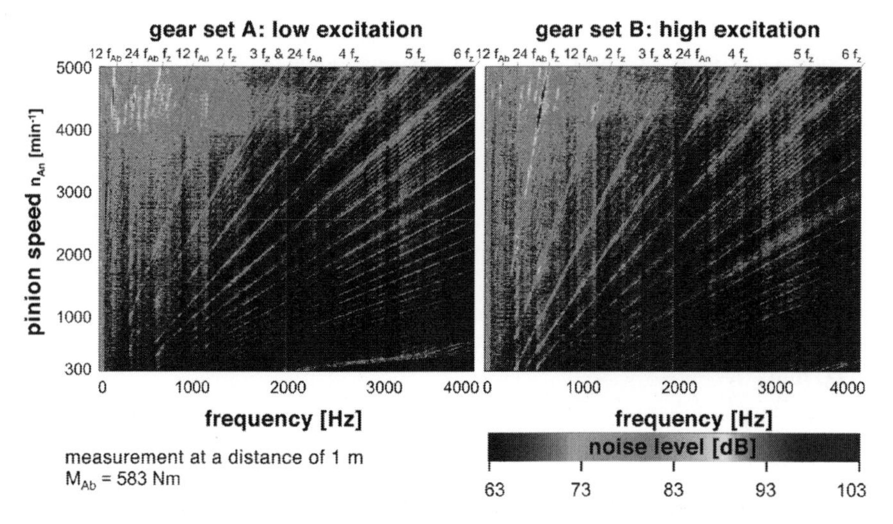

Fig. 5.39 Speed run-up: air-borne noise on the output bearing side with a torque load of 583 Nm

A comparison of the Campbell diagrams for gear sets A and B shows that gear set B produces higher noise emission. Air-borne and structure-borne noise are not analysed at individual operating points, but necessarily over the entire operating range. However, it is sufficient to examine the excitation behavior by means of the transmission error at a small number of static operating points.

References

[BAGH73] Bagh, P.: Über die Zahnfußtragfähigkeit spiralverzahnter Kegelräder. Dissertation, RWTH Aachen (1973)

[BECK99] Becker, J.: Analyse und Optimierung des Lauf- und Geräuschverhaltens bogenverzahnter Kegelräder. Dissertation, RWTH Aachen (1999)

[BOSC65] Bosch, M.: Über das dynamische Verhalten von Stirnrad-getrieben unter besonderer Berücksichtigung der Verzahnungsgenauigkeit. Dissertation, RWTH Aachen (1965)

[BREC07a] Brecher, C., Gacka, A.: Simulationsmodell zur Abbildung des dynamischen Verhaltens von Kegelradgetrieben – Berücksichtigung weiterer Anregungsmechanismen. Tagungsband zur 48. Arbeitstagung „Zahnrad- und Getriebeuntersuchungen", RWTH Aachen (2007)

[BREC07b] Brecher, C., Hesse, J.: Untersuchung des Geräuschverhaltens flankenmodifizierter Radsätze. Tagungsband zur 48. Arbeitstagung „Zahnrad- und Getriebeuntersuchungen", RWTH Aachen (2007)

[BREM92] Bremer, H., Pfeiffer, F.: Elastische Mehrkörpersysteme. Teubner, Stuttgart (1992)

[DALE87] Dale, A.K.: Gear noise and the sideband phenomenon. In: Gear Technology, Heft Jan./Feb., S. 26–33 (1987)

[DIET99] Dietz, P., Gummersbach, F.: Lärmarm Konstruieren XVIII, Systematische Zusammenstellung maschinenakustischer Konstruktionsbeispiele. Schriftenreihe der Bundesanstalt für Arbeitsschutz und Arbeitsmedizin. Bundesanstalt für Arbeitsschutz und Arbeitsmedizin, Dortmund (1999)

[FAUL68] Faulstich, H.I.: Zusammenhänge zwischen Einzelfehlern, kinematischem Einflanken-Wälzfehler und Tragbildlage evolventenverzahnter Stirnräder. Dissertation, RWTH Aachen (1968)

[GEIS02] Geiser, H.: Grundlagen zur Beurteilung des Schwingungsverhaltens von Stirnrädern. Dissertation, TU München (2002)

[GERB84] Gerber, H.: Innere dynamische Zusatzkräfte bei Stirnradgetrieben. Dissertation, TU München (1984)

[HAGE71] Hager, D.: Verzahnungsgenauigkeit und Laufruhe von Kegelradgetrieben. Dissertation, RWTH Aachen (1971)

[LAND03] Landvogt, A.: Einfluss der Hartfeinbearbeitung und der Flankentopographieauslegung auf das Lauf- und Geräuschverhalten von Hypoidverzahnungen mit bogenförmiger Flankenlinie. Dissertation, RWTH Aachen (2003)

[MÖLL82] Möllers, W.: Parametererregte Schwingungen in einstufigen Zylinderradgetrieben – Einfluss von Verzahnungsabweichungen und Verzahnungssteifigkeitsspektren. Dissertation, RWTH Aachen (1982)

[NAUM69] Naumann, D.: Untersuchung über den Einfluss von Einzelfehlern auf den Einflanken-Wälzfehler und die Tragbildlage oktoidenverzahnter Kegelräder. Dissertation, RWTH Aachen (1969)

[NEUP83] Neupert, B.: Berechnung der Zahnkräfte, Pressungen und Spannungen von Stirn- und Kegelradgetrieben. Dissertation, RWTH Aachen (1983)

[OPIT69] Opitz, H.; Faulstich, H.-I.: Zusammenhänge zwischen Einzel-fehlern, Einflanken-Wälzfehler und Tragbild evolventen-verzahnter Stirnräder. In: Brandt, L. (Hrsg.) Forschungsberichte des Landes NRW, Nr. 2058, 1. Aufl. Westdeutscher, Köln (1969)

[REIC60] Reichardt, W.: Grundlagen der Elektroakustik. 3. Aufl. Akademische Verlagsgesellschaft Geest & Portig KG, Leipzig (1960)

[SCHW94] Schweicher, M.: Rechnerische Analyse und Optimierung des Beanspruchungsverhaltens bogenverzahnter Kegelräder. Dissertation, RWTH Aachen (1994)

[SMIT85] Smith, R.E.: Identification of gear noise with single flank composite measurement. In: Technical Paper, no. 85FTM13, AGMA Fall Technical Meeting, San Francisco, 14.–16. Oct 1985. AGMA, Alexandria, VA (1985)

[TESC69] Tesch, F.: Der fehlerhafte Zahneingriff und seine Auswirkungen auf die Geräuschabstrahlung. Dissertation, RWTH Aachen (1969)

[THOM80] Thompson, A.: Fourier analysis of gear errors. In: Tagungs-band zur Konferenz NELEX 80, Glasgow, 7.–9. Oktober 1980. Eigendruck National Engineering Laboratory East Kilbride, Glasgow, Paper Nr. 3.5, S. 1–21 (1980)

[TOPP66] Toppe, A.: Untersuchungen über die Geräuschanregung bei Stirnrädern unter besonderer Berücksichtigung der Fertigungsgenauigkeit. Dissertation, RWTH Aachen (1966)

[TOSH61] Toshime, T.: Der Stoß des Flankenpaares der Zähne als Ursache des Geräusches. In: VDI Berichte 47, S. 99–101 (1961)

[VDI90] Richtlinie VDI 3720; Blatt 9.1; (Januar 1990) Lärmarm Konstruieren-Leistungsgetriebe-Minderung der Körperschallanregung im Zahneingriff (1990)

[WECK92] Weck, M., Bartsch, G., Reuter, W., Stadtfeld, H.J.: Verzah-nungsmessung auf Mehrkoordinaten-Meßgeräten. In: Weck, M. (Hrsg.) Moderne Leistungsgetriebe. 1. Aufl. Springer, Berlin, S. 190–212 (1992)

[WECK99] Weck, M., Landvogt, A.: Untersuchung des Einflusses von Oberflächenrauheit und Einlaufkorrekturen höherer Ordnung auf die Laufeigenschaften und das Geräusch achsversetzter Kegelräder. Abschlussbericht zum FVA Forschungsvorhaben 282 „Kegelradgetriebegeräusche“. Forschungsvereinigung Antriebstechnik e.V., Frankfurt a.M., Heft Nr. 589 (1999)

[WECK00] Weck, M., Hohle, A.: Untersuchung des Einflusses von topologischen Abweichungen, Oberflächenstrukturen und Rauheiten auf das Geräuschverhalten von Zylinderrädern. Abschlussbericht zum FVA Forschungsvorhaben 285 „Oberflächenbeschaffenheit von Zylinderrädern“. Forschungsvereinigung Antriebstechnik e.V., Frankfurt a.M., Heft Nr. 601 (2000)

[WECK02] Weck, M., Landvogt, A.: Bestimmung der von Kegelradverzahnungen ohne eine Beeinträchtigung des Geräuschverhaltens tolerierbaren Fertigungs- und Montageabweichungen. Abschlussbericht zum FVA Forschungsvorhaben 337 „Kegelradgeräuschtoleranz". Forschungsvereinigung Antriebstechnik e.V., Frankfurt a.M., Heft Nr. 668 (2002)

Chapter 6
Manufacturing Process

6.1 Introduction

Bevel gear manufacturing processes may be divided into non-cutting and cutting operations (Table 6.1).

Table 6.1 Bevel gear manufacturing processes

Non-cutting processes	Cutting processes
Casting	Planing
Sintering	Milling
Extrusion	Hard skiving
Die forging	Grinding
Tumble forging	Lapping
	Honing

Non-cutting Processes A major difficulty in non-cutting processes is ejecting the work piece from the mould or die, severely limiting or even excluding their applicability to bevel gears, especially spiral bevel gears.

Casting is not used in the field of power transmission gears, but only for mass production of bevel gears with low requirements, for example plastic gears.

Sintered bevel gears are used in relatively large numbers in hand tools such as angle grinders. Apart from the limitations noted above in relation to ejection, sintering, unlike other non-cutting processes, poses problems of compacting the material evenly in order to obtain a sufficiently homogeneous structure [BART06]. The dies are also very expensive to produce and modifications on the gears are costly if the dies have to be altered.

Extrusion is used predominantly for straight bevel gears.

Along with cutting processes, straight bevel gears for vehicle differentials are produced mainly by forging. Requirements in terms of tolerances for pitch and tooth flank topography are lower than on rotating power gears. Forging processes also allow certain gear blank geometries which cannot be realized in cutting operations. Apart from these advantages, the chief drawbacks of a forging process

J. Klingelnberg (ed.), *Bevel Gear*, DOI 10.1007/978-3-662-43893-0_6

are the high cost of dies and the complex process design. Any modification of the gear, for example contact pattern corrections, which in a cutting process can be realized by changing the machine settings, will demand a new and costly die.

Cutting Processes This topic applies solely to planing (shaping) and milling of straight and skew bevel gears. Production processes for spiral bevel gears are described in Sects. 6.2, 6.4 and 6.5.

In the past, the planing of straight or skew bevel gears was widely used. The best known method in Germany was developed by Heidenreich-Harbeck [DEGN02] and has in the meantime been replaced by more productive methods. It is now used to only a limited extent in single and spare part production.

Straight Bevel Gear Milling The operations currently used to produce straight bevel gears are milling and broaching.

Three methods, distinguished only by the tools employed, are used for milling with a generating motion. These are known as Coniflex® (Gleason), Konvoid (Modul) and Sferoid (Klingelnberg).

Typical for all three methods are:

tooth shape:	tapered depth
manufacturing:	single-indexing method;
	both pinion and wheel are generated
tool:	two intermeshing cutters with radial blades
profile crowning:	by machine kinematics
lengthwise crowning:	curved blade paths

The axes of two disc-shaped cutters, one for the left tooth flank, one for the right tooth flank, are at an angle to each other, causing the blades to intermesh alternately in such a way that their primary cutting edges form a trapezoidal profile. Since the cutting edges do not lie precisely in the plane of rotation, but at a small inclination angle (see Fig. 6.1), the teeth receive a fixed lengthwise crowning and the bottom of the tooth slot is not straight but arc-shaped. The angle between the cutters is given by the flank angles of the blades and the tapered slot width of the bevel gear.

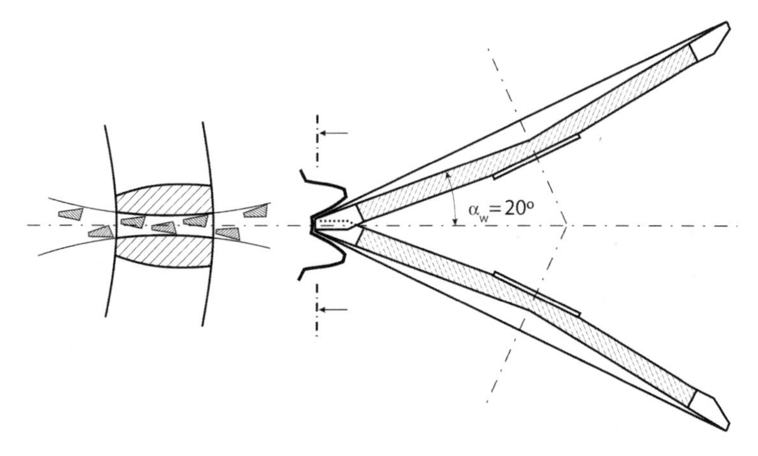

Fig. 6.1 Principle of straight bevel gear milling using intermeshing cutters

The broaching process known as Revacycle® is the most productive cutting process to manufacture straight bevel gears. As a specialized tool is needed for each transmission ratio, it is suited only to mass production. The tool is disc-shaped in diameters of approximately 530–635 mm. As there is no generating motion, it has a large number of differently profiled form blades on its circumference. Each slot is produced by a single revolution of the tool. The first segment of the circumference consists of roughing blades, a little radially offset to each other to be able to cut. This is followed by segments with finishing blades and then a gap in which the work piece is rotated by one pitch. The finishing blades have a concave arc profile, which is reproduced in the work gear during a linear shift of the centre point of the tool, thus creating a straight bottom slot.

Typical for the broaching process is:

tooth shape:	tapered depth
manufacturing:	single indexing method; both pinion and wheel are form cut
tool:	broaching cutter with many different blades on the circumference
profile crowning:	by the tool
lengthwise crowning:	by the tool

6.2 Cutting of Spiral Bevel Gears

6.2.1 Development History

The first spiral bevel gear generator, which operated using the single indexing method, was developed in the US in 1913. The tool was a face cutter head for the 5-cut method (see Sect. 2.1).

The first continuous indexing generator for spiral bevel gears was presented in Germany in 1923 using a conical hob (see Sect. 1.1). This gear cutting method is named Palloid® (see Sect. 2.1). Two hobs with opposing threads are required to produce a pinion and a wheel.

In 1946, a face hobbing machine with a cutter head for the continuous indexing method was developed for the first time in Switzerland. Tooth depth was constant and the tooth lengthwise shape was an elongated epicycloid.

Since then, on the one hand, machines with a one-part tool spindle were developed, on which lengthwise crowning is generated either by means of special cutter designs or by tilting the tool axis (see Sect. 2.1). On the other hand, machines with a two-part tool spindle for two-part cutter heads were also developed. On these machines one spindle centre could be offset relative to the other, allowing an appropriate difference in tool radii for lengthwise crowning. One of these manufacturing methods is termed Zyklo-Palloid® and was introduced in 1955 (see Sect. 2.5). The other is known under the name of Kurvex.

The first stick blade cutter heads for high volume production were introduced in 1967 (see Fig. 6.7).

For decades, all these gear cutting machines had extremely complex mechanics and mechanical drive chains. Since the middle of the 1980s, it has been possible to replace these high-precision drive chains by numerically controlled drives, known as "electronic gearboxes". Later, a coordinate transformation was used to convert the up to ten setup- and motion-axes of the mechanical machines into the kinematics of a CNC machine with six axes. Generally speaking, all six axes move simultaneously in a timed relationship when cutting spiral bevel gears (see Sect. 3.2.2).

For the manufacturing methods explained in Sect. 2.1, there were originally numerous machines working on different mechanical principles, depending on which motions were required. Specific manufacturing methods consequently became assigned to specific machine manufacturers. Progress in CNC technology made these specialized machines obsolete.

Nowadays only six-axes-machines are produced which, given suitable tools, can perform nearly all known manufacturing methods for spiral bevel gears.

6.2.2 Development Trends

Production engineering has evolved in parallel with the gear cutting machines. Since the end of the 1990s, dry cutting has steadily replaced the previously dominant wet cutting which used cooling lubricants. An almost exclusive use is now made of cemented carbide tools, generally designed in the form of coated stick blades. Increased cutting speeds also reduce machining times substantially. Today's machine concepts are specially adapted for the needs of dry cutting, where particular attention must be given to the optimum removal of the hot metal chips. More and more direct drives with high rotation speeds are being used to achieve high cutting speeds even on smaller spiral bevel gears and hence using smaller tool diameters. Work piece automatic loading into the gear cutting machines is steadily expanded. Efforts to integrate additional processes into the machines, such as deburring, are observed.

6.2.3 Tools

6.2.3.1 Technological Angles in the Cutting Edge Geometry

The geometry of the bevel gear tooth flank is defined solely by the form and position of the cutting edge of the tool. Aside from the form, the technological angles of the cutting wedge are important for the manufacturing process. Figure 6.2 defines these angles. For the sake of simplicity, a cutting wedge is shown with a plane front face and plane relief surfaces, and without tool edge radii.

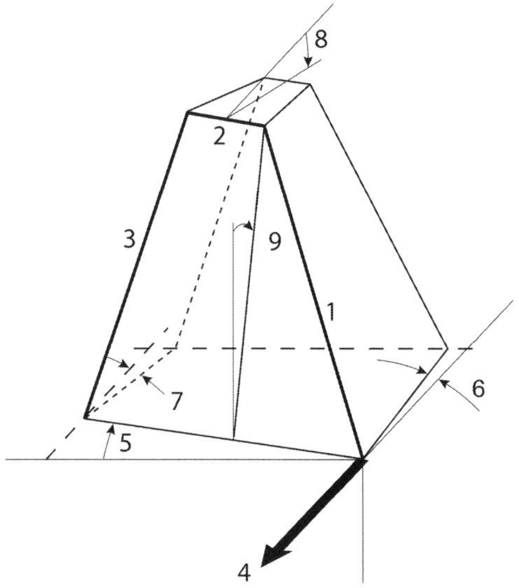

Fig. 6.2 Technological angles on the cutting wedge

The cutting wedge consists of three sections: the primary cutting edge (1), the tip cutting edge (2) and the secondary cutting edge (3). The front face is the plane which is made by these three sections of the cutting wedge. It is inclined by the side rake angle (5) in relation to the plane normal to the direction of primary motion (4). The relief surface behind the primary cutting edge is set back against the direction of primary motion by the side relief angle (6), and the relief surface behind the secondary cutting edge is similarly set back by the side relief angle (7). The tip relief is inclined against the direction of primary motion by the tip relief angle (8) and the front face is additionally inclined to the plane normal to the direction of primary motion by the hook angle (9).

These technological angles always relate to the shape of the cutting wedge irrespective of the geometry of the cutting edge holder.

6.2.3.2 Profile Blades and Cutter Heads

Relief ground profile blades are reground on the front face only. The amount of regrinding is towards the back of the blade (see Fig. 6.3), such that the blade profile remains unchanged. The sharpening or regrinding can be repeated until the remaining cross-section of the blade becomes too weak. With most types of tools, the relief ground blades remain in the cutter head for regrinding purposes (see Sect. 6.2.3.5). The stock amount S_A to be removed must be bigger than the crater depth in the front face and bigger than the width of wear on the relief surfaces. The blade displaying the largest amount of wear determines the amount of regrinding for all the blades on a cutter head, since all profile blades must have the same tip height after re-sharpening.

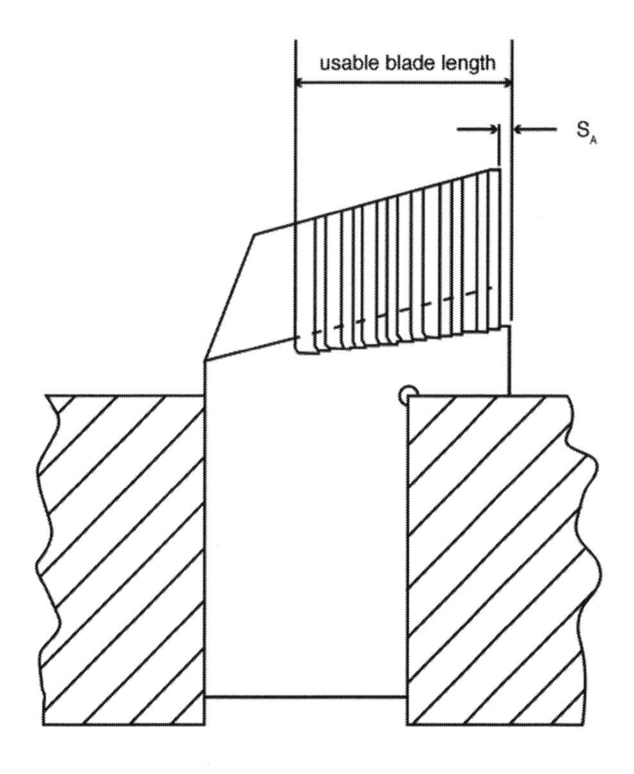

Fig. 6.3 Principle of a relief ground profile blade

As a result of the stock amount S_A removed from the front face of the blade, there is a reduction Δ_H in the blade tip height according to the tip relief angle γ_K where $\Delta_H = S_A \tan \gamma_K$. After blade re-sharpening, it is therefore necessary to adjust the depth setting on the cutting machine by Δ_H so that tooth depth is not altered.

Standardized profile blades are available in stepped ranges according to module, pressure angle, point width, tool edge radius and protuberance. They are now almost exclusively produced with coated surfaces. After regrinding, only the relief surfaces remain coated; the coating on the front face is removed.

Advantages of profile blades:
- less sharpening effort on the front face of the blades;
- simple, low-cost surface grinding machines are used for regrinding;
- the low mounting accuracy of the blades in the cutter head is adequate for roughing;
- for finishing, the blades can be adjusted radially in the cutter head without affecting tip height;
- profile blades can be reground in the cutter head

Disadvantages of profile blades:
- cost of the profile blades;
- small number of possible regrinds compared to stick blades;
- high effort for radial adjustment of the blades;
- only the flanks remain coated, not the front face;
- no cemented carbide blades available for dry cutting;
- no modification of the cutting edge geometry is possible in order to optimize the contact pattern

Cutter head with profile blades for the N-method The tool for this cutting method (see Sect. 2.1) is a highly rigid single-piece cutter head holding three HSS blades per group, comprising a pre-cutter, inner cutter and outer cutter blade. These profile blades are standardized and for each cutter head size, they are designed for pressure angles of 17.5°/20°/22.5° or 25° and for three different module ranges.

There are four different cutter head designs available, in which the angle between two consecutive blades differs. Different lengthwise crowning can be achieved by selecting adequate cutter heads for the pinion and wheel. Changing the angle between consecutive blades alters the tooth thickness. To compensate this, the relevant blade is shifted radially. This generates a different lengthwise curvature of the tooth flank and, in combination with the mating flank, a changed lengthwise crowning. The radial settings of the individual blades are adjusted by means of wedge plates and a setscrew, and the tip height of the blades is adjusted by another setscrew.

Zyklo-Palloid® cutter head In this case, a two-part cutter head with HSS profile blades is used (see Sect. 2.1). In general, these cutter heads are applicable for pinions and wheels, irrespective of the hand of spiral. They carry three or four blades per group with one or two middle blades. The radial position of the blades is adjusted using parallel plates and one screw. For reasons of space, a maximum of five blade groups can be mounted on a two-part cutter head (see Fig. 6.4).

Tooth thickness and lengthwise crowning can be varied flexibly via the eccentricity Ex_B of the two-part machine spindle for the Zyklo-Palloid cutter head. The inner part of the cutter head with the inner blades is located on the first, inner tool spindle. The outer part with the outer blades is fixed on the second, radially adjustable outer spindle, driven at an exact angle to the first spindle by means of an Oldham coupling.

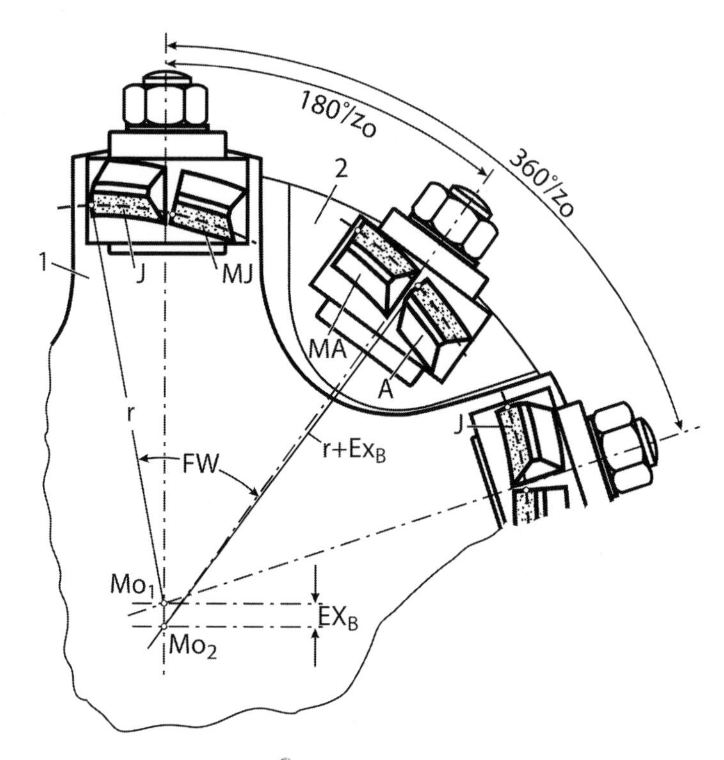

Fig. 6.4 Geometry of a Zyklo-Palloid® cutter head.

1	inner cutter head	2	outer cutter head	J	inner blade
MJ	inner middle blade	MA	outer middle blade	A	outer blade
r	radius of the inner blades	$r + Ex_B$	radius of the outer blades		
Ex_B	crowning eccentricity	FW	angle of blade sequence		

FM Cutter Heads with Radially Non-adjustable Profile Blades Various HSS blade types exist for these cutter heads which are used for the Face Milling (FM) method. The blades are clamped in the cutter head using either T-shaped clamping pieces (RIDG-AC®), as shown in Fig. 6.5, or clamping wedges (WEDG-AC®). Depending on the version, there will be a different number of blades for the same tool diameter. Two or three blades per group can be used, positioned by parallel plates.

Fig. 6.5 RIDG-AC® cutter head with inserted profile blades

FM Cutter Heads with Radially Adjustable Profile Blades For greater precision, the HARD-AC® cutter head allows a very precise radial adjustment of the HSS blades. The blades are aligned in reference to the center using a gauge, wedge plates and setscrews, up to an accuracy of 1 μm. Two or three blades per group are used, depending on the type.

FM Finish Cutter Head with Radially Adjustable Profile Blades This HSS tool (SINGLE CYCLE®) for the plunge cutting of wheels has five outer and inner blades alternately arranged in the cutter head in such a way that a sector remains without blades. When a tooth slot is machined, chip removal is similar to that in broaching because the blades are laterally offset one to another from the first to the last, fifth, blade group. This allows a tooth slot to be finish cut in a single revolution of the cutter head and the work piece to be rotated by one pitch in the sector without blades. So, there is no need to retract the constantly rotating cutter head from its position while cutting all teeth of a gear.

Compact Tool with Radially Non-adjustable Blades The cutter head and blades are machined from a single piece of tool steel (SOLID cutter). Therefore, the cutting edges cannot be adjusted radially and are only be usable for a given point width. The entire cutter head is consumed once the regrindable part of the blades has been used up.

6.2.3.3 Palloid® hob

The only tool for cutting spiral bevel gears which is not a face cutter head, is the Palloid hob. The conical single-start hob (see Sect. 2.1) has a pitch cone angle of

30°. On its conical body is a single-start worm with a lead of $\pi \cdot m_0$, which worm is provided with several straight flutes to form the teeth of the hob (see Fig. 6.6). This tool allows continuous indexing; however, the generating motion requires a superimposed pivoting motion of the hob around the axis of the virtual crown gear. Depending on the required cutting length SF, which results from the face width of the bevel gear to be cut, a hob from series A, B or C is selected. These hobs are also graded according to module size and tooth thickness modifications, and differ in their cutting direction and hand of lead.

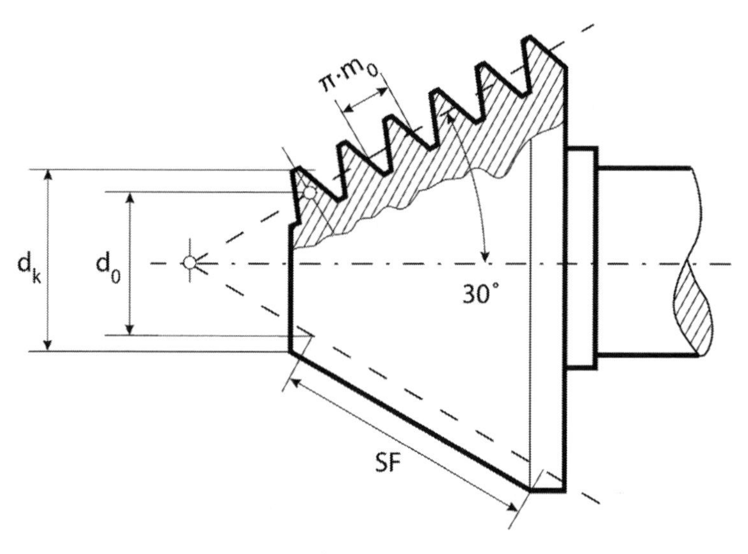

Fig. 6.6 Simplified geometry of a Palloid® hob.
d_k tip diameter of the first full tooth
d_0 pitch diameter of the first full tooth
m_0 hob module
SF cutting length

The cutting teeth of a Palloid® hob are relief-ground and only its front faces need regrinding in straight flutes. So, the geometry of the individual hob tooth is preserved except for a slight reduction in the hob diameter. To prolong tool life, the Palloid® hob can be coated and after regrinding even recoated.

6.2.3.4 Stick Blades and Cutter Heads

Characteristics of Stick Blade Cutter Heads These cutter heads were formerly made from two parts: a disk–shaped inner part had precisely ground slots on the circumference which were enclosed by an outer ring that was either bolted, welded or shrunk fixed to the inner part. Because of the higher rigidity, integral cutter heads with precision fit eroded chambers for the stick blades are now state of the art (see Fig. 6.10). Another characteristic is the cross-section geometry of the blade shank. Most designs are rectangular and can be adjusted flexibly to the required tool radius using parallel plates. There are also shanks with a pentagonal cross-section, for which no parallel plates can be used. The number of fixing screws per stick blade depends on the supplier.

Cutter heads with different numbers of blade groups can be used for the same nominal tool radius. In this case the proper cutter head is chosen according to the module and the expected chip volume. If there is not enough space between consecutive blades the chips will jam. The avoidance of a common divider between the number of blade groups and the number of teeth of the work piece may also be the deciding factor. The width of the blade shank is chosen on the basis of the required profile height and the stiffness of its cross-section. Blade shanks which are too wide have an unused shoulder and increase the cost of the blank, incurring additional grinding costs and unnecessary wear on the grinding wheel.

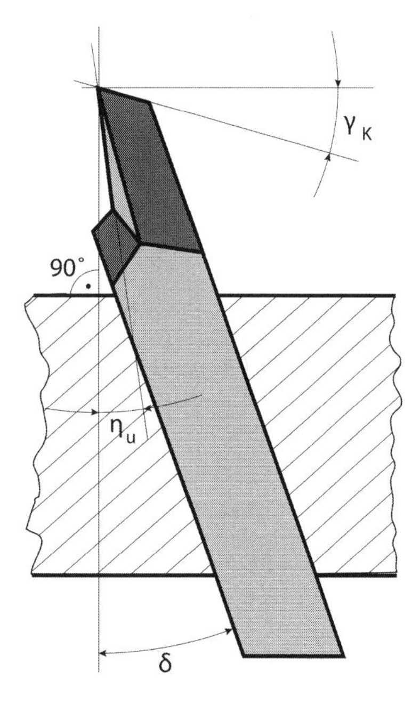

Fig. 6.7 Geometry of a stick blade.
γ_K top relief angle
η_u hook angle
δ inclination angle of the stick blade in the cutter head

On a stick blade, the surfaces which form the blade profile are ground to match the blade shank (see Fig. 6.7). For regrinding, the stick blades are removed from the cutter head and the entire blade profile is reground in the direction of the blade shank. These stick blades are mounted in the cutter head again and adjusted to the defined point height using a cutter head setup device. In contrast to profile blades, no compensation of the cutting depth is necessary. The cutting edge geometries of stick blades are differentiated according to the number of profile surfaces to be reground: "two-face grinding" (2F) or "three-face grinding" (3F).

Advantages of stick blades:

- high-performance tool due to the large number of stick blades in the cutter head compared with profile blades;
- large usable length for regrinding in the shank direction;
- flexible design of the cutting edge geometry and technological angles;
- simple blade coating techniques;
- simple low-cost procurement of HSS or cemented carbide blanks for the blades;
- favorable geometry for the use of cemented carbide

Disadvantages of stick blades:

- specialized blade grinding machines required;
- at least two functional surfaces need to be ground;
- high accuracy requirements for the stick blade geometry

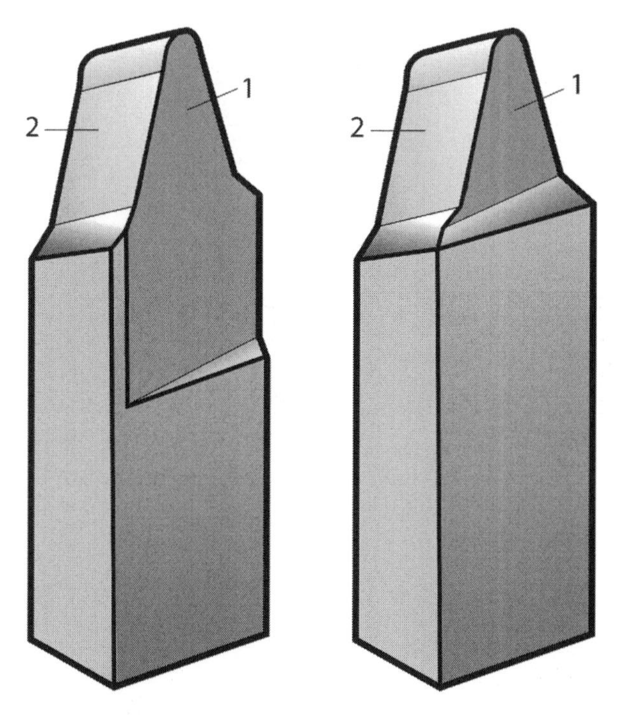

Fig. 6.8 Different stick blade versions:
left for two-face grinding (2F)
right for three-face grinding (3F)
1 front face, 2 side relief surface

Special Features of Two-Face Grinding (2F) Only the two relief surfaces are ground when profiling and regrinding; the front face has already been built along the usable stick length (see Fig. 6.8). Thus, in conjunction with the inclination of the stick blade in the cutter head, the effective rake angle for the process is necessarily fixed. Subsequently, this cannot be modified or optimized, but in return it is only necessary to coat the front face once. The relief surfaces of the primary and secondary cutting edges are not recoated after regrinding, and are exposed to increased wear.

This 2F concept was designed for use with HSS stick blades. Because of typical crater wear, coating the front face resulted in a significant increase in tool life. This concept proved not to be optimal for cemented carbide blades because in this case typically flank wear patterns appear. A disadvantage of the 2F concept is that the quality of the once-only deposited coating changes during dry cutting. As a result, the more often the blade is reground, the more marked is the scatter in the service time that can be obtained between two re-sharpening operations.

Special Features of Three-Face Grinding (3F) When profiling and regrinding, three surfaces are ground, namely the two relief surfaces and the front face. Depending on the work piece involved, the face is ground at differing angles and positions which are optimized for the cutting process (see Fig. 6.9). Since changes to the technological angles entail a change in the form of the cutting edge,

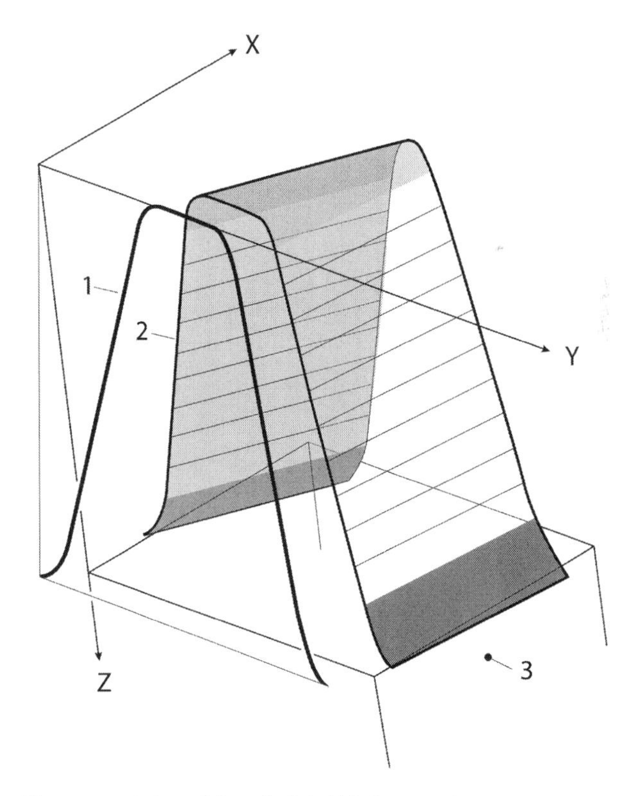

Fig. 6.9 Graphic representation of the calculated blade geometry.
1 desired profile in cutting plane
2 form of the cutting edge at the blade
3 blade shank

specialized software is required to optimize the technological angles and the form of the cutting edge independently of each other.

Given the many advantages in dry cutting operations, the 3F concept has prevailed in wide areas of bevel gear manufacture. It uses cemented carbide blades requiring a new coating after each re-sharpening operation. The blade is therefore always in brand-new condition such that there is no scatter in the service times. The coating completely encloses the cemented carbide blade, protecting it against cobalt wash-out and thermal influences and significantly reducing wear on the relief surfaces. Also new fully coated 2F blades achieve 2–3 times higher tool life before re-sharpening in comparison to only front face coated blades. The thin hard layer of the full coating causes light rounding of the cutting edge, diminishing the risk of micro-chipping.

Stick Blade Cutter Heads with 2 Blades per Group for Single Indexing These cutter heads for face milling (FM) operate without roughing or middle blades (see Fig. 6.10). They are used mainly for cutting gear sets that will subsequently be ground. Stick blades, arranged on two concentric circles in the cutter head, alternately cut the convex and concave tooth flanks. The sticks are predominantly made of cemented carbide; an HSS material is also possible, but is becoming less important.

Fig. 6.10 Stick blade cutter head with two blades per group for the single indexing method

Variant Designs of Stick Blade Geometry The target for the design of bevel gear cutting tools is to distribute the cutting work in each tooth slot between the outer blades (OB) cutting the concave tooth flanks and the inner blades (IB) cutting the convex tooth flanks. The example shown in Fig. 6.11 is for a hypoid gear with different pressure angles on the two flanks of the work piece.

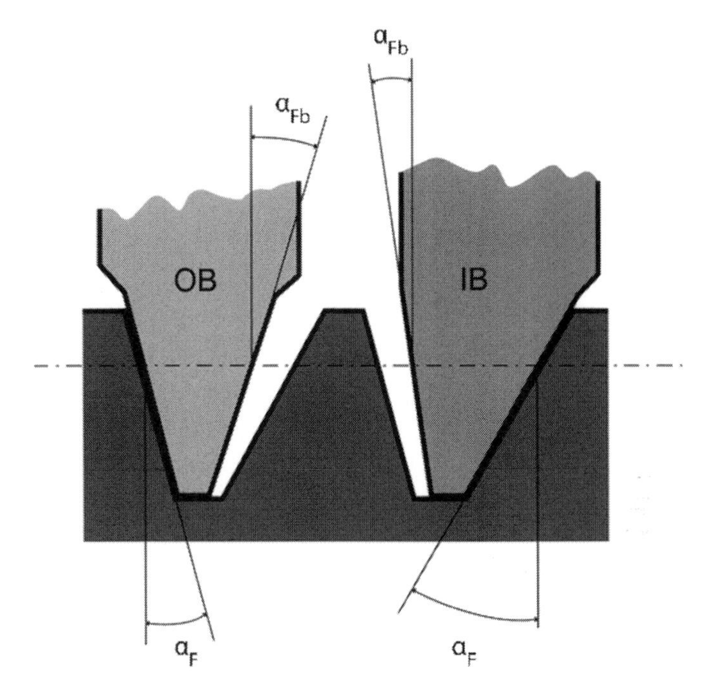

Fig. 6.11 Schematic of outer and inner stick blades (shown here in different tooth slots). *OB* outer blade, *IB* inner blade

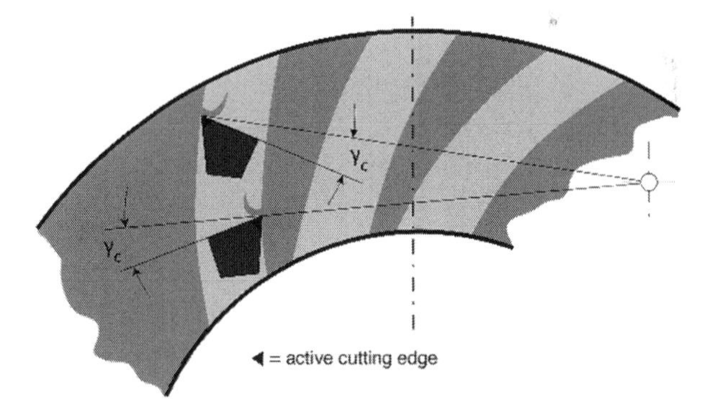

Fig. 6.12 Chip formation with two succeeding outer and inner blades

Normally, a positive side rake angle γ_C is selected for the primary cutting edge on each blade (Fig. 6.12). The secondary cutting edge then necessarily has a negative rake angle, but due to a lower point width s_{a0} than the slot width e_{fn} and a smaller flank angle $\alpha_{Fb} < \alpha_F$ practically it plays no part in the cutting process. With this concept, attention must be paid in the blade tip region to the transition from the primary to the secondary cutting edge. If there is insufficient overlapping of the two blade profiles, unfavorable wear behavior occurs.

More recent stick blade concepts for the Completing method (see Sect. 2.1) are designed such that the blade profile covers the entire cross-section of the tooth slot as, for instance, the TwinBlade concept does. A blade of this kind has the maximum possible point width $s_{a0} = e_{fn}$, giving it optimum stiffness (Fig. 6.13).

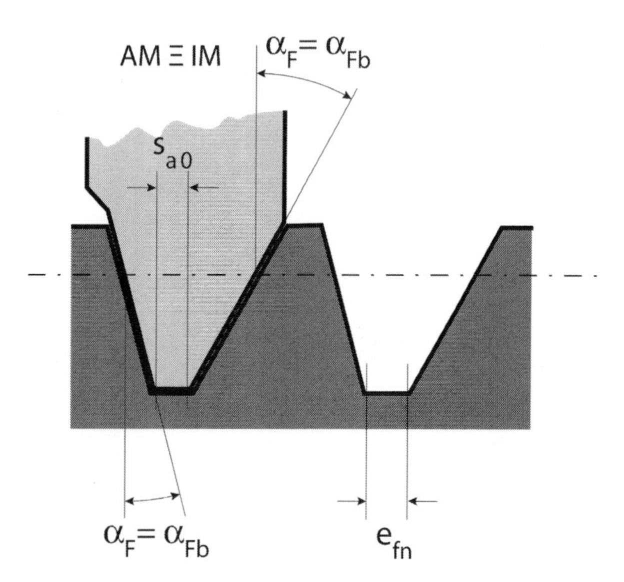

Fig. 6.13 Schematic view of a stick blade cutting over the full depth

Blades which cut the full profile depth necessarily have a side rake angle close to 0°. Although this angle is not optimal, the advantage of double cutting performance is more important. Moreover, the chips do not slide sideways over the front face, as in the conventional concept, but in the axial direction of the blade (Fig. 6.14), preventing both jamming of the chips in the tooth slot between two blades and micro-chipping on the cutting edges resulting from jamming.

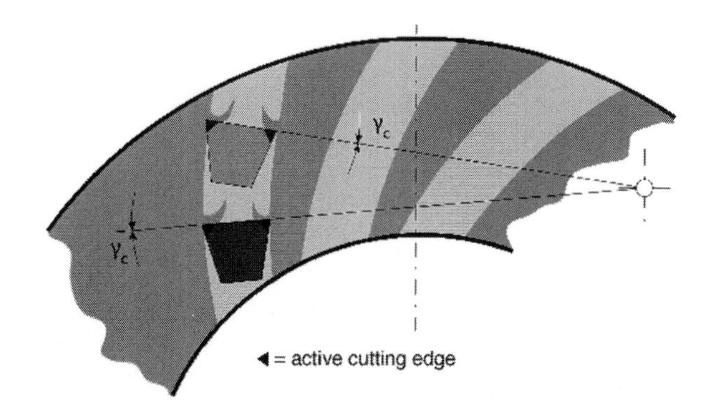

Fig. 6.14 TwinBlade chip formation

Stick Blade Cutter Heads with 2 Blades per Group for Continuous Indexing These cutter heads with a high number of starts for the face hobbing method work without roughing or middle blades (see Fig. 6.15). The outer and inner blades alternate. The chambers in the cutter head are oriented not towards the centre, but tangentially to the roll circle of the epicycloid (see Fig. 2.4). The pitch plane points of the outer and inner blades lie on the same radius and lengthwise crowning is achieved by tool tilt. Parallel plates are not used in this system. Due to the required tooth flank surface quality, tool material is always cemented carbide.

Fig. 6.15 Stick blade cutter head for continuous indexing

Stick Blade Cutter Heads with 3 Blades per Group for Face Hobbing These cutter head variants are used for the Oerlikon FN, FH and FS gear cutting methods. The additional third blade compared to the cutter head in Fig. 6.15 does a roughing operation and is not involved in cutting the finished tooth flank geometry. The roughing blade provides an optimum support for the two flank blades.

The FN cutter head series has the same diameter grades and numbers of blade groups as the cutter heads with profile blades used for the N method (see Sect. 6.2.3.2).

FH cutter heads were the first stick blade cutter heads used for the continuous indexing method in large production series (see Sect. 6.2.1). Lengthwise crowning is no longer generated by means of different angular blade sequence but by tilting the cutter head on the machine (see Sect. 3.2.1).

On FS cutter heads, the blades are closer together than on the two previous variants and the blade sequence is different: on the pinion cutter head the sequence is roughing blade, outer blade and inner blade; on the wheel cutter head, the sequence is outer blade, roughing blade and inner blade.

All three variants use the 3F cutting edge geometry.

6.2.3.5 Tool Conditioning

Processes for the Conditioning of Profile Blades Relief ground blades are usually reground by clamping the entire cutter head on a special machine and positioning a grinding wheel parallel to the front face. Each face is re-sharpened in a pendulum grinding operation. An alternative is to clamp each blade individually in a fixture. It is important to maintain all the profile blades of one cutter head in the same state for regrinding. The blade with the highest wear determines the amount to be ground for the entire set of blades. Excessive wear on one single blade can consequently lead to high tool costs.

Processes for the Conditioning of Stick Blades Originally, HSS stick blades were clamped in an adjustable fixture and re-sharpened by a pendulum grinding operation using a profiled grinding wheel. The fixture was repositioned to grind the different surfaces or the blades were re-clamped in different positions. Another variant is to clamp the blades in a ring-shaped fixture and grind them using a cup wheel. Profile crowning and protuberances are then the same on all blades. In order to obtain greater flexibility in the profile form, new CNC machines working in a "point-by-point contour grinding" have been developed which allow cemented carbide stick blades to be reground. On these machines, the stick blades are placed individually in a rotating clamping head, usually by a loading robot. In this process, it is important that the blades are positioned with the same surface in the clamping fixture and in the cutter head.

When cemented carbide stick blades are point-by-point contour ground, usually electroplated or vitrified-bonded diamond wheels which cannot be profiled are used. The profile form of the blade cutting edge is generated point by point in a CNC-cycle by the diamond wheel. Doing so, the grinding wheel is always oriented in space such that the desired tool relief angle and surface clearance are created at the same time. Geometrical variables like profile curvature, protuberance, tip relief, tool edge radius and profile length can be designed differently for the primary and secondary cutting edges (2F) as well as for the front face (3F). In order to make grinding efficient, a pendulum grinding step for roughing is followed by a finishing contour grinding step which accurately produces only a narrow facet. This results in lower grinding pressure and less thermal influence. Roughness attained in point-by-point contour grinding hovers around $Ra \approx 0.2–0.3\ \mu m$.

To increase grinding accuracy, especially when the machine is running cold or still warming up, the position of the grinding wheel in relation to the clamping fixture can be determined at regular intervals by contactless measurement, and compensated appropriately. Stick blade regrinding machines are a necessary pre-requisite to achieve stick blade profile deviations of only a few micrometres. On 3D measuring devices, it is possible to accurately measure and record blade profiles using a high point density along the entire length of the profile. Any deviation can be corrected in closed loop between the measuring device and the grinding machine.

6.2.4 *Blade Materials*

Blade materials used to cut bevel gears are mainly high speed steels (HSS) and cemented carbide substrates. Due to their complex manufacturing, profile blades (see Sect. 6.2.3.2) are made exclusively in HSS, nowadays predominantly in powder metallurgical variants. Matters are different with stick blades (see Sect. 6.2.3.4), which are much simpler to manufacture from usually rectangular semi-finished sticks, which can be obtained in both HSS and cemented carbide.

Until 1997, mainly powder metallurgical HSS stick blades were used. Since then, the increase in dry cutting of bevel gears required cemented carbide stick blade material. Two cemented carbide grades predominate. Initially, the preferred material belonged to the finest grained K10 cemented carbide classification with approximately 6 % cobalt content (grain size 0.6 . . . 0.8 μm), or the K30 grade with about 10 % cobalt content. In order to improve the combination of fracture strength, fracture toughness and wear resistance, new micro-grain substrates (grain size 0.3 . . . 0.5 μm) with a rather higher cobalt content were developed. The substrate choice should be matched to the particular machining task. All cemented carbide substrates are subject to diffusion wear when cutting case hardened steels, in which the tungsten carbide dissolves. It is absolutely essential to coat the stick blades, or at least their front faces; full coating is better, significantly prolonging tool life.

6.2.5 *Manufacturing Technology*

6.2.5.1 Characteristics of the Single Indexing Method

Spiral bevel gears can be milled most economically with the Completing method (see Sect. 2.1) in a single working cycle. The advantages of the tool, on which all blades are arranged successively in a circular arc, are evident: the blade cutting edges can be set in good coincident positions for milling, and the inactive regions of the cutting edge never make contact. However, the development of the contact pattern of such gear sets is very difficult without computer support. So, in the past, most of these bevel gears were designed for a multi-cut process where the two flanks of each tooth could be corrected separately. Nowadays, most new bevel gear sets are designed by computer simulation in such a way that each bevel gear can be finished in a single cut.

A multi-cut process may, however, be needed for various reasons, for example when there is a desire to reduce the number of variants of the necessary tool versions, or when the tool cannot be tilted in order to keep the tooth root linear in the axial section (see Fig. 3.2). Such specifications often come from the aircraft industry.

A further reason may be the need that simple manual corrections are possible on the gear cutting machine to improve tooth flank topography. A multi-cut process can then be used on both generated and plunge cut gears. Formerly, the 5-cut method was employed for this purpose (see Sect. 2.1); only three cuts are now required with modern tool systems in which the wheel is finished in one cut while, on the pinion, the first cut finishes one tooth flank and the second tooth flank is cut with a different tool and machine settings.

6.2.5.2 Characteristics of the Continuous Indexing Method

This method is suited for finishing in a single cut. The teeth have constant height and the taper of the space width is the same as that of tooth thickness (see Fig. 2.4), such that the pinion and wheel can always run together correctly. However, the paths of the outer and inner blades cross and a complete overlap of the blade point widths cannot be ensured. Regions of the cutting edge which are not involved in generating the final contour may still come into partial contact beforehand. This can occur only in the generating process and, because of the unfavourable side rake angle, leads to severe blade wear which needs to be accounted for when designing the process. When applying the continuous indexing method, vehicle gear sets are usually manufactured with the forming method (see Fig. 3.6).

6.2.5.3 Plunge Cutting Process in the Forming Method

In this process, the tool profile is reproduced exactly in the tooth slot of the wheel and chip width increases with plunge depth. The cradle position where plunge cutting has to be made is that at which the finished tooth slot will pass through the middle of the face width. Otherwise deviations in pressure angle will occur. The amount of feed is chosen to maintain chip thickness at the optimum value for the blade material. Plunge feed should only be reduced if either the machine loading escalates too much or the resulting chip becomes so large that the space between two adjacent blades is no longer sufficient to accommodate it, creating a risk that the blade will break.

The cutting direction is predominantly from toe to heel. One reason is that the burrs occur on the outside of the bevel gear where removal is easier. Another reason is that the resulting cutting forces are directed towards the work piece fixture. Cutting speeds range from 45 to 70 m/min for HSS tools, and from 160 to 280 m/min for cemented carbide tools. The shorter face width, the higher the cutting speed, that may be used since thermal stressing of the cutting edge increases with increased cutting length. The material properties of the bevel gear blank likewise have a significant effect on cutting edge load.

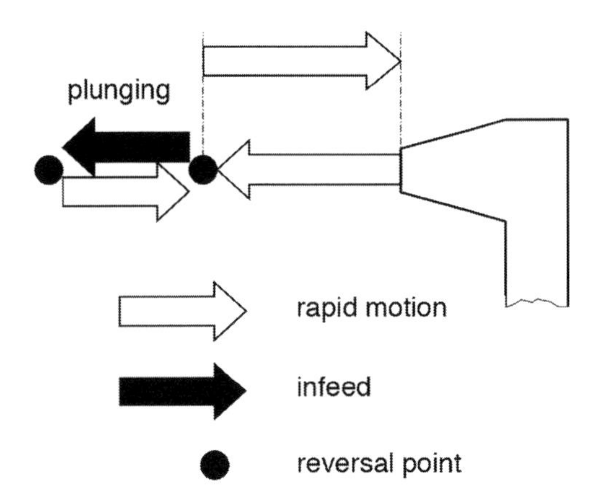

Fig. 6.16 Elements of the plunge cutting process

The plunge cutting process can be designed in various ways. Chip thickness and cutting speeds can be changed as a function of plunge depth on modern CNC machines. A clean cut at the bottom of the tooth slot is also usually possible in the reversal point. Figure 6.16 is a schematic diagram of different machine motions for the plunge cutting process used by the forming method.

6.2.5.4 Generating Process with Prior Plunge Cutting

The generating process of spiral bevel gears is often preceded by a plunge cutting process. The cost effectiveness of plunge cutting is used to remove as much material as possible from the tooth slot. The cradle angle at which the plunge cut is performed does not influence the final geometry of the work piece, and can therefore be chosen to cause the least possible wear to the tool. The inner and outer cutter blades are loaded to roughly the same extent.

In the single indexing method the plunge cut operation usually starts at the heel since, because of the tapered tooth slot, it is there that the largest volume of stock material is to be removed. The final plunge position is usually the start position for the generating process. The machine then generates till the end roll position, the tool is withdrawn from the tooth slot and returned to the starting position for another plunge. Meanwhile the machine does one indexing turn and is ready for the next tooth slot (Fig. 6.17).

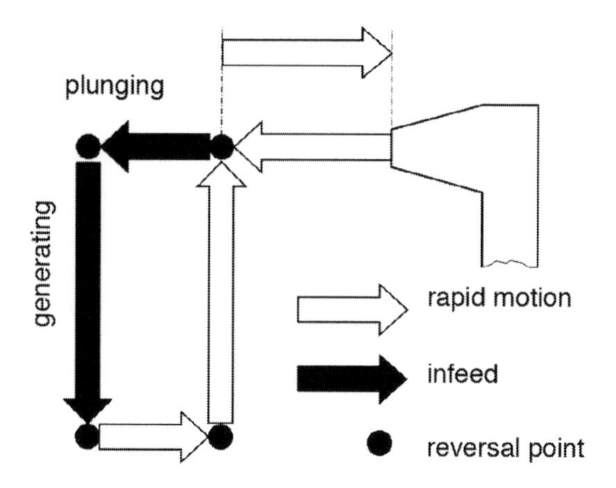

Fig. 6.17 Elements of combined plunge cutting and face hobbing

On bevel gears cut with the continuous indexing method, the tool which is employed determines the best plunge position. With a two-blade-per-group cutter head, plunging is made near the middle of the generating zone since it is there that the inner and outer blades are cutting under the same load and, if at all, to a slight extent with the secondary cutting edge (Fig. 6.2) which has unfavorable cutting properties. On a three-blade-per-group cutter head, plunge cutting preferentially occurs at the heel to let the pre-cutter remove as much stock as possible from the tooth slot. The pre-cutter is usually the first blade in the group to enter the tooth slot, but is somewhat shortened in height such that it cannot work effectively if it is plunged in another position. In the continuous indexing method, the final plunge cutting position is often not the first generating position which must then be reached by rotating the cradle. In compensation however, contrary to the single indexing method, all the tooth slots have been finished at the end of the generating process and the machine can return in rapid traverse to the initial position in order to change work piece.

There are, of course, different ways to design the plunge cutting process. On CNC machines, chip thickness and cutting speed can be altered depending on the depth position. Clean cutting at the end of the plunge is also possible. The same is true for the generating process.

6.2.5.5 Generating Process Without Plunge Cutting

This process for generated spiral bevel gears comprises at least one complete generating cycle which is necessary to achieve the correct final geometry. In order to obtain a good tooth flank surface, climb milling is recommended. Depending on the tool cutting direction, climb milling dictates the direction of roll. For the preferred "toe to heel" cutting direction, generation proceeds from heel to toe. In any generating process, very different relative positions occur between the work piece and the tool along the path of generation, Therefore, quite different chips are produced as all points of the cutting edge successively come into contact with the work piece during the generating process, the top of the blades being the most solicited.

In the single indexing method, the tool rapidly cuts the blank to full tooth depth, at the start-of-roll position, and then performs the generating process until the end-of-roll position is reached. The tool is then withdrawn from the finished tooth slot and is traversed back to the starting position. After the indexing operation the next tooth slot is generated (Fig. 6.18). This sequence can be shortened on modern machines by simultaneously moving a number of axes.

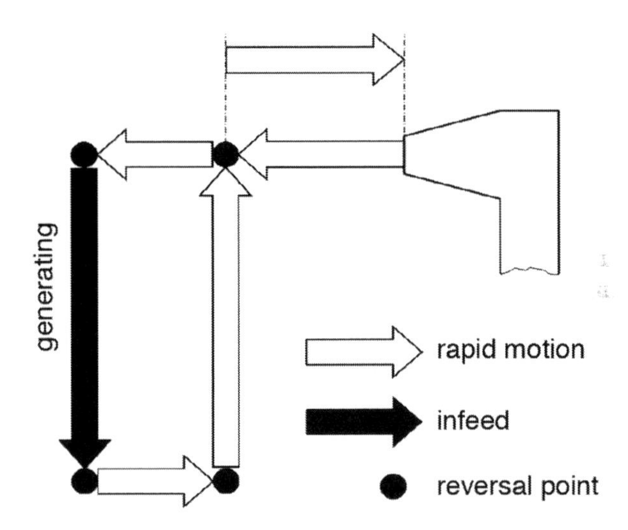

Fig. 6.18 Elements of a generating process without plunge cutting

With the continuous indexing method, the start-of-roll position is the same as before, but once the generating process is completed, not only one but all the tooth slots are finished and the machine can return to its start-of-roll position. Depending on the tool and the required tooth flank surface quality, a single generating cycle may not be sufficient. For example, coarse chips may scratch the tooth flanks and another generating cycle with a small finishing tool feed may be required.

Essentially, the criteria to determine cutting speeds are the same as those in the plunge cutting process. Since the chips in a generating process are thinner and narrower than those in a plunge cutting process, and since the cutting length does not always extend across the entire face width, somewhat higher cutting speeds may be selected for generating than for plunge cutting. The generating speed depends on the number of required generating cuts. In the case of a finishing process, the number of cuts must be chosen such as to ensure that the desired tooth flank envelope produces smooth running. More generating cuts are needed for a highly curved pinion tooth flank than for a lightly curved wheel tooth flank. In the case of a roughing cut, the generating speed can be determined entirely according to tool load.

6.2.5.6 Pitch Compensation with the Single Indexing Method

Modern dry cutting at high cutting speeds produces continuous heating and expansion of the gear body. In the single indexing method, the positions of the tooth slots in the gear change during the machining operation, resulting in pitch deviations. This is most evident when comparing the first and last slot being produced. Since this behavior can be reproduced, it is possible to counteract it. On CNC machines, it is possible to modify the position of each individual tooth slot by altering the depth position of the tool. When there is a difference between the pressure angles on the two flanks, a change in the angular position of the work piece must be superimposed to the change in depth. With the continuous indexing method, the effect of work piece heating on the tooth pitch is smaller, such that the usual temperature compensation on the cutting machine is adequate.

6.2.5.7 Compensation of Hardening Distortions

To obtain the most-cost effective production of spiral bevel gears, it is necessary to pay special attention to the influence of the hardening process on the final tooth flank topography. This is particularly important when producing lapped gear sets because the hardening process has a negative effect on the contact parameters. Corrective lapping aimed at improving the contact pattern frequently leads to rougher running. On ground gear sets, an irregular grinding allowance, increasing the risk of grinding burn, occurs if hardening distortions are not accounted for.

Using modern calculation methods, hardening distortions in the intended spiral bevel gear can be considered at the design stage where machine kinematics and tool design are modified as a function of the size and direction of the assumed distortions (see Sect. 7.1.5.3).

6.2.5.8 Effects of Cutting Quality on the Finish Grinding Process

In the automotive industry, most bevel gears using the single indexing method are subsequently ground. Low wear in the tip region of the full-profile blades produces highly constant tooth depths and fillets. These areas are important for the subsequent grinding process: if they are not accurate enough, excessive stock may be removed by grinding. On the one hand, this may cause a disadvantageous component in the grinding force; on the other hand there may be a significantly increased risk of grinding burn in the tooth root, where it is particularly critical. This risk can only be diminished by a slower and hence more expensive grinding process.

6.3 Heat Treatment

6.3.1 Fundamentals of Hardening

The initial state of steels used to make gears is either acceptable for machining or optimized for later operation in the gearbox. The requirements for load capacity and suitable wear properties in gears during operation impose a hard contact surface zone, requiring locally sufficient high carbon content. However, steel can be formed and economically machined only if it combines low strength with a homogeneous structure. Heat treatment therefore serves to alter the material properties of the steel for economic workability prior to, and for gear-typical strength, after heat treatment.

The crystalline structure of iron is always cubic, with allotropic modifications depending on the temperature. At room temperature, pure iron shows a body centered cubic (bcc) structure, known as ferrite. At a temperature of 911 °C there is a transformation into a face centered cubic (fcc) crystal cell, known as austenite. The volume of the bcc structured cell is less than the volume of the fcc structured cell.

Since pure iron is not suitable for gear applications, use is made of alloys with which it is possible to combine the opposing demands for machinability and load capacity. Instead of an iron atom, a foreign atom of roughly the same size is substituted in the iron lattice. Some possible foreign atoms include chromium, nickel, molybdenum, vanadium, silicon, boron, manganese, tungsten and titanium. With smaller foreign atoms, an interstitial solid solution is generated. Of all reasonable alloying elements, carbon has the greatest importance for gear applications.

In their initial state, case hardening steels show a phase mixture of ferrite and iron carbide in their microstructure. The phase components are proportional to the carbon content in the alloy. The cubic bcc ferrite and carbide are transformed into austenite starting at a temperature of 723 °C. The fcc structure of austenite allows dissolving a large amount of carbon. When cooled slowly, austenite is transformed completely into perlite. Perlite is a stratified mixture of ferrite and iron carbide (Fe_3C) which is called cementite. The stability range in regard to temperature and concentration in ferrite and austenite can be varied within wide limits using several different alloying elements.

If austenite with high carbon content is quenched sufficiently fast, no further perlite structure can be formed by diffusion-controlled processes. Instead martensite is created. The martensitic transformation performs a lattice shearing resulting in a tetragonal distortion of the bcc-structure. The solid solution of carbon within the iron lattice is "frozen". The higher the carbon content in the martensite, the higher the hardness will be since the residual stresses due to the distortion of the lattice increase. Hardness can be increased by increasing the mass-% carbon ratio to approximately 0.6. Higher carbon contents lead to higher amounts of retained austenite preventing further hardness increase.

Hardenability is a specific property of various steels. It is influenced by different alloy compositions. A distinction is drawn between hardness increase and hardness penetration. The first depends mainly on the carbon content, the second on the additional alloying components in the steel.

Tempering at low temperatures has proved useful after martensitic transformation. It reduces the brittleness of the material and reduces the risk of crack initiation. The process sequence for hardening thus generally consists of the steps austenitizing, quenching with martensite formation and tempering [SIZ05], [GIES05].

6.3.2 *Heat Treatment Processes*

Heat treatments can be differentiated between thermal and thermo-chemical processes, as listed in Table 6.2.

Table 6.2 Thermal and thermo-chemical processes

Hardening processes for gear parts	
Thermal hardening processes	**Thermo-chemical hardening processes**
Sectional heating	Case hardening
Induction hardening	Carbonitriding
Flame hardening	Nitriding
Laser hardening	Nitrocarburizing
Volume heating	Boronising
Quench and temper	

Thermal processes require a suitable hardenability of the base material. Depending on the temperature profile, surface hardening or through hardening are possible. Only certain alloying systems are suitable for through hardening. Hardening the entire cross-section of the component is in any case an exception in gears. In practice, induction and flame hardening are relevant, along with laser hardening. These processes are characterized by relatively light equipment, simple operation and high output.

In the case of thermo-chemical processes, the chemical composition of the material in the surface layer can be designed according to the stress profile of the gears, such as wear resistance on the surface and toughness in the core. The wear resistance on the surface is achieved by hardening; the strength and toughness of the core by using a suitable alloy composition.

6.3.3 *Thermal Processes*

6.3.3.1 Induction Hardening

The heat is generated in the material by means of AC induction at different frequencies. The frequency of the eddy current and the distance of the inductor from the surface determine the depth of penetration and hence the attainable surface layer hardness depth. The form of the inductors and their motion relative to the work piece determine the characteristic of the hardened surface layer. Processes with short cycle times like, for example, progressive hardening and spin hardening, generate an uneven heat input due to the shape of the tooth, and hence an uneven hardness in the surface layer. On gears, this leads to a risk of overheated teeth at the tip followed by simultaneously inadequate hardening at the bottom of the slot. This difficulty can be reduced by the use of specially shaped inductors.

6.3.3.2 Flame Hardening

In flame hardening, the energy needed for austenitization is supplied externally to the regions which are to be hardened. Special burner systems are used for this purpose. The necessary high energy density of the flame on the work piece surface and its great directionality during energy input to the work piece set the limits for the application of this process. If the surface of the work piece is exposed too long to the flame, undesired burning-off will result. If exposure to the flame is too short, the necessary temperatures are not reached. Flame hardening can consequently be used only for low quality standards and shallow hardening depths.

6.3.3.3 Laser and Electron Beam Hardening

In laser and electron beam hardening, only a spot of the surface is heated. The necessary cooling happens by rapid dissipation of the heat into the work piece (self-quenching). The limitation of this method is the accessibility of the surface to the laser or electron beam. An evenly hardened layer can be obtained over the entire tooth flank only if the beam hits the surface perpendicularly or at least at a constant angle. When compared to laser hardening, electron beam hardening has a higher energy density, but vacuum is required. Due to these restrictions, laser or electron beam hardening can be used only to a very limited extent for gear hardening.

6.3.4 Thermo-chemical Processes

6.3.4.1 Advantages of the Processes

Gear-typical stresses on the material are essentially the contact stress on the tooth flank and the bending stress in the root. Their influence on material choice was discussed in Sect. 3.5.3, with the apparently conflicting result that gears require a material which is hard, tough and strong.

By using thermo-chemical hardening processes it is possible to achieve these contradictory material properties, with the hardness, ductility and strength curves changing gradually from the surface to the core of the work piece. Different chemical compositions in the surface layer and in the core of the gear teeth allow the creation of differentiated material conditions for tooth flank load capacity and bending strength.

Several different diffusion processes can be used for this purpose, allowing the alloying elements to penetrate the base material under heat, and thus changing its composition and the associated properties. Apart from a few niche applications, these processes require concentration gradients for the diffusion component and heating of the entire work piece.

Figure 6.19 represents the relationship between residual and bending stresses. The residual stresses are shown along the ordinate. The dashed line (curve 2) shows the residual stress curve, the dotted line (curve 3) indicates the equivalent load-dependent stress, and curve 1 indicates the resulting cumulative stress in section A-A. It is evident that the compressive residual stresses induced by heat treatment reduce the tensile stresses in the root on the loaded side of the tooth.

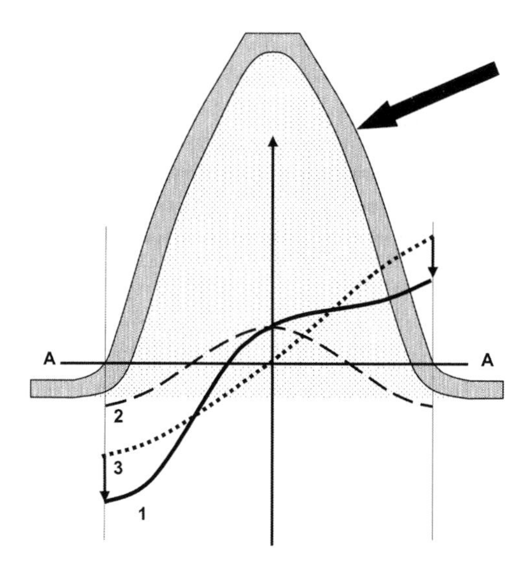

Fig. 6.19 Residual and bending stresses in the tooth root

The form of such residual stress profiles can be adjusted by the distribution of alloying elements introduced in the surface layer and the concentration towards the core. These can be controlled and slightly adapted by the process parameters of the diffusion treatment. Nitrogen and carbon have proved to be suitable enriching elements for gears in many process variants.

6.3.4.2 Nitriding

Nitriding processes can always be applied when thin, very hard, wear-resistant nitrided and nitrocarburized layers tend best to meet the expected stresses.

Only very slight distortions are expected because of the low process temperatures. However, the hardness curve in the layer and its small thickness limit the amount of stock which can be removed in a subsequent hard finishing operation. In practice, nitrided gears receive their final tooth flank geometry and surface during soft machining, such that the gear is finished after nitriding.

Enriching the surface layer of the material with elemental nitrogen increases the hardness and strength of the base material. Unlike martensitic hardening, this process relies on the mechanism of precipitation hardening where strength increases with hardness in the nitrogen-enriched surface layer. In addition, as a rule compressive residual stresses appear in the surface layer. Ductility is inversely proportional to hardness.

This latter characteristic indicates what the main problem is with nitrided gears: as a result of their high hardness and inadequate ductility, micro-cracks propagate very quickly when there is local overload. There is a high probability that the type of damage after a sudden overload will be a fracture in the root. Nitriding layers are therefore preferably used where the main focus is protection against adhesive and abrasive wear and corrosion. If carbon as well as nitrogen is added to the surface in a nitrocarburizing process, a surface layer with a very small coefficient of friction is produced. This increases efficiency in the gearbox along with tooth flank load capacity.

The temperature range of the process is 500–580 °C. The nitrogen source is primarily ammoniac. The atmosphere is normal pressure in gas nitriding; partial vacuum processes with and without plasma support through a glow discharge are also used. Salt bath processes with liquid reaction partners in the form of cyanides as the suppliers of nitrogen and carbon are used as well.

Depending on the base material, nitriding depths of up to 1 mm are possible. The tendency to crack occurrence can be reduced to a certain level by starting from a quenched and tempered base material. Using specially alloyed nitriding steels will not necessarily solve this problem. Elements with an affinity to nitrogen increase the surface hardness but they also steepen the hardness gradient toward the core, which is even more critical for crack occurrence. In addition, when using alloying elements which form carbides and nitrides, carbon is replaced by absorbed nitrogen and the carbon precipitates at the grain boundaries which reduces material toughness.

6.3.4.3 Carburizing

For heavily stressed gear components, case hardening using special case hardening steels is the optimum combination. The material can be formed and machined well prior to heat treatment, and provides the best property profile for typical gear stresses after heat treatment. Case hardening is a thermo-chemical diffusion process with many possible variants.

The carbon needed for hardening is added to case hardening steels after soft machining. These steels have low initial carbon content, and are alloyed to improve hardenability or core toughness, depending on the type of load. In this way, it is possible to modulate the mechanical properties to meet the specifications.

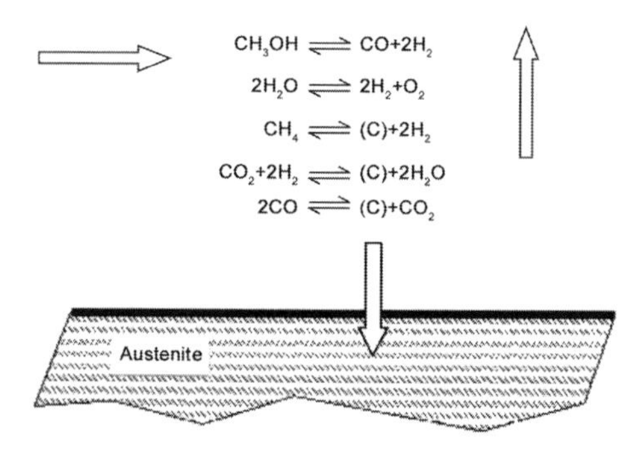

Fig. 6.20 Reaction diagram of carbon transition

Carburizing uses liquid or gaseous carbon carriers and operates at temperatures allowing a high diffusion rate in an austenitic structure. In austenitic structures the solubility of carbon in the iron lattice is more than one order of magnitude higher than that in ferritic structures. The mechanism of carbon transition is well understood and, thanks to its reaction kinetics, it offers a large series of process indicators which can be used to evaluate and control the process.

The thermal decomposition results in the reaction shown in Fig. 6.20. The reaction products CO, H_2, C, CO_2, H_2O, and O_2 can all be measured. Partial oxygen pressure, infrared absorption or dew point can be measured. All these techniques indicate the prevailing carbon activity in the gas atmosphere, with varying degrees of accuracy and reproducibility. If more than one measuring method is used, the reliability and reproducibility of the carburizing process are improved.

Due to the high concentration gradients between the atmosphere and the work piece and the high process temperature, enrichment of the surface layer with carbon at first proceeds very efficiently. Intelligent control algorithms allow highly precise process control, permitting carburizing depths of more than 5 mm to be reached without undesirable side-effects influencing the surface layer [WUEN68], [WEIS94], [WYSS95]. The precondition for this is steelmaking with high quality melt metallurgy, to ensure low scatter of the alloying elements and high fine-grain stability. However, ever higher case hardening depths are desirable, particularly for marine drives. Since conventional process design is not feasible for this approach, special micro-alloying systems are used to improve grain size stability for carburizing at even higher temperatures and in shorter times. However, this process design is currently feasible only for small dimensions in laboratory conditions.

Further potential to shorten process times is offered by modern low pressure processes which use oxygen-free carbon carriers to work at substantially higher carbon concentrations and without inter-granular oxidation. This process can additionally be supported using plasma. Unfortunately, at this time there are no reliable sensor systems available for reaction products without oxygen, so that process control has to be done on empirically determined parameters. The empirical factors are determined by varying the parameters over several test batches. This process is an interesting alternative for mass production, as low-pressure carburization followed by high-pressure gas quenching is very cost-effective and sufficiently proven. As there are no consumables with emissions to contaminate the parts, this technology is ready for integration in production lines.

The combination of nitrogen and carbon is also possible in a carburizing operation. The process, known as carbonitriding, generates a material composition with significantly greater hardenability in the surface layer. This is achieved by a much more sluggish transformation of austenite into martensite, but with a greater tendency to higher residual austenite content, and hence lower surface hardness. If the treatment temperature is reduced in order to counteract this tendency, there is a risk that the core structure will consist of a heterogeneous mixture of martensite and ferrite, a state which is referred to as under-hardened. A suitable process control of the nitrogen feed rate is therefore important in carbonitriding.

6.3.5 Temperature Profiles in Case Hardening

For martensite formation, a sufficiently high cooling rate from the austenitization temperature to the martensite initiation temperature must be guaranteed. The martensite initiation temperature depends on the chemical composition of the alloy. The optimum cooling rate should be just high enough to ensure that the maximum possible volume of martensite is created.

Many properties on the work piece can be influenced using different temperature profiles. For example, undesirable residual stress curves resulting from unfavorable changes in sectional area in the work piece can be improved by stepwise quenching.

The currently prevailing process combination is carburization using carbon providing atmospheres followed by quenching in liquid media. Depending on the time sequence and temperature profile, these are referred to as direct hardening or as single hardening, as shown in Fig. 6.21. A variant form of single hardening is hardening after isothermal transformation. The double austenite-ferrite transformation ensures a fine austenite grain size prior to hardening.

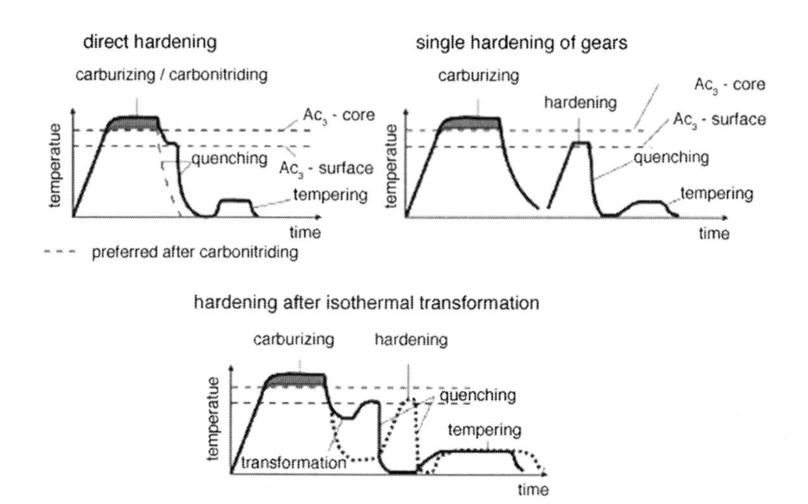

Fig. 6.21 Direct hardening, single hardening and hardening after isothermal transformation [DIN-EN10052]

Direct hardening is the most economical option in terms of expected changes in size and shape. The risk of unacceptable microstructures occurs only at case hardening depths of more than 3 mm. In this case, a more complex form of process management becomes necessary, thus increasing the costs.

Salt melts, mineral oils and synthetic oils are possible quenching media, along with aqueous polymer solutions. When choosing the medium, a compromise must always be found between cooling rate, risk of cracking, fire risk, susceptibility to distortion, work safety, ecological restrictions, productivity and ergonomy.

6.3.6 Hardening Distortions

Dimensional changes to the work piece always occur during heat treatment. These distortions may be regarded as the sum of all geometrical changes in the work piece. They occur as linear dimensional changes and as non-linear plastic and elastic deflections. Figure 6.22 classifies the distortion components. Linear dimensional changes are caused by volume increase, due to martensitic transformation, which entails a growth in volume of roughly 1 %. They are classified as unavoidable changes in size and shape.

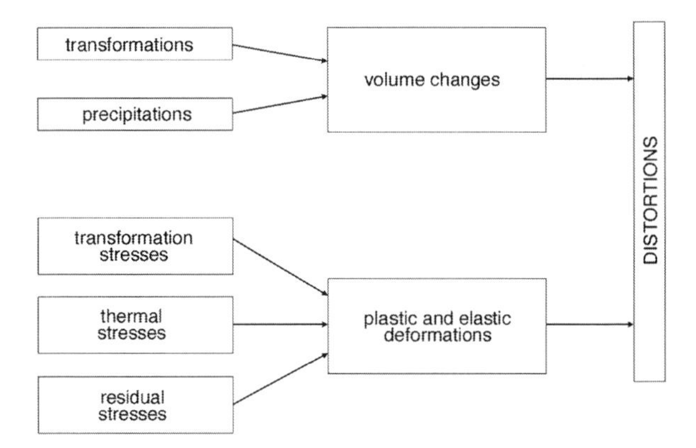

Fig. 6.22 Classification of hardening distortions

On the other hand, strictly speaking, non-linear changes are avoidable since they are caused by plastic and elastic deformations. They arise as a result of residual stresses, non-homogeneity in the material, asymmetrical shape of the work piece, asymmetrical forging or uneven temperature profiles during heat treatment.

Fundamentally, small distortions can be expected only if the work piece has a low, homogeneous residual stress level. High residual stresses also increase the risk of crack formation and provide the worst possible preconditions for the hardening process.

Avoidable distortions are caused chiefly by thermal stresses and by residual and processing stresses. In order to control these, requirements must be imposed on the material, on the bevel gear design and on the heat treatment itself.

Minimal residual stresses will occur only in an original material which is as homogeneous as possible. Non-homogeneity, for example segregation, leads to stresses which can be released during cutting or heat treatment. Residual stresses can also be induced in the work piece through unsuitable cutting parameters or blunt tools, and always lead to distortions if they are released by heat treatment.

Apart from the stress state of the bevel gear, the blank form has a significant effect. A design optimized for low distortion will envisage an even mass distribution and gradual variations in cross section.

When work pieces are installed in a rack for heat treatment and hardening, force may under certain circumstances be exerted on them. Since the high temperature yield strength of case hardening steel is low at the necessary process temperatures, deformations of the support rack may be reproduced irreversibly on the work piece. Other factors which increase distortion are steep temperature gradients inside the work piece which lead to time-shifted and locally-shifted thermal effects if heating rates are too high.

Also of importance are the packing density and mass distribution on the rack, which influence the flow of the process and quenching media around the work pieces. This is an especially critical factor where bevel gears are concerned, as there are widely different geometrical conditions for flows around work pieces at tooth tip and root.

Quenching in vaporizing media like oils or aqueous polymer solutions implies considerable distortion potential. All liquid quenching media have a boiling point significantly below the work piece temperature at the beginning of the quenching process. The quenching medium does not wet the surface of the work piece, but rather surrounds it with a layer of steam of low thermal conductivity which keeps the medium away from the surface. This so-called Leidenfrost effect locally produces very different cooling rates, when departure from film boiling occurs. Figure 6.23 represents different phases in the quenching process. At the start of quenching, the entire surface of the work piece is insulated from the quenching medium by the vapor film. As the temperature falls, the boiling phase forms. In this phase both the quenching medium itself and small steam bubbles cool the work piece further. Once the work piece temperature falls below the boiling point, the convection phase begins, further cooling the work piece through the flow of the quenching medium over its surface.

As the re-wetting front progresses over the surface, the point of maximum volume change likewise migrates as a result of thermal contraction. The boundary may be located at several points simultaneously. The three stages of heat transfer, with their different cooling rates, consequently generate an extremely inhomogeneous stress state in the bevel gear. Moreover, the teeth disturb an even circulation around the surface and steam bubbles may cling in the tooth slots and delay the course of martensite formation. A remedy here may come from agitation systems in the quenching tank which help the steam skin to break down without forcing the pace of heat transfer excessively through too strong a flow.

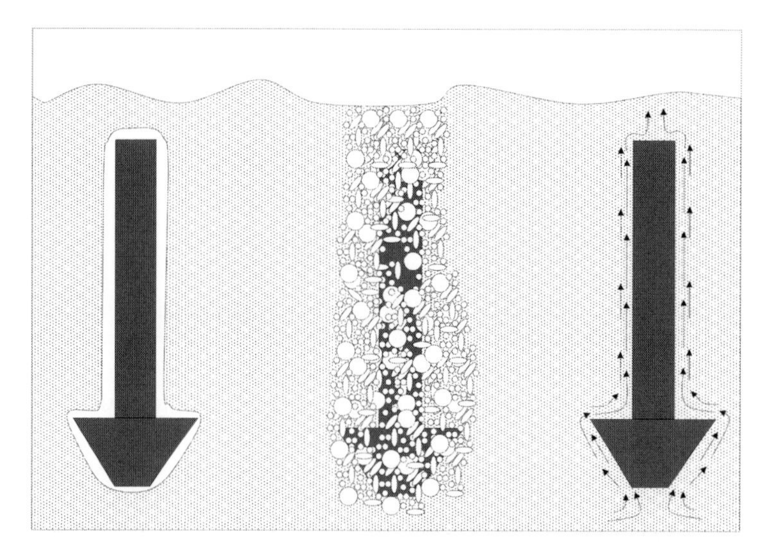

Fig. 6.23 Steam shell, boiling and convection phase of the quenching process

These problems do not arise with non-vaporizing media such as gases and molten salts or metals which have the advantage of being easily controlled. For example, in the case of high-pressure gas quenching, the possible work piece mass is limited due to poor heat conductivity, even with hydrogen or helium as the quenching media. The use of salt melts or metallic melts provides a favorable quenching process because of their greater thermal conductivity, but is questionable in terms of the environmental impacts.

6.3.7 Fixture Hardening

The increasing demand in decreasing bevel gear mass is in conflict with the conditions needed for hardening to be as distortion-free as possible. Especially in the case of wheels, which are kept very flat for weight reasons, substantial distortions are to be expected. These distortions manifest themselves principally in the form of axial run-out and non-concentric bore. Sufficiently accurate axial run-out and bore concentricity are an indispensable condition for the lapping process on bevel gear sets which are not ground or hard skived.

This can be achieved only with press or fixture hardening, where the wheel is fixed at several points during quenching to keep axial run-out and non-concentricity as small as possible.

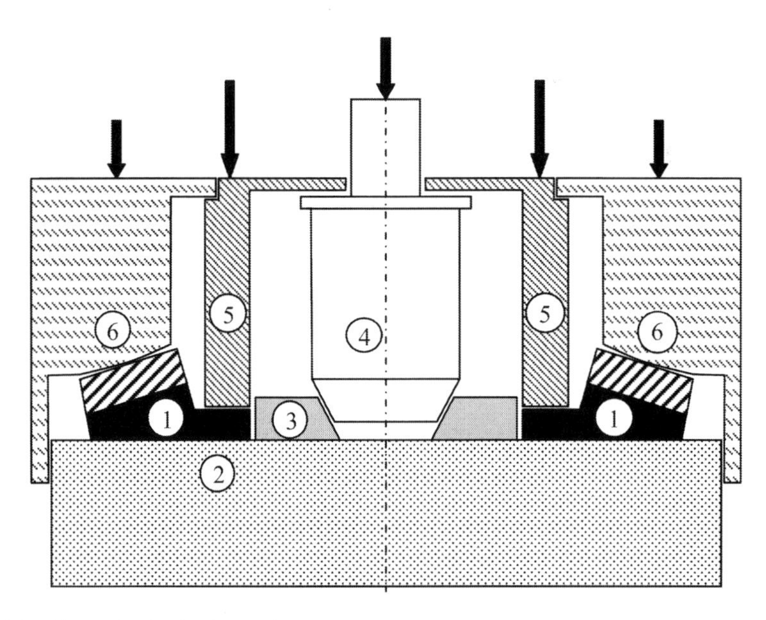

Fig. 6.24 Example of a hardening fixture

Figure 6.24 shows a hardening fixture for wheels. Prior to hardening, the fixture is opened and circular sections 4, 5 and 6 are raised. Wheel 1, which has been heated to hardening temperature, is then placed, preferably by a handling robot, over expanding mandrel 3 on base plate 2, and the fixture is closed. In the process, tapered arbor 4 widens expanding mandrel 3 to align the bevel gear, around which the quenching medium begins to flow. Defined forces are applied to arbor 4 and the two flat rings 5 and 6. The circular expanding mandrel 3 locates the wheel bore such that the axial run-out of the wheel, in the tooth tip zone and on the mounting flange, is minimized. In this manner, any uncontrolled dimensional changes in the wheel during the quenching process are largely excluded. The temporal and local interaction between thermal stresses and transformation stresses must be accounted for in this process. Specialized know-how is involved both in the geometry-related choice of supporting points at which force is exerted, in time control of the force and in the design of the fixture tool. The latter must include suitable channels for the quenching medium which must, above all, allow even, sufficiently rapid quenching. The fixture must also be simple to set up.

Hardening fixtures for shaft-shaped pinions are much more complex. The straightness of the pinion shaft can be ensured only by means of a rotational movement of the shaft over a number of force- and path-controlled rolls during the quenching process.

6.4 Hard Skiving

The terms hard skiving or skive hobbing designate a machining process in which special cutting tools are used to re-hob already hardened tooth flanks. Its special advantage is that it can be used to eliminate hardening distortions from the gear. The same machines and tools are used as for soft machining, except that the cutting edges consist of a different material. In practice, polycrystalline cubic boron nitride (CBN) has proven effective as a tool material. The initially feared detrimental re-hardening zones do not occur. The surface roughness attained in hard skiving is usually lower than that found in ground tooth flanks.

Hard skiving is at present the only process which can be used to hard finish spiral bevel gears by the continuous indexing method. The now customary large gears, with up to 2,300 mm outside diameter, are, without exception, roughed by the Zyklo-Palloid® method and hard skived after heat treatment. The continuous process has the advantage of high pitch accuracy, and the universally usable tool system allows cost-effective manufacture even with a batch size of one. The sole prerequisite for hard skiving is a generated wheel. Plunge-cut wheels cannot be hard-skived because of the length and width of the chips.

Tools for hard skiving do not differ substantially from Zyklo-Palloid® blades on the HPG-S method (see Sect. 2.1). Instead of the profile blade (see Fig. 6.4), a cutter bar carrier is placed in a clamping device mounted on the surface of the cutter head. This clamping device is tilted relative to the surface of the cutter head by the required tip clearance angle. The radial setting of the cutting edge is adjusted with distance plates. Each time the CBN cutting edge is re-ground at the cutting face, the cutter bar carrier is shifted in the clamping holder such that it is at the correct height.

6.5 Grinding Spiral Bevel Gears

6.5.1 Development History

The first bevel gear grinding machine was developed in the USA in the 1930s, followed shortly by a pinion machine with a tilting axis for the tool [HOFM39]. The grinding machines developed later were almost exclusively used for generated spiral bevel gears in the aircraft industry and, despite their age, machines are still found in some manufacturing companies. Because of the lightweight construction of the gear housing and the large resulting deflections, the highest requirements are imposed on the flank topography and the tooth pitch of aircraft gears. Grinding is therefore almost essential as a hard finishing process for these gear types.

Attempts to use more modern grinding machines in the automotive industry during the 1970s and 1980s failed because their long machining and set-up times were not economically viable. The use of these machines, on which the tool could no longer be tilted, was restricted to the 5-cut method.

The breakthrough into large series production came only with the advent of CNC machines and flexible numerical computation processes. In 1982, the pioneer of CNC bevel gear grinding, Wiener, developed a grinding machine for Formate® wheels which had a grinding spindle possessing a second, eccentrically arranged, driven bearing sleeve (see Sect. 6.5.5.2). This allows avoiding full faced contact between the grinding wheel and the tooth flank.

A machine for pinion grinding followed later the same year. The grinding wheels were not yet path-controlled; it was possible to alter the flank angle, but not profile crowning. At roughly the same time, attempts were made to grind spiral bevel gears using CBN-coated grinding wheels on modified mechanical cutting machines. However, the coating of these wheels lacked the necessary consistency and the resulting surface quality was still inadequate, such that this method was used only for rough grinding in preparation for subsequent structural lapping.

In 1985, Wiener developed a bevel gear grinding machine with a path-controlled dressing device, thereby achieving much greater flexibility. The machine featured a two-spindle design such that work pieces designed for the 5-cut method could be machined in a single clamping.

In 1989, a newly developed CNC grinding machine came to the market in the USA. The particular feature in its design concept was that the traditional mechanics for tool swivelling and tilting were made obsolete through the swivelling motion of the machine tailstock. This machine technology, the more sophisticated computation processes and the development of stable ceramic-bonded grinding wheels improved the cost-effectiveness of bevel gear grinding in high volume production.

From 1988 to 1990, an attempt was made in Switzerland to grind spiral bevel gears using the continuous indexing method [STAD90]. The idea was to use a gear type tool with tooth flanks conjugate to the work piece, and with a large hypoid offset, to be meshed with the work piece. High rotational speeds of the pinion resulted in high sliding velocities which allowed bevel gear tooth flank grinding. However, the timed relationship of the tool and the work piece was not stable enough at these speeds, and the manufacture of the tool proved so costly that the attempt failed.

Since the turn of the millennium, new machine concepts have been launched where the motional axes needed for grinding are arranged on a central column, eliminating the former machine bed. The machine dynamics were also improved by the use of direct drives. A concept from another manufacturer features a vertical grinding spindle above the work piece. This ensures significantly better chip disposal, the drive and measuring systems are located above the working area and the workspace is easily accessible. Both machine concepts have gained acceptance in the automotive industry where they are frequently integrated with automated loading in complete production lines.

Since 2005 grinding machines are also being manufactured for the machining of larger bevel gears with diameters up to 1,100 mm.

6.5.2 Development Trends

Grinding as a hard finishing process for spiral and hypoid bevel gears is now established in several application fields. In the aircraft industry, virtually all bevel gears have traditionally been ground, and the number of ground gears is increasing in general gear applications. In the automotive industry, depending on the manufacturer, bevel gear grinding is responsible for a varying percentage of bevel gear production as a whole. The equipment in the machine park, the quality standards and the operating personnel, along with company know-how, are all given as reasons for the manufacturing method to be used, although it is hard to recognize a general trend. Economic aspects and the advantages of ground gears in optimizing load capacity are reasons of the rise in importance given to grinding, especially for high torque applications.

6.5.3 Tools

Bevel gear grinding is exclusively done with the single indexing process, using cup or cone-shaped grinding wheels. Because of their price to performance ratio, vitrified bonded wheels with sintered corundum abrasives are the main type used, with CBN grinding wheels for some applications. Vitrified bonded wheels usually have a base plate allowing them to be adapted to the machine.

Cup Grinding Wheels Cup shaped grinding wheels are used for plunge and generation grinding (Fig. 6.25). The rated diameter of the grinding wheel is defined in the design, and is determined mainly by the size of the gear and the desired displacement behavior. The grinding wheel diameter is identical to the cutter head diameter used in soft machining. For most applications, the diameter is graduated in inches (according to the FM cutter head scale). In the former Soviet Union and China, metric tool systems are also used.

Geometrically, cup grinding wheels might be regarded as hollow cylinders rotating round their axes. The resulting centrifugal forces may cause the bodies of vitrified bonded cup grinding wheels to burst beyond certain rotation speeds. The permitted circumferential speed of vitrified bonded cup grinding wheels is accordingly limited to 32 m/s by the FEPA standard. Higher circumferential speeds are permitted only if the manufacturer has conducted a bursting test and suitability is certified.

Fig. 6.25 Cup grinding wheel for bevel gear grinding

Flared Cup Grinding Wheels The flared cup grinding process introduced by Krenzer in 1986 [KREN91] uses grinding wheels in the form of a truncated cone to hard finish spiral bevel gears. The main characteristic of this process is the CNC motion used to generate a wheel whose geometry is identical to that obtained with the plunge cutting process. Figure 6.26 shows the geometry of the grinding wheel used in the process. The process has meanwhile largely been replaced by plunge grinding, which is more cost effective.

Fig. 6.26 Flared cup grinding wheel

6.5.4 *Abrasives*

6.5.4.1 Grinding Wheel Materials

The following materials are generally applicable:

- Corundum, semi-friable aluminium oxide, aluminium oxide
- sintered corundum
- silicon carbide
- cubic boron nitride (CBN)

The abrasives described below are widely used in application because of their positive properties.

Aluminium Oxide This universally applicable, easily splintered abrasive is available in various compositions. Aluminium oxide is suitable for machining hardened steels, but removal rates are limited. Aluminium oxides are well suited for grinding unhardened materials.

Applications: universal machining of unhardened and hardened steels
grinding speed: 15–30 m/s
depth infeed: 0.05–0.15 mm
dressing speed: 20–250 mm/min
dressing removal: 0.1–0.2 mm

Sintered Corundum This is Al_2O_3 generated from aluminium hydroxide. After drying and crushing, it is sintered in a number of operations. Sintered corundum has an extremely small crystallite of 0.0002–0.0005 mm, leading to a self-sharpening effect. Crystallite fragments of varying sizes splinter off during grinding, exposing underlying sharp grit cutting edges. Sintered corundum consequently attains significantly longer tool lives. Surface roughness can be improved systematically if the dressing removal amount is lowered or the speed is reduced. Sintered corundum wheels are frequently supplied with an open-pored bond, allowing more cooling lubricant to come into contact and increasing grinding performance.

Applications: high power grinding of hardened steels
grinding speed: 15–35 m/s
depth infeed: 0.1–0.5 mm
dressing speed: 100–400 mm/min
dressing removal: 0.03–0.1 mm

Silicon Carbide Silicon carbide has higher hardness and lower toughness when compared to aluminium oxides. Pale green silicon carbide, with extremely good cutting ability, and dark silicon carbide which is extremely hard and less brittle, can both be applied. Silicon carbide has high thermal conductivity and is therefore used especially for grinding heat-sensitive materials.

Applications: high-alloyed and tough materials, titanium alloys
cutting speed: 15–35 m/s
depth infeed: 0.1–0.5 mm
dressing speed: 20–300 mm/min
dressing removal: 0.08–0.15 mm

CBN Cubic boron nitride (CBN) came to the market as an abrasive in 1969. CBN is synthesized and is supplied in monocrystalline and microcrystalline forms. Advantages include its great intrinsic hardness of 4,700 N/mm^2 according to Knoop (by comparison, diamond hardness is 7,000 N/mm^2) and the long resulting tool life and high thermal stability.

CBN is vitrified bonded for dressable grinding wheels. Dressing removal rates are significantly smaller than for sintered corundum grinding wheels, and the operation is referred to as conditioning. Profile changes can therefore be realized only to a limited extent. An alternative is to electroplate steel bodies with a CBN-layer. The steel substrate provides the profile which is then coated in the positive electroplating process. The cutting ability of CBN changes over the operational tool life, and thus the geometry of the grinding wheel changes because of wear. Grinding wheels manufactured in the reverse process (negative coating) are not yet in use.

Applications: high power grinding of hardened steels
grinding speed: 30–70 m/s
depth infeed: 0.1–0.5 mm
dressing speed: 100–300 mm/min (with vitrified bonded CBN)
dressing removal: 0.005–0.01 mm (with vitrified bonded CBN)

6.5.4.2 Abrasive Parameters

Grit size Grit sizes are given according to the FEPA standard [FEPA06] or in US mesh. The determining factors in the choice of grain size are the required removal rate and surface quality. The rule of thumb is: the larger the grit, the higher the removal rate and the longer the wheel life. However, this is only partially true since the dressing parameters also play a crucial role. In case of bevel gear grinding, grit sizes of 60 (grit diameter 250–297 μm) up to 120 (grit diameter 88–105 μm) are generally used. Different grit mixtures combining the advantages of coarse and fine grain sizes are more often used.

Bond The bond ensures the cohesion between all the constituents of the grinding wheel. The wheel is pressed as a green compact which is then sintered in a furnace. The bond acquires its final properties during this process.

Possible bond types:
Bonds for CBN and diamond

- vitrified low-firing bond (V)
- resinoid bond (B)
- brittle bronze bond (sintered metal bond)
- electroplated bond

Bonds for corundum and silicon carbide

- vitrified bond (V)
- resinoid bond (B)
- fiber-reinforced resinoid bond (BF)
- rubber bond (R)
- polyurethane bond (P)

The bond type used almost exclusively for dressable CBN, corundum, silicon carbide and sintered corundum is the vitrified bond, which has the following crucial advantages:

- it is the only type of bond with a controllable structure
- given an appropriate choice of bond, it assures the best chip transport and very good cooling lubricant feed
- bond hardness, pore size (structure) and pore volume can be influenced precisely
- higher removal rates are possible than with resinoid bonds

Hardness, Structure and Porosity The force required to break a grit out of the bond is the measure of static hardness which is determined by grain fraction, bond mass, structure or porosity, any pore forming agents which are present and the manufacturing method. Static hardness is defined according to the ISO or FEPA standard, as given in Table 6.3.

The dynamic hardness, which is based on static hardness, but is additionally influenced by grinding speed, is crucial to the process. Dynamic hardness rises with

Table 6.3 Static hardness according to ISO or FEPA

Code letter	Hardness
A–D	Extremely soft
E–F	Very soft
G–I	Soft
J–M	Medium hard
N–Q	Hard
R–T	Very hard
U–Z	Extremely hard

increasing grinding speed. Grinding wheels graded at static hardness from F to K are generally used to grind bevel gears.

Grinding wheels with a variety of structures are used. A fine structure is employed when low surface roughness is required. A porous structure allows a higher removal rate, as it allows more material to be removed from the contact area, and the grinding wheel can absorb more cooling lubricant. Porous grinding wheels are usually employed for power grinding.

Grinding Wheel Codes Manufacturers use different systems for the classification of grinding wheels. In most cases, the following properties are specified:

Abrasive, granulation, hardness, structure, bond, porosity

Example: 93A 60H 15 VH [WINT11]

93A	abrasive	sintered corundum
60	granulation	medium
H	hardness	soft
15	structure	porous structure
V	bond	vitrified
H	porosity	high porosity

6.5.5 Grinding Technology

6.5.5.1 Calculating the Specific Material Removal Rate

The specific material removal rate is an important variable in the analysis of a grinding process. It specifies the volume of stock removed per unit time and millimetre of contact width, making it possible to compare different grinding methods and processes.

The volume of stock removed V_w is a measure of the volume of metal removed from the work piece by grinding, and is calculated from the area of the tooth flank and the removed grinding allowance. When compared to a standard grinding process like flat grinding, the influencing variables for the specific material removal rate in a bevel gear grinding process, such as the feed a_e and the grinding wheel width b, are not constant. More analytically, the differences in feed along the profile height and the changing contact line along the face width must be accounted for. This results in a specific material removal rate which can only be calculated point-by-point and is represented over the whole tooth area. Calculations of this kind for cylindrical gears are performed by [KLOC05].

In a simplified approach, for spiral bevel gears, an attempt can be made to approximate conditions to those in a standard grinding process (Table 6.4).

Table 6.4 Simplified calculation of the specific volume removal rate for a tooth flank

Designation	Formula	No.
Tooth flank area excluding the bottom of the tooth slot	$A_z = h_m \cdot b_t / \cos\ \alpha_n$	(6.1)
	where: b_t = face width as arc length $b_t = \frac{r_{co} \cdot \pi}{90} \cdot \arcsin\left(\frac{b}{\cos \beta_m \cdot 2 \cdot r_{co}}\right)$ α_n = pressure angle according to Fig. 2.20	(6.2)
Removed volume	$V_w = A_z \cdot a_e$	(6.3)
	where a_e = flank allowance perpendicular to the flank	
Chip volume per unit time	$Q_w = \frac{dV_w}{dt}$	(6.4)
At constant generating speed v_w (cradle speed of rotation) and generating interval $\Delta\alpha$	$Q_w = \frac{V_w \cdot v_w}{\Delta\alpha}$	(6.5)
At constant plunge cutting speed v_t	$Q_w = \frac{V_w \cdot v_t \cdot \sin \alpha_n}{a_e}$	(6.6)
Specific material removal rate for generation grinding	$Q'_w = \frac{Q_w}{h_m \cdot \sin \alpha_n} = \frac{V_w \cdot v_w}{\Delta\alpha \cdot h_m \cdot \sin \alpha_n}$	(6.7)
Specific material removal rate for plunge grinding	$Q'_w = \frac{V_w \cdot v_t}{a_e \cdot h_m}$	(6.8)

6.5.5.2 Process Parameters

Generating Speed, Radial Feed These variables determine the specific material removal rate for the generating and plunge cutting processes.

When a tooth flank (slot) is being generated, the cradle and the work piece perform a combined motion whose result is an octoid tooth flank profile. The rotational speed of the cradle is called the generating speed whose value is determined by the module of the gear and feed per cycle (guide value: 5–20° per second).

The plunge feed defines the speed at which the tool plunges, into the tooth slot, along its axial direction and is determined by the module of the gear and the depth feed per cycle (guide value: 10–100 mm/min).

Possible problems caused by an excessive generating speed or plunge feed are grinding burns, surface defects (in generation grinding) and uneven tool wear.

Infeed The combination of stock allowance which is to be removed by gear grinding, the existing hardening distortions and radial and axial run-out determine the necessary depthwise infeed. For power grinding of gears with medium surface quality requirements, the depth feed is made in one turn (flank allowance: 0.10–0.12 mm). If the allowance is larger, it is divided in several cycles.

Guide values for the infeed per cycle are 0.4–0.6 mm in plunge grinding and 0.05–0.3 mm in generation grinding.

Grinding Speed The circumferential speed of the grinding wheel at its active diameter is referred to as the grinding speed. Increasing the grinding speed has the following effects:

- The chip cross-section and the number of kinematic cutting edges are reduced, resulting in lower cutting forces.
- The reduction in the chip cross-section lowers the load on the grinding wheel, diminishing wear. However, increased speeds increase frictional power.
- The roughness of the ground surface falls due to the smaller chip cross-section, despite the falling number of kinematic cutting edges.
- The thermal load is increased as a result of the rise in grinding speed, and more of the cooling lubricant fed to the wheel is spun off due to increased centrifugal forces.

Guide values for the grinding speed are 18–25 m/s in generation grinding and 16–20 m/s in plunge grinding.

Eccentric Motion According to Waguri In 1967, the development of an additional oscillating motion was published in Japan. The grinding spindle bearing is mounted in a separately driven eccentric bushing. The degree of eccentricity cannot be varied. If the eccentric drive is activated, the grinding wheel performs an oscillating motion in addition to its main rotation (grinding motion), interrupting the wheel-to-work piece contact. This allows an improved cooling lubricant feed; this is a particular advantage in plunge grinding. In generation grinding, it is possible to influence the surface structure with the aid of the eccentric motion. The ratio between the grinding wheel and eccentric bushing speeds, termed the eccentric factor, is variable and is chosen as a function of the grinding wheel diameter (see Table 6.5). It is necessary to take into account that the oscillation of the grinding wheel must not excite the grinding machine at an unfavourable frequency as the resulting grinding quality will deteriorate.

Table 6.5 Guide values for the eccentric factor

Grinding wheel diameter [mm]	Eccentric factor
50–70	0.2–0.35
70–100	0.35–0.8
100–150	0.5–1.3
150–220	0.5–1.8
220–300	0.7–2.5
Above 300	1.1–3.0

Direction of Rotation of the Grinding Wheel The direction of rotation of the grinding wheel should in general be opposed to the generating direction to avoid material jam. If the gear geometry makes this impossible, or in a plunge grinding operation, the direction of rotation which allows better grinding oil supply to the contact point should be chosen.

Distribution and Reduction of the Grinding Stock In order to minimize grinding time and the risk of grinding burn, it is important to distribute grinding stock removal evenly between the concave and convex flanks. On Duplex gears, this is achieved by means of a slantwise infeed motion.

The grinding stock can only be reduced by optimizing the entire process chain of soft machining → hardening → ID/OD grinding → gear tooth grinding. The minimum allowance depends on the gearing system and the size of the gear. In an ideal case, stock allowance can be reduced to 80–120 μm for usual applications in the automotive industry (mean normal module 3.5–10 mm).

Grinding Power Grinding power as a function of the specific removal rate can be monitored and analysed by the gear grinding machine. Appropriate machine software can trigger an emergency response to protect the tool and the work piece if the grinding power limit is exceeded during the process.

Some of the measures which can be used to reduce grinding power are lower infeed per cycle, reduced generating/plunging speed, a more suitable grinding wheel, a higher dressing speed for a rougher grinding wheel and checking or correcting the grinding oil supply.

Wear Compensation Particularly in operations involving high power grinding in a single cycle, grinding wheel wear can be observed between the first and last ground slot of a gear, leading to tooth-to-tooth pitch deviation (see Sect. 5.3.1). The software on bevel gear grinding machines can compensate for such wear. Normally, a combination of infeed along the grinding wheel axis and a rotation of the work piece are required. The required compensation values can be calculated from the pitch diagram of a measuring machine.

6.5.5.3 Cooling Lubricant

Mineral oil based grinding oils, semi-synthetic and fully synthetic oils can be used in bevel gear grinding. In order to meet the demands of power grinding, high power grinding oils with special additives must be used. The market also supplies alternatives to oil based cooling lubricants in the form of water based grinding emulsions. Cost-effective use of these emulsions in gear grinding is not possible because of the chip removal rates involved.

A wide range of grinding oils is sold by various manufacturers. Grinding oil usability is determined by the chemical and physical properties. The recommendations of the manufacturer and the oil supplier should be considered when choosing the grinding oil.

Conditioning Proper supply of the grinding oil is a crucial part in process optimization. A high power grinding process requires the cooling lubricant to be supplied in sufficient quantity, at a constant temperature and at the right place. A separate filtration unit is used to clean and cool the grinding oil. In the unit, the oil is allowed to settle (such that air may be released), filtered and cooled to the necessary feed temperature.

In general, the oil is supplied via a volume circulation and a separate high pressure circulation. The volume flow is fed via pipes and nozzles to points in front of and behind the contact zone. The high pressure circulation feeds a V-shaped cleaning nozzle placed above the active profile of the wheel. This flushes material deposits from the pores of the grinding wheel, enabling the absorption of fresh cooling oil.

It has proved effective when designing pipes and nozzles for cooling lubricant to choose an oil discharge velocity which roughly matches the circumferential speed of the grinding wheel. If the oil discharge velocity is too small or too high, the oil is not fed effectively to the grinding zone (Table 6.6).

Table 6.6 Calculating the oil discharge speed velocity v_{oil}

Designation	Formula	No.
Oil discharge velocity	$v_{oil} = \varphi \cdot \sqrt{2 \cdot \frac{p}{\rho}}$	(6.9)
	φ = coefficient of friction of the liquid	
	ρ = density of the oil	
	p = pressure in the supply line	

According to this formula, the cooling oil supply pressure needed for an oil discharge velocity and circumferential speed of 20 m/s is approximately 6–8 bars.

6.5.5.4 Dressing

Dressing the grinding wheel is necessary for the following reasons:

- The grinding wheel has become worn in the process; dressing restores the profile accuracy of the wheel.
- The grinding wheel becomes loaded with removed material during the process, and the abrasive grits become blunted. Dressing sharpens the grit cutting edges and the removed material is flushed from the grinding wheel.
- The grinding wheel and, hence, the surface roughness of the tooth flanks which is achieved, can be influenced by varying the dressing parameters.

Dressing is carried out once or twice per work piece during the process. To improve surface quality and pitch accuracy, the wheel is not dressed at the very last or the next to last few cycles in a multi-cycle operation.

Dressing Tool Vitrified bonded grinding wheels in CNC controlled bevel gear grinding machines are profiled almost exclusively with a rotating dresser roll. As an alternative, a stationary dressing tool (monocrystalline diamond) can be employed. The diamond attains better surface qualities owing to the low contact ratio. However, the higher wear on the monocrystalline diamond, inherent to the system, means that form deviations occur on the profiled grinding wheel after a short period of operation.

Diamond dresser rolls with the geometry shown in Fig. 6.27 are generally used in bevel gear grinding. The diameter, d_R, of the dresser roll depends on the diameter of the grinding wheel which is to be dressed; diameters, d_R, from 35 to 100 mm are common. The edge radius, ρ_R, of the diamond roll is determined by the individual application, and is usually between 0.15 mm and 6 mm.

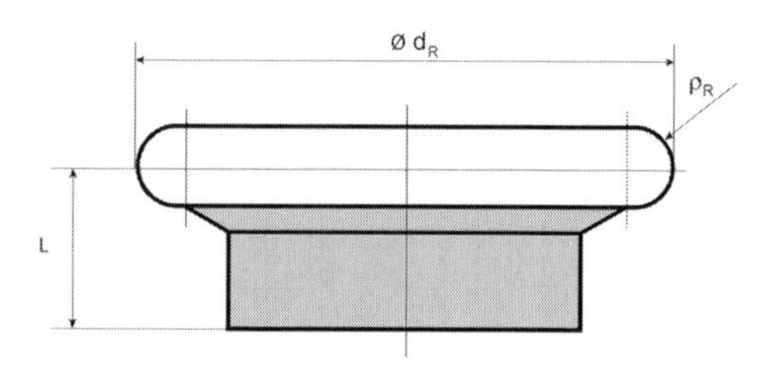

Fig. 6.27 Geometry of a dresser roll

The types of diamond used in the coating differ fundamentally in structure, and may be classified as follows (Table 6.7):

Table 6.7 Classification of diamond dresser rolls

	Natural diamond	MCD[a]	With bonding phase	Without bonding phase
Manufacture	Natural	High pressure synthesis	Sintering process	Gas phase deposition
Hardness [Knoop]	6,000–9,000	8,000–9,000	4,000–5,500	8,500–9,000
Application	All diamond tools	Stationary dressers	Corner reinforcement for form rolls wear protection	Form rolls with small radii and angles

[a]Mono-cristalline diamond

Diamond dresser rolls are manufactured using a direct (positive) process or a reverse (negative) one. In the positive process, a profiled body is coated with diamonds, usually arranged irregularly. The bond material is nickel. An advantage of tools manufactured in the positive process is that they can be re-coated.

In the negative process, high precision moulds of metal or graphite are used, in which the diamonds are incorporated in a single layer. A nickel film is deposited in an electroplating process, serving as a bond for the diamonds and joining them to the steel body. An alternative to electroplating is sintering where the diamonds are placed or scattered into a graphite mould. The cavity between the diamond coating and the body is filled with a wear-protective powder containing tungsten, and is sintered at a high temperature. Steps must be taken to prevent temperature-induced distortions [LIER03].

Dressing Process The dressing process and the resulting grinding wheel structure are defined by:

- the ratio between the circumferential speeds of the grinding wheel and the dresser roll (dressing ratio)
- the dressing removal amount
- the dressing speed and resulting contact ratio

Dressing Ratio q_d The ratio between the circumferential speeds of the dresser roll v_R and the grinding wheel v_S is referred to as the dressing ratio q_d.

$$q_d = \frac{v_R}{v_S} \tag{6.10}$$

Dressing can be performed as climb dressing (positive dressing ratio) or counter rotational dressing (negative dressing ratio). The dressing ratio decisively affects the way in which the cutting paths of the individual dressing diamonds are configured and hence the effective roughness depth of the grinding wheel. The two overlapping rotary motions means that the path curves of the individual dressing diamonds assume a quasi-cycloid shape. A positive ratio results in steeper effective path flanks, such that counter rotational dressing tends to produce lower grinding wheel roughness [KLOC05], [WESS07].

Guide values for the dresser ratio q_d are 0.4 to 1.6 in climb dressing, and −0.5 to −1.2 in counter rotational dressing.

Dressing Removal Amount a_d The dressing removal amount a_d determines the amount of material removed in the axial direction of the grinding wheel, corresponding to the amount by which the grinding wheel is shortened in one dressing operation.

Dresser Infeed a_{ed} The dresser infeed a_{ed} is a function of the dressing removal amount a_d and the flank angle φ_s of the concerned grinding wheel flank. It determines the actual dressing removal amount on the flank in the normal direction.

$$a_{ed} = a_d \cdot \sin \varphi_S \tag{6.11}$$

Guide values for the dressing removal amount are 0.05–0.15 mm for vitrified bonded sintered corundum grinding wheels and 0.005–0.010 mm for vitrified bonded CBN grinding wheels (Table 6.8).

Table 6.8 Legend to Figs. 6.28, 6.29 and 6.30

Symbol	Designation	Unit
a_d	Dressing removal amount	mm
φ_S	Profile angle of the grinding wheel	degree
d_S	Grinding wheel diameter	mm
d_R	Dresser roll diameter	mm
v_S	Grinding wheel circumferential speed	m/s
v_R	Dresser roll circumferential speed	m/s
v_{fd}	Dressing feed rate	mm/min
ρ_R	Dresser roll rounding radius	mm
$W_{S\ theo}$	Theoretical grinding wheel waviness	mm
n_S	Speed of rotation of the grinding wheel	r.p.m.

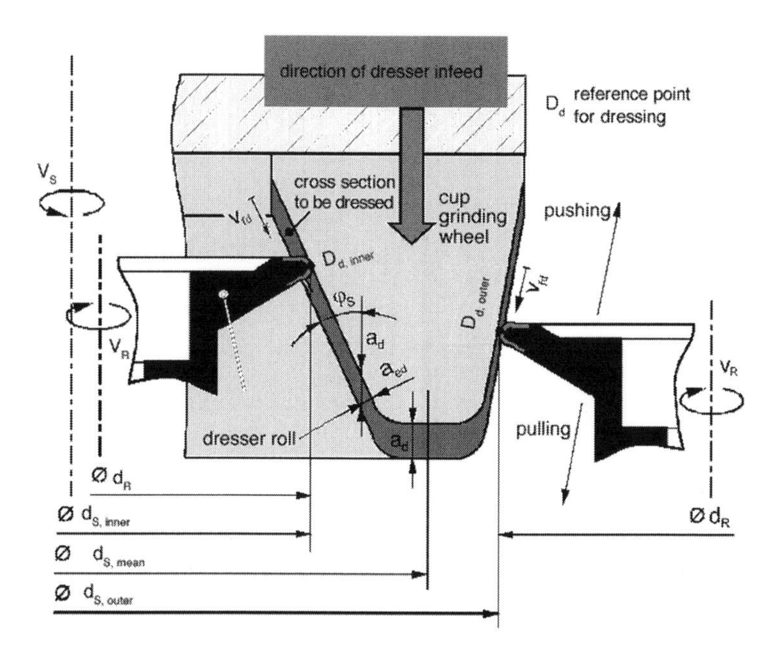

Fig. 6.28 Dressing removal rate and dresser infeed [WESS07]

Contact Ratio u_d and Dressing Speed v_{fd} Contact ratio u_d is defined as the ratio between the contact width a_{pd} of the dresser roll and the dressing feed rate s_d (see Fig. 6.29). The contact ratio is a universal variable for dressing processes, and permits a comparison between different methods and processes. The contact ratio describes how closely the individual spiral feed paths created in dressing are spaced [TUER02] (Table 6.9).

Table 6.9 Calculating the contact ratio

Designation	Formula	No.
Contact width	$a_{pd} = \frac{b_d + s_d}{2}$	(6.12)
Effective width	$b_d = 2\sqrt{\rho_R{}^2 - (\rho_R - a_d \cdot \sin\varphi_S)^2}$	(6.13)
Dressing feed rate	$s_d = \frac{v_{fd}}{n_S}$	(6.14)
Contact ratio	$u_d = \frac{a_{pd}}{s_d}$	(6.15)

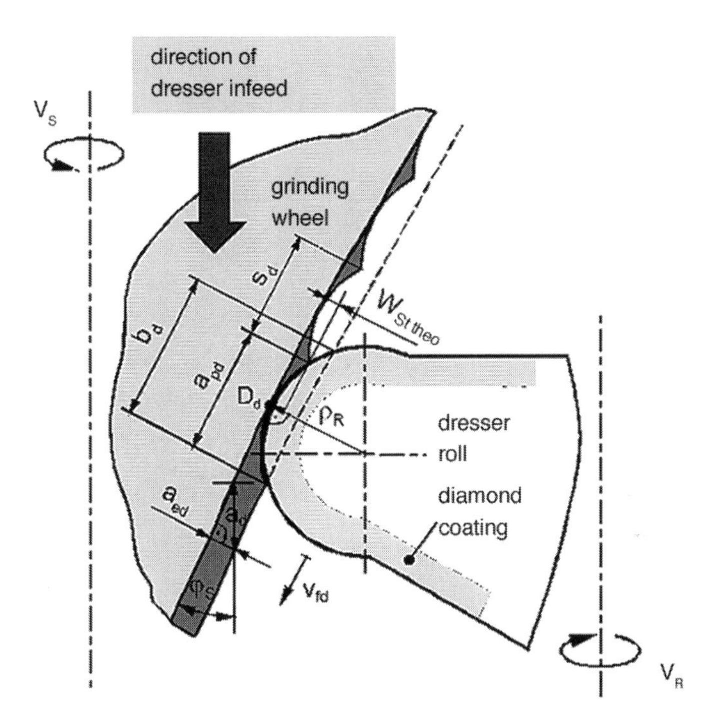

Fig. 6.29 Contact ratio in dressing [WESS07]

The dressing speed v_{fd} is adjustable, and may differ for the inner and outer sides and for the tip of the grinding wheel. To ensure an even grinding wheel surface, the dressing speed should be chosen to make the contact ratio $u_d > 1$.

A special aspect results from the different contact conditions on the inner and outer sides of the grinding wheel when dressing cup wheels. Figure 6.30 illustrates the way in which the contact arc, when dressing the inner side, is significantly longer than that when dressing the outer side. These conditions are similar to internal and external cylindrical grinding. In this case, the equivalent grinding wheel diameter is used to compare the processes [CZEN00].

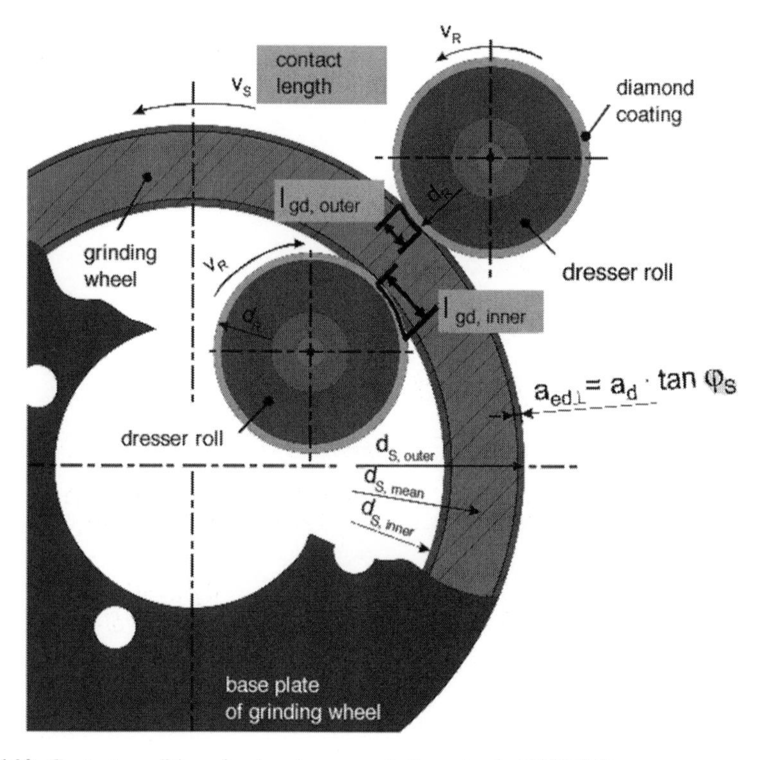

Fig. 6.30 Contact conditions for dressing cup grinding wheels [WESS07]

Similarly, it is possible to calculate the equivalent dresser roll diameter d_{eqd}, which combined to the tangential dresser infeed a_{ed} determines the contact length l_{gd} between the dresser roll and the grinding wheel. The difference in contact length produces different theoretical grinding wheel wavinesses on the two grinding wheel flanks (Table 6.10).

Table 6.10 Calculating the contact length

Designation	Formula	No.
Equivalent dresser roll diameter outer flank	$d_{eqd} = \dfrac{d_S \cdot d_R}{d_S + d_R}$	(6.16)
Equivalent dresser roll diameter inner flank	$d_{eqd} = \dfrac{d_S \cdot d_R}{d_S - d_R}$	(6.17)
Contact length	$l_{gd} = \sqrt{a_d \cdot \tan \varphi_S \cdot d_{eqd}}$	(6.18)

The different flank angles of the grinding wheel have a significant influence on the achievable surface roughness of the bevel gear. The chip thickness is smaller when the outside of the grinding wheel is dressed at the smaller flank angle. As a result, the grinding wheel surface is more finely profiled and a measurably finer surface is ground than on the inside with the larger flank angle.

This means that different dressing parameters have to be used in order to generate the same surface roughness on both flanks. This can be done by varying the contact ratio with guide values ($u_d = 1 \ldots 6$).

6.6 Lapping

6.6.1 Development History

In the 1930s, lapping processes for bevel gears were developed in Germany and in the US. Based on the first devices which drove the bevel gear set and rolled the gears against one another while lapping compound was added, lapping machines which also allowed the pinion shaft to oscillate during rotation were developed shortly afterwards [KRUM41]. This significantly reduced the formation of lapping edges at profile height. A further improvement came through the superimposition of a slight swing movement of the gear shaft, and subsequently also of the pinion shaft (swing lapping). This allowed moving the contact zone between pinion and wheel along the tooth face width, and hence to lap the tooth flanks variably along the face width [PRRE60]. Within certain limits, it became possible for the first time to influence crowning and thus the position of the contact pattern.

After the Second World War, the use of hypoid gears in car rear axles increased. Higher sliding velocities created conditions in which the productivity of the lapping process could be significantly increased. In the 1950s, oscillatory lapping was simplified in the US to the point where only the shaft angle changed as a result of an oscillating pinion shaft (swinging pinion cone: SPC). This process was superseded when V and H lapping was developed in Switzerland which allows the contact zone to be moved along the entire tooth flank by use of a suitable periodic change in the mounting distances of the pinion and wheel and in the hypoid offset (see Sect. 3.4.3). Using this process, it was possible to improve significantly the control of lapping removal along the tooth flank, with the result that the method is still in use today. In the 1970s, the initially hydraulic and template controlled axial path systems were replaced by electronic controls. A considerable sophistication in process control was achieved in Switzerland in 1988 through the first CNC lapping machine.

6.6.2 Description of the Process

Lapping is a hard finishing process using a geometrically undefined cutting edge, and is employed in bevel gear production as a means to improve the running and noise behavior of the gear set after hardening.

In bevel gear lapping, the pinion and wheel are in mesh at low torque and high rotation speed such that abrasive contained in the lapping compound may remove material from the tooth flanks in the contact zone (Fig. 6.31). During the lapping process, the contact zone is displaced by changing the relative positions of the pinion and wheel such that the complete tooth flank is lapped, possibly with increased material removal at selected points.

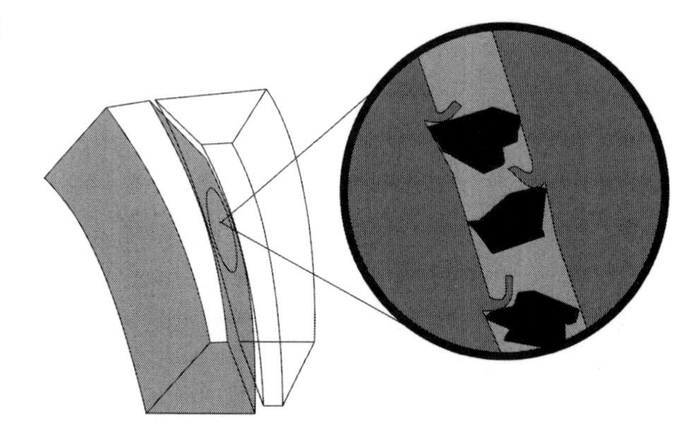

Fig. 6.31 Material removal by means of lapping

Contact pressure between the meshing tooth flanks is generated by an applied torque which acts with the sliding velocity to remove material.

6.6.3 Lapping Media

6.6.3.1 Lapping Compound

The task of the lapping compound is to transport the lapping abrasive into the tooth contact zone and provide sufficient lubricating effect to prevent scuffing of the gear set. The lapping compound should also retard settling of the lapping abrasive in the lapping medium tank, and still be easy to wash off the gear set after the lapping operation. Use of oil with thixotropic properties is recommended to keep the lapping abrasive in suspension.

6.6.3.2 Lapping Abrasive

The lapping abrasive has a decisive effect on material removal. In practice, silicon carbide has proved suitable because of hard, sharp-edged grains combined with its good cost-performance ratio. Less usual are lapping grains of aluminium oxide or boron nitride.

The chosen grain size depends on the application. Basically, a higher material removal can be obtained with a large grain, but only at the expense of greater roughness on the tooth flanks. A grain size of 280 for a module equal to 2.5 mm, or 240 for modules larger than 5 mm, is generally used in the automotive industry. The mass mix ratio of lapping compound to abrasive is usually 1 part of oil to 1.4 or 1.5 parts of abrasive.

6.6.4 *Process Parameters*

6.6.4.1 Feeding the Lapping Abrasives to the Gear Set

The jet of lapping medium is not usually fed directly into the meshing point but, depending on the direction in which the gear set is rotating, ahead of it. By directing the lapping jet at the toe of the wheel, the spinning effect distributes the lapping compound conveniently over the face width.

The lapping nozzles are adjusted radially such that two-thirds of their cross-section overlaps with the face width. The distance from the nozzles to the gear teeth should be about 10–15 mm, in the direction of the wheel shaft (Fig. 6.32).

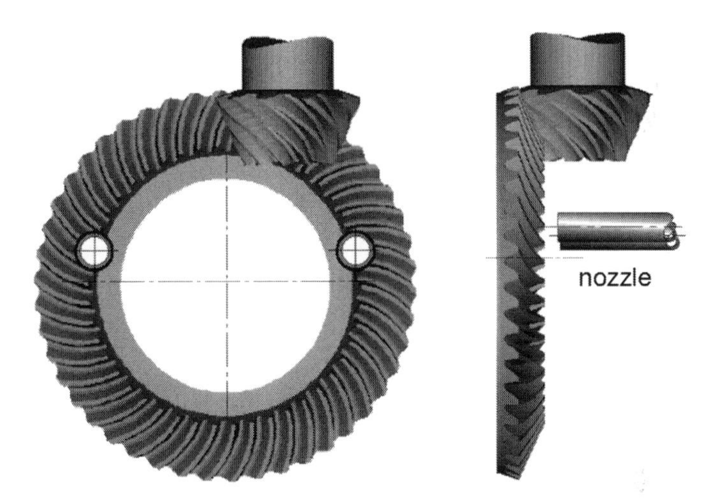

Fig. 6.32 Position of the lapping nozzles

6.6.4.2 Influences of Gear Set Design

As established in Sect. 3.4.4, the tool radius crucially affects the load-free displacement behavior of a gear set. For lapping, this means that with a large tool radius the contact zone is easy to change, whereas with a relatively small tool radius this is possible to only a limited extent.

The hypoid offset influences the local sliding velocities which vary over the tooth flank (see Sect. 2.4.3). Since pure rolling takes place on the pitch cone (Fig. 6.33) of a spiral bevel gear set without offset, scarcely any lapping removal is possible about this area such that if lapping time is too long, a clear elevation will appear about the pitch cone.

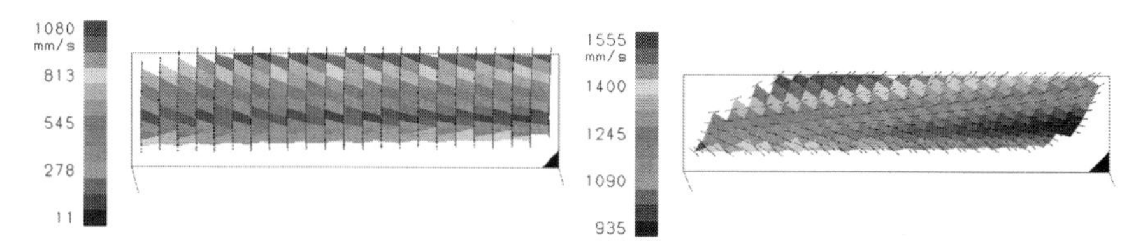

Fig. 6.33 Sliding velocities for spiral and hypoid bevel gears

Because material removal on the pinion increases with transmission ratio, the tool used to mill the pinion is usually equipped with a protuberance to remove more material at the root of the tooth and thus avoid a lapping edge being formed in this area of the tooth. The height of the protuberance is chosen such as to leave a small unlapped area at the root of the pinion tooth, thus preventing premature contact with the tip of the wheel.

Of course, a greater surface hardness on the pinion counteracts the higher lapping removal.

Figure 6.34 describes the desirable contact pattern positions before and after lapping for different tool radii and hypoid offsets. The contact patterns after lapping must be adapted to the environment of the gear set, and are shown here only as examples (see Sect. 3.4.5).

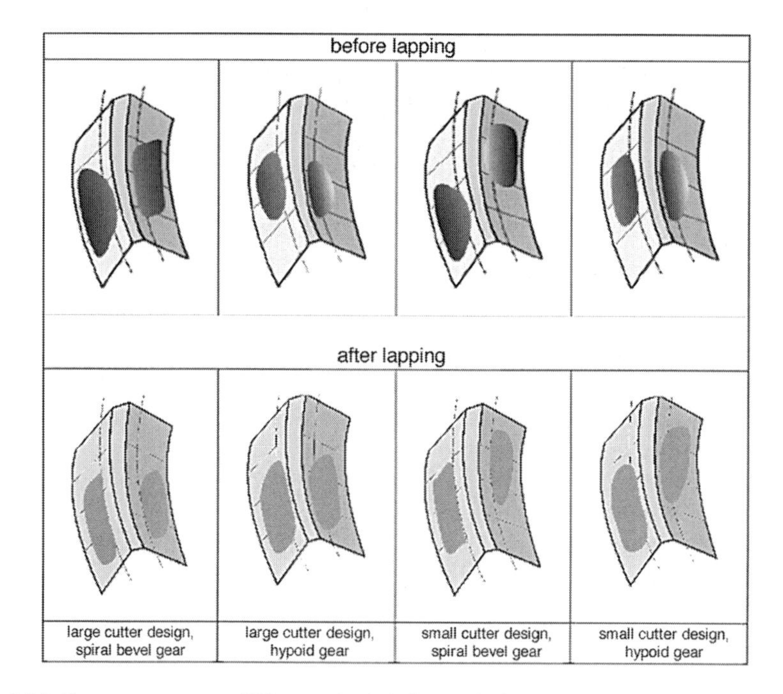

Fig. 6.34 Contact patterns on different wheels before and after lapping

6.6.4.3 Lapping Machine Parameters

The setting parameters for the lapping machine are the speed of rotation, the braking torque, the lapping time and the lapping motion. The speed of rotation is determined from the desired circumferential speed for the particular mean diameter of the wheel. The permissible circumferential speed will be from 3 to 4 m/s on older mechanical machines to about 7–10 m/s on modern CNC machines.

Torque, like the speed of rotation, depends not only on the type of machine, but also on the pitch qualities of the pinion and wheel which may excite rotary oscillations. For CNC machines, values in the region of 2.5–4 Nm per 100 mm wheel diameter are customary. To influence the lapping result, both speed of rotation and torque can be varied during the lapping process if technically possible on the lapping machine.

The lapping motion has a great influence on the lapping process. It is made up of a series of pinion and wheel displacements and the sequence in which they are realized. Apart from displacement speed, the dwell time spent in the individual positions affects the result.

The axial shifts along V and H determine the displacement of the contact zone from the start position to the extreme positions (see Sect. 3.4.3). The V and H values are determined mainly by the V-H check, and are fixed with the desired displacement of the contact pattern and desired bias of the gear set after lapping.

Backlash is reduced during lapping to prevent unlapped regions at the root of wheel teeth to come into contact with the mating pinion tooth tips. This is realized by shifting the wheel axially, which typically reduces backlash by 30 %.

The lapping time "per half round" describes the time taken to move the contact zone from the starting position to an extreme position and back. It can be entered separately for each of the extreme positions. Reference values of 1.5–2 s per 0.1 mm of V displacement can be used for gears cut with medium to large tool radii, or 0.8–1.5 s for small tool radii. Dwell time can be included at any desired point to increase material removal.

6.6.5 *Using Lapping to Change Running Properties*

The tooth flank topography of the pinion and wheel changes due to uneven material removal. In a typical automotive lapping process, lengthwise crowning is reduced by 30–40 % and profile crowning by 50–70 %. Besides reducing flank crowning lapping typically leads to bias-in behavior being advantageous for the running properties (see Sect. 3.4.3).

Lapping usually reduces transmission error by 60–90 %. At the middle of the tooth, material removal is about 5–30 μm per tooth flank, which enlarges the contact pattern.

Roughness attained by lapping fundamentally depends on grain size and lapping time. As a reference, roughness R_z in the order of 6–8 μm can be attained using 280 grain. Surface structure is oriented in the sliding direction, which can have advantages for efficiency.

References

[BART06] Barth, W.: Fräsen, Fließpressen und Sintern von Kegelrädern für Power Tools – Ein Wettbewerb der Technologien. Innovation rund ums Kegelrad (2006)

[CZEN00] Czenkusch, C.: Technologische Untersuchungen und Prozessmodelle zum Rundschleifen. Dissertation, Universität Hannover (2000)

[DEGN02] Degner, W.: Spanende Formung. Theorie, Berechnung, Richtwerte. Carl Hanser Verlag, München (2002)

[DIN-EN10052] DIN-EN 10052, Ausgabe 1994-01: Begriffe der Wärmebehandlung von Eisenwerkstoffen (1994)

[FEPA06] FEPA: FEPA-Standard 42-1. www.fepea-abrasives.org (2006)

[GIES05] Gießmann, H.: Wärmebehandlung von Verzahnungsteilen. Expert, Renningen ISBN3-8169-1928-6 (2005)

[HOFM39] Hofmannn, F.: Gleason Spiralkegelräder. Springer, Berlin (1939)

[KLOC05] Klocke, F.; König, W.: Fertigungsverfahren Band 2 – Schleifen, Honen, Läppen. 4. Aufl. VDI, Düsseldorf (2005)

[KREN91] Krenzer, T.: CNC Bevel gear generators and flared cup formate gear grinding. AGMA Technical Paper 91 FTM 1 (1991)

[KRUM41] Krumme, W.: Klingelnberg Palloid-Spiralkegelräder. Verlag Julius Springer, Berlin (1941)

[LIER03] Lierse, T.: Abrichten von Schleifwerkzeugen für die Verzahnungsbearbeitung. Seminar: Feinbearbeitung von Stirnrädern in Serie 3./4.12.2003 Aachen

[PRRE60] Preuger, E.; Reindl, R.: Technisches Hilfsbuch Klingelnberg, 14. Auflage Verlag Julius Springer, Berlin

[SIZ05] Stahl-Informations-Zentrum: Merkblatt 450, Postfach 10 48 42, 40039 Düsseldorf (2005)

[STAD90] Stadtfeld, H.J., Kotthaus, E.: Das kontinuierliche Schleifen gehärteter Kegelräder mit bogenförmiger Flankenlinie. In: Bartz, W.J. (Hrsg.): Kegelradgetriebe. 1. Aufl., S. 145–163. Expert, Ehningen (1990)

[TUER02] Türich, A.: Werkzeug-Profilerzeugung für das Verzahnungsschleifen. Dissertation, Universität Hannover (2002)

[WEIS94] Weissohn, K.H.: Die Technik der C-Pegel-Regelung, HTM 49 (1994)

[WESS07] Wessels, N.: Analyse des Abrichtprozesses von Korund-Topfscheiben für das Kegelradschleifen. 48. Arbeitsstagung „Zahnrad- und Getriebeuntersuchungen" WZL, Aachen 23./24.05.2007

[WINT11] Winterthur Technology Group: Precision Grinding Wheels, 2011 Catalogue, Winterthur Schleiftechnik AG, Winterthur, Switzerland (2011)

[WUEN68] Wüning, J.: Weiterentwicklung der Gasaufkohlungstechnik, HTM 23 (1968)

[WYSS95] Wyss, U.: Regelung des Härteverlaufs in der aufgekohlten Werkstück-Randschicht, HTM 50 (1995)

Chapter 7
Quality Assurance

7.1 Measurement and Correction

7.1.1 Measuring Tasks

Two tasks are foremost when testing bevel and hypoid gears. The first is to assure the quality specifications of the production job in the certification framework according to the regulations laid down in ISO 9000 ff. and ISO 14000 ff. This task is becoming increasingly important for the manufacturer, as the regulations stipulate that the quality of the products must be documented on a continuous basis. This means that measurements must be made much more frequently than has been the case in recent years. Bevel gear measurement plays an important part in machine capability testing, which represents a decisive criterion for production machines of all kinds.

The second task of bevel gear metrology is to determine the geometry of the tooth flanks as actually manufactured in order to guarantee that the previously calculated and developed form of the tooth flanks, including all modifications, has been achieved in the production process. If tolerances are shown to have been exceeded, the setting parameters of the production machines must be corrected. In the past, such corrections were generally made on the basis of contact pattern tests and then set manually on the machines. Since the 1990s, however, production has been transformed by the quality assurance closed loop (see Sect. 7.1.5), in which corrective machine setting values are calculated automatically from the results of measurement and are then transferred to the machine. In the spiral bevel gear manufacturing sector, this has made bevel gear production much more flexible and progressive than that of other gear types.

J. Klingelnberg (ed.), *Bevel Gear*, DOI 10.1007/978-3-662-43893-0_7

7.1.2 Pitch Measurement

In principle, circular pitch measurement on bevel gears follows exactly the same procedures as on cylindrical gears. A measurement of this kind could be made using simple pitch testing devices without any detailed knowledge of the design. However, it has become customary to combine pitch measurement to the study of tooth flank topography using a 3D gear measuring machine. The pitch is measured in the transverse section, generally in the middle of the tooth flank. Depending on the chosen standard, the assessment is converted to the point at which the relevant tolerances relate.

For example, DIN 3965 [DIN3965] is used for the pitch quality of bevel gears. The tables contained in DIN 3965 refer to the corresponding cylindrical gear standard [DIN3962], but allow bevel gears a "bonus" accuracy grade. This means that a bevel gear is graded one accuracy grade higher than a cylindrical gear of the same size with exactly the same pitch deviation. As the DIN tolerances are production-oriented rather than function-oriented, this accounts for the enhanced manufacturing effort which has gone into bevel gear production with the same deviations as corresponding cylindrical gears. As in all older gear standards, the tables are graded by module and diameter. When comparing similar bevel gears, a jump in accuracy grades can result. Since DIN 3965 refers expressly to DIN 3961 [DIN3961], the pitch tables apply to measurements on the mean pitch diameter and, naturally, in the transverse section alone. However, this fact is not spelled out explicitly in DIN 3965 such that it is repeatedly observed in practice that pitch results are converted to the normal section and compared to the cylindrical gear values contained in the tables. Using this method, which is quite unacceptable, one easily "jumps up" one accuracy grade.

Since 2006, an international standard for bevel gear tolerances, overcoming the shortcomings of DIN, has been published in the form of [ISO17485]. The ISO standard employs formulae to calculate the permitted tolerance values in which there are no longer jumps from grade to grade by way of the module and diameter. The assessment is based on what is referred to as the tolerance diameter (see Table 7.1), located at mid- face width and at half the effective whole depth, and can therefore always be used as the direct measuring point. ISO 17485 is based relatively closely on the tables in [ISO1328] for cylindrical gears and, like DIN, grades bevel gears one accuracy class higher. The calculated tolerances correlate extremely well with those in DIN 3965.

Each individual pitch deviation f_{pt} represents the difference between the actual, measured position of a tooth flank and the theoretical circular pitch. A positive

Table 7.1 Tolerance diameter according to [ISO17485]

Designation	Formula	No.
Tolerance diameter, pinion	$d_{T1} = d_{m1} + 2 \cdot (0,5h_{mw} - h_{am2}) \cos\delta_1 = d_{m1} + 2 \cdot (h_{am1} - h_{am2}) \cos\delta_1$	(7.1)
Tolerance diameter, wheel	$d_{T2} = d_{m2} + 2 \cdot (0,5h_{mw} - h_{am2}) \cos\delta_1 = d_{m2} + 2 \cdot (h_{am2} - h_{am1}) \cos\delta_2$	(7.2)

deviation results if the measured value is larger than the desired value, and vice versa (see Fig. 7.1). The differences, measured from tooth to tooth, are plotted on a chart, starting with deviation 0 on tooth 1. Measurement data determine the single pitch deviation f_{pt} of the bevel gear, which is the maximum amount of all positive and negative single values in µm, and the cumulated pitch deviation F_p, which is the difference between the maximum and minimum of the summed values (see Fig. 7.2 and also [DIN3960] or [ISO1328]).

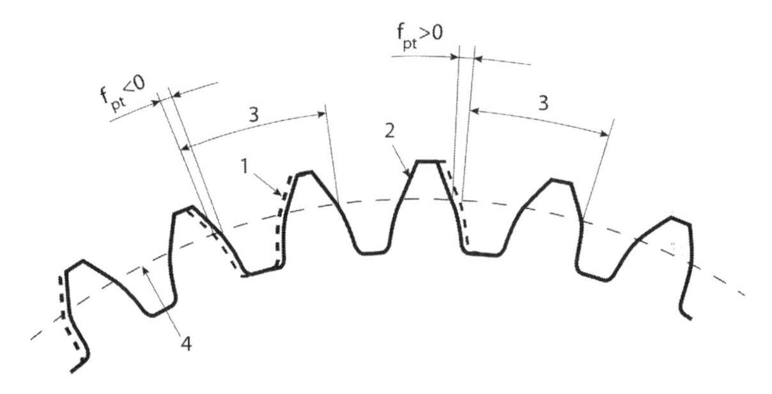

Fig. 7.1 Definition of pitch deviation.
1 theoretical tooth flank position
2 actual tooth flank
3 theoretical circular pitch
4 tolerance diameter

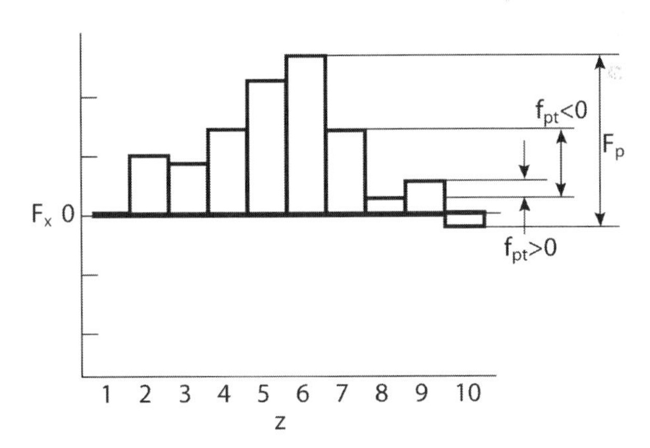

Fig. 7.2 Determination of single and cumulative pitch deviations.
z tooth sequence number
f_{pt} single pitch deviation
F_x pitch deviation
F_p total cumulative pitch deviation

The pitch tolerance may be determined for a quality class designated B using the formulae from Table 7.2. The factor between classes is $\sqrt{2}$. The following rules, to round up or down, are employed:

- If the result is larger than 10 μm, it is rounded to the nearest whole number.
- If the result is larger than 5 μm, but smaller than 10 μm, it is rounded to the nearest 0.5 μm.
- If the result is smaller than 5 μm, it is rounded to the nearest 0.1 μm.

Table 7.2 Pitch tolerances according to [ISO17485]

Designation	Formula	No.
Single pitch tolerance	$f_{ptT} = (0.003d_T + 0.3m_{mn} + 5)\left(\sqrt{2}\right)^{(B-4)}$	(7.3)
Total cumulative pitch tolerance	$F_{pT} = (0.025d_T + 0.3m_{mn} + 19)\left(\sqrt{2}\right)^{(B-4)}$	(7.4)
with	$2 \leq B \leq 11$ $1.0\ \text{mm} \leq m_{\text{mn}} \leq 50\ \text{mm}$ $5 \leq z \leq 400$ $5\,\text{mm} \leq d_{\text{T}} \leq 2{,}500\ \text{mm}$ For other m_{mn}, z and d_{T}, the above formulae may be applied in the same sense, but the results are outside the ISO standard. The same applies to accuracy grades 1 and 12	

7.1.3 Flank Form Measurements

7.1.3.1 Fundamentals

On cylindrical gears, the tooth profile is generally measured against a desired reference involute, which is represented in the test chart as a straight line. In the same way, the actual tooth trace is compared to the desired tooth trace, likewise represented as a straight line. On crowned cylindrical gears, deviations are toleranced by means of templates (K-profiles). Even when ignoring the fact that crowning can be in any direction, bevel gears do not have an involute tooth profile and, at least in the case of spiral bevel gears, no straight tooth trace. Therefore, bevel gear tooth flanks cannot be measured like those of cylindrical gears. Some approaches attempt to proceed in a similar manner with bevel gears, measuring individual profile curves and tooth traces against the calculated conjugate mating tooth flank. However, since the results of such measurements simultaneously represent the attained crowning and deviation, the information conveyed is minimal. For these reasons, with bevel gears one rather uses 3D measurement of tooth flank topography compared to that of the desired tooth flank.

7.1.3.2 Flank Form Data and Measuring Grid

Using a tooth flank generator (see Sect. 3.3.1) which accounts for tooth flank modifications, it is possible to calculate the exact 3D coordinates and the corresponding normal vector at any point on a tooth flank. The coordinates of points on a desired measuring grid are calculated, along with tooth thickness which is given in the form of a tooth thickness angle (see Fig. 7.3). The resulting target grid data is fed to a 3D gear measuring machine in a specific file format.

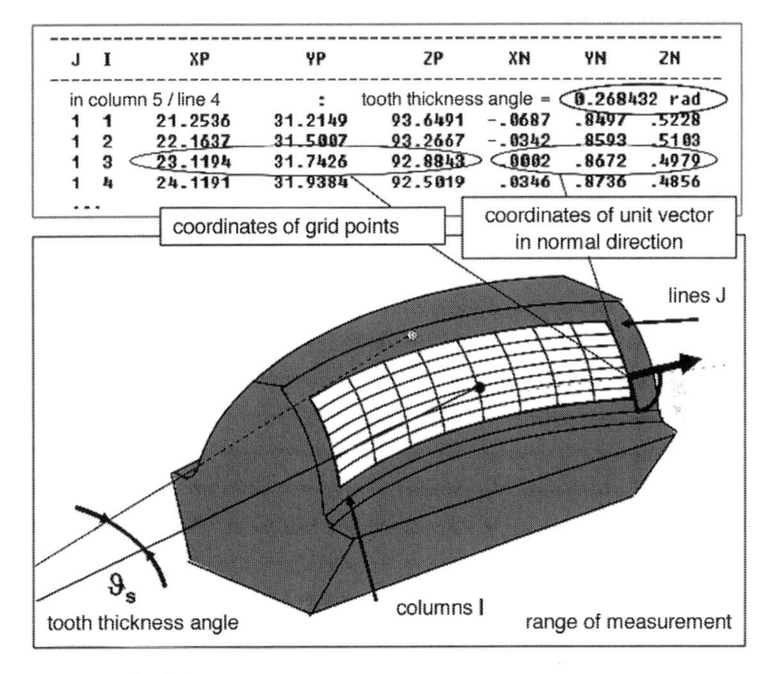

J	I	XP	YP	ZP	XN	YN	ZN
in column 5 / line 4			:	tooth thickness angle =	0.268432 rad		
1	1	21.2536	31.2149	93.6491	-.0687	.8497	.5228
1	2	22.1637	31.5007	93.2667	-.0342	.8593	.5103
1	3	23.1194	31.7426	92.8843	.0002	.8672	.4979
1	4	24.1191	31.9384	92.5019	.0346	.8736	.4856
...							

Fig. 7.3 Nominal flank data

In order to obtain sufficient information to analyze the measured results, a measurement grid must be sufficiently fine and should cover most of the tooth flank area. The following measurement grid recommendations are generally used:

- A measurement grid with five lines and nine columns is often adequate for machine capability tests and production checks.
- More precise analyses can be performed using measurement grids with 39 × 39 points, naturally requiring longer measuring times. There are normally no benefits in using higher point densities.
- Odd numbers of columns and lines are preferred to establish an unambiguous centre point. This centre point should be immediately adjacent to the tolerance diameter d_T (see Table 7.1) and should also be used to determine the actual tooth thickness angle.
- The projection of the tooth flank in the axial plane is normally neither rectangular nor trapezoidal since blank modifications such as cylindrical segments, tip shortening, etc. may be present. To prevent the probe from straying outside of the actual tooth flank, or to guard against losing too much information as a result

of exaggerated toe, heel, tip and root margins, modern calculation softwares apply margins to the actual blank shape (see Fig. 7.4).

- Tip, root, toe and heel margins must be large enough to avoid erroneous measurements in edge bending, chamfers or burr areas. The edge margins of the measurement grid should not be too large and should normally be selected as half the diameter of the probe, plus chamfer width. The margin at tooth root is usually large enough to ensure that measurement will in all cases occur outside the fillet and any protuberance area.

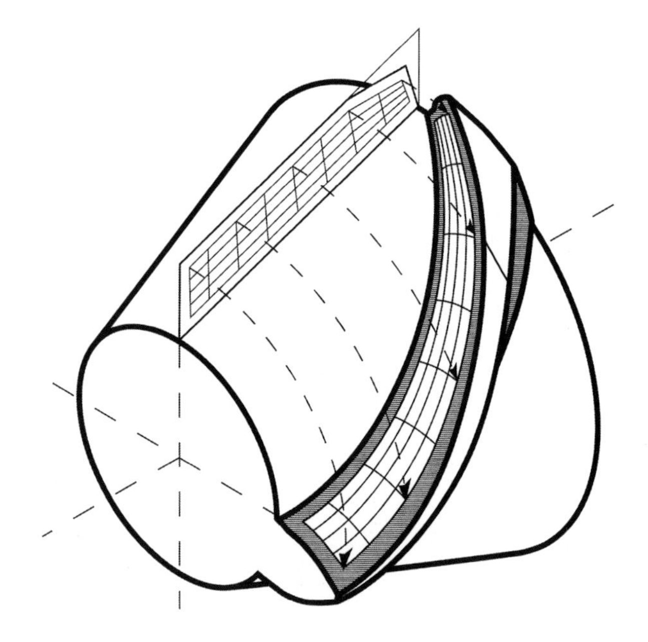

Fig. 7.4 Projection of the measurement grid on the tooth flank

7.1.3.3 Measurement and Analysis

Tooth flanks are measured at the exact points defined in the target measurement grid, using 3D gear measuring machines and accounting for probe radius in the direction of the tooth flank normal vectors. The 3D measuring machine initially searches for a tooth slot and sets the deviation to zero at a reference point, often the central point of the target grid. The target measurement points are then traversed, and the deviations in the normal direction are determined in relation to the reference point. The actual coordinates of the tooth flank points are saved in the same way as the target measurement grid data, and are then available for the analysis of actual contact patterns and for the calculation of machine corrections (see Sect. 7.1.5.2). It is also possible to display the desired tooth flank as a plane and show the measured deviations in a direction normal to this plane (see Fig. 7.5). Typically, three to four teeth evenly distributed around the circumference are measured and the mean measurement values are used to calculate corrections.

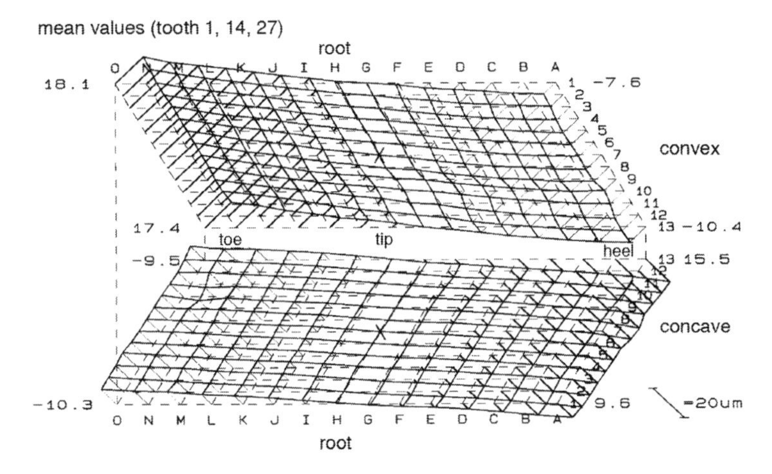

Fig. 7.5 Typical results of tooth flank form measurement

For quantitative analysis, it is possible to output the five parameters used to evaluate the ease-off (see Fig. 3.15), i.e. pressure and spiral angle errors, profile and lengthwise crowning, and bias, these parameters relating to the deviations measured between the desired and actual tooth flanks rather than to the ease-off. Neither DIN 3965 nor ISO 17485 defines tolerances for the tooth flank form of bevel gears.

Another result obtained from tooth flank form measurements on bevel gears is tooth thickness deviation in the transverse section, which stems from the difference between the theoretical (see Fig. 7.3) and measured tooth thickness angle. This difference may be converted to give deviations in the values normally used, i.e. normal circular tooth thickness s_{mn} or chordal tooth thickness s_{mnc}.

7.1.4 Additional Measuring Tasks

7.1.4.1 Face and Root Cone

In addition to the desired measuring tooth flank data, coordinates to measure points on tooth tip, or the face cone, and tooth root, or the root cone, are calculated and supplied to the measuring device.

The result of the face cone measurement does not depend on the gear cutting process and is only a check on the form of the blank, which may, however, be the cause of other deviations.

On the other hand, the result of the root cone measurement depends on gear cutting and can consequently be used to check this process. It should be noted that older calculation programs use the geometry calculations described in Sect. 2.2 to determine the root measuring points along the theoretical root cone. In the case of gear teeth manufactured using machine settings such as tilt and helical motion, this approach leads to misinterpretation. The desired points should therefore always be calculated on the tooth root curve resulting from machine motions (see Fig. 3.2).

7.1.4.2 Whole Depth and Tooth Depth

The tooth whole depth is measured indirectly from the difference in measurements on the face and root cones, e.g. at a point on tooth tip and a corresponding point on the root. Since tooth tip measurement depends solely on the accuracy of the blank, the measured whole depth provides no information about the cutting process. The whole depth actually attained is, therefore, determined using a distance measurement, e.g. from the center point of the tooth root curve to the axis of the bevel gear. This measurement can provide a useful result only if the desired measuring point includes all tooth root modifications. The tooth depth measurement can, amongst other things, be used to judge whether the gear cutting tool has the correct point height.

7.1.4.3 Run-Out

According to DIN, run-out measurements on cylindrical gears are defined as direct measurements of the radial depth position of a specimen such as a ball or sphere in a tooth slot (see Fig. 7.6).

Fig. 7.6 Runout measurement

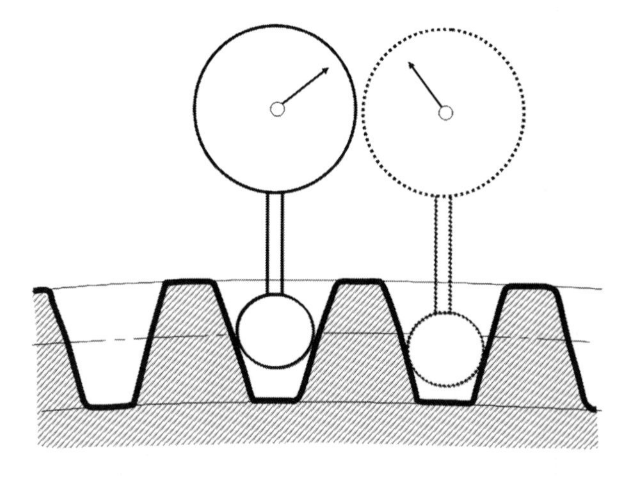

On bevel gears, this measurement is made perpendicular to the pitch cone. In practice, this run-out measurement is only a first check performed directly next to the gear cutting machine in order to obtain quick information about the current tooth slot width or tooth thickness, or about any faulty clamping of the blank. Quantitative run-out assessments are usually calculated from the results of pitch measurement. Permissible tolerances, to determine run-out quality, are specified both in DIN 3965 and in ISO 17485 (see Table 7.3).

Table 7.3 Run-out tolerance according to [ISO17485]

Designation	Formula	No.
Run-out tolerance	$F_{rT} = 0.8(0.025d_T + 0.3m_{mn} + 19)\left(\sqrt{2}\right)^{(B-4)}$	(7.5)

Radial run-out F_r, the difference between the maximum and minimum values in the run-out measurement, is a variable which is always determined using both tooth flanks. It is therefore a measure of the variation in tooth slot width. Occasionally, individual run-out for either the right or left tooth flank is also obtained. There are no standard run-out tolerance values which are calculated from the appropriate pitch measurement and, consequently, a quality rating cannot be assigned to it; given careful analysis, however, conclusions on the running behavior of a bevel gear set can be obtained from it, as from the pitch itself.

7.1.5 Closed Loop Production

7.1.5.1 Fundamentals

Series production of spiral bevel gears now takes place in a closed loop control system. This technology aims to ensure uniform quality in the manufactured components such that a finished bevel gear will match the computer design within close tolerances. Only an exact adherence to the tooth flank micro-geometry optimized during the design phase can guarantee the validity of calculated predictions relating to tooth contact, including load capacity, operational life and noise emissions [TRAP02], [LAND04].

In order to produce bevel gears within the tolerances, it is not enough to monitor the last step in the production process. All the process steps—from roughing through hardening to finishing—need to be controlled in terms of the quality of the semi-finished results in comparison to the desired specifications. In the case of grinding, for example, it is important that the intended grinding allowance should actually be present and that it should be distributed evenly. The accuracy required for the individual steps will naturally differ. The milling of a gear which is subsequently ground will certainly not require the same accuracy as a milling operation prior to lapping.

Closed loop production means that the production process is monitored on the basis of desired specifications and tolerances for the individual steps, and that corrective interventions are made as needed. It should be remembered in this context that lapping, as a hard finishing process with undefined geometry, cannot be closed-loop controlled in the manner defined above.

7.1.5.2 Machine Corrections

Figure 7.7 illustrates the realization of the closed loop principle for gear cutting processes on milling or grinding machines: the starting point is given by the machine and tool settings resulting from design calculations, along with the appropriate theoretical target values.

An ideal gear manufacturing machine would use the specified settings to produce a bevel gear corresponding exactly with the desired theoretical data. The manufacturing process on a real gear cutting machine will, however, lead to deviations from the desired values. The reason for this is, firstly, that every real gear cutting machine is itself subject to faults as a result of tolerances and inherent deflections. Secondly, the manufacturing process is subject to dynamic and thermal effects which generally depend on technological parameters like feed rates and generating speeds.

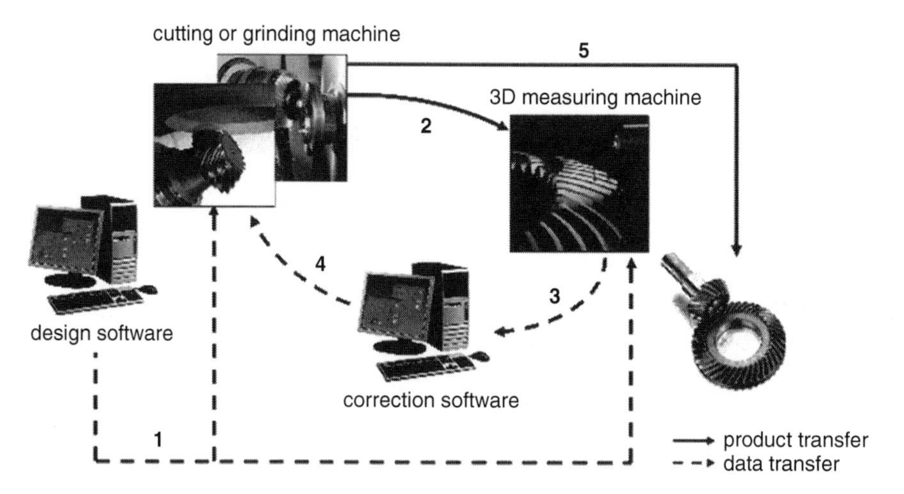

Fig. 7.7 Bevel gear production in a closed loop.
1 theoretical data
2 uncorrected bevel gear
3 actual measured data
4 corrective settings
5 bevel gears complying with the tolerances

The deviations of a generated gear member from the desired data are determined by measurement on a 3D measuring machine. Monitoring by means of manufacturing tolerances determines whether the bevel gear satisfies the quality requirements or whether machine corrections are necessary.

The task is to determine corrective settings for the production machine and if necessary also for the tool settings, which have just been used. The aim is to use the new, modified settings to generate a gear whose deviations from the desired theoretical data are smaller than before—or better—do not exceed the specified tolerances.

When computing corrective settings, certain corrective values are specific to the machine and to the process, i.e. they are valid only for a particular gear cutting machine and only for the process parameters used in production. On the other hand, if corrected machine and tool settings are transferred to another gear cutting machine of the same type while retaining the process parameters, only a slight re-correction will be necessary to compensate the altered basic errors of the machine.

Typical machine corrections include the correction of topography and tooth thickness deviations. If a correct tooth depth measurement can be performed, an automatic tooth depth correction is also feasible but is rarely performed in practice. The correction of pitch deviations also falls into the realm of machine corrections, and is often referred to as pitch compensation (see Sect. 6.2.5.6).

7.1.5.3 Hardness Pre-correction

When bevel gears are hardened, deflections known as hardening distortions occur (see Sect. 6.3.6). These alter the starting geometry for hard finishing, sometimes significantly, and must not be neglected in controlling the complete manufacturing process.

It is hard to predict by calculation the type and size of the hardening distortions which may occur. Hardening distortions of individual parts are often scattered about a mean value. In practice, these mean hardening distortions are determined by measuring a number of hardened bevel gears on 3D measuring devices. It should be noted that bevel gears are manufactured according to desired data in the preceding soft machining process. Only in this way can deviations resulting from hardening distortion be kept separate from machine-dependent production deviations, and compensated by means of a so-called hardening pre-correction.

As Fig. 7.8 shows, hardening pre-correction also takes place in a closed loop process. The starting-point is formed by the designed machine and tool settings with the corresponding target values, according to which the bevel gears are machined and then hardened in the machine closed loop. Subsequent measurement and averaging provides the mean hardening distortion in the form of deviations. On the basis of these deviations, the machine and tool settings can be modified in such a way that the bevel gear which theoretically results from the new settings exhibits almost the original desired geometry after hardening.

Adaptation of the cutting tool may be necessary in order to compensate larger hardening distortions. In a milling process, such a modification of the tool geometry can take place only during the actual process development phase. In this phase, hardening pre-correction is usually passed through only once. Slight re-corrections using modified machine settings then take place as needed in the running process.

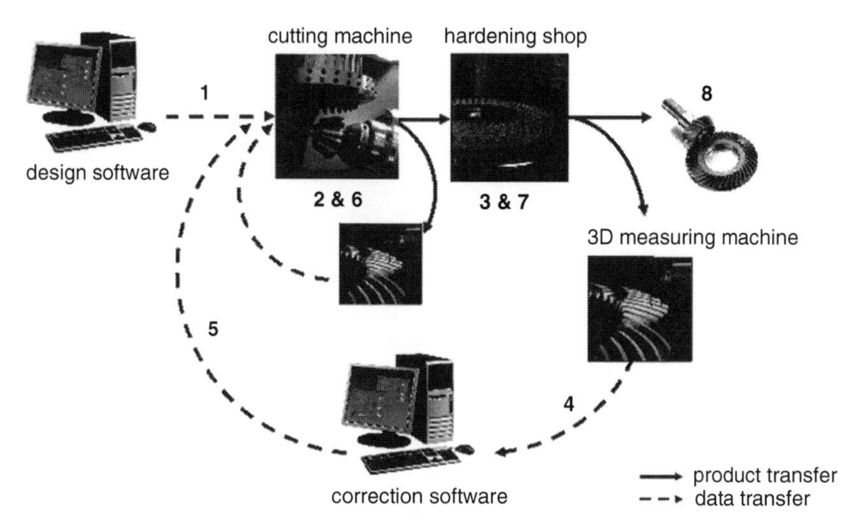

Fig. 7.8 Hardening pre-correction.
1 theoretical desired data
2 closed loop cutting
3 hardening
4 actual measured data of the hardened bevel gear
5 modified nominal data compensating the hardening distortion
6 closed loop cutting with modified nominal data
7 hardening of the compensated bevel gear
8 bevel gears complying with the tolerances

7.1.5.4 Calculating Corrections

Calculation of machine and hardening pre-corrections is very closely related to the calculation of flank modifications, for example for the purpose of contact pattern development. The effects of additional machine motions on the tooth flank form (see Sect. 3.3.4) indicate the fundamental options for kinematic shaping of the flank form.

The basic principle underlying all correction calculations is to compensate, with corrected machine—and possibly tool—settings, the deviations from the nominal data which result from manufacture. The necessary corrections must be determined in such a way that they will approximate the geometry as closely as possible to a reference geometry obtained by mirroring the measured deviations on the desired theoretical geometry. For tooth flank topography, for example, mirroring can take place point wise along the respective tooth flank normal vectors, transferring each measured deviation in the opposite direction along the normal vectors. The same procedure can be followed for tooth depth. For tooth thickness, the tooth thickness angle error is mirrored.

Solving the inverse problem, that is to determine the corrected machine and tool settings for a prescribed reference geometry, i.e. the systematic calculation of flank modifications, is not trivial. Whereas experienced engineers and machine operators can generally manually correct geometry errors on the machine by modifying one or two machine settings, computer technology may be required for complex deviations resulting from machine dynamics, where a combination of several machine settings and several of the additional motions referred to above may provide the solution. It should also be noted that certain deviations simply cannot be corrected.

Modern computer programs to calculate corrections make use of derivative-based numerical non-linear optimization procedures to minimize the deviation from the (virtual) reference geometry and hence the manufacturing deviations [VOGE07]. The derivatives used in the process are also termed sensitivities, and describe the quantitative influence of individual machine and tool settings on the reference variables.

Whether deviations can be corrected depends crucially on which machine and tool settings can be varied. In a milling process, for example, the tool settings in ongoing production are not available for correction. Correction options on purely mechanical machine tools are significantly restricted when compared to those on modern CNC machines, which at least theoretically possess all the kinematic degrees of freedom. The result of a correction calculation is a prediction of the expected manufacturing deviations after the correction has been applied. As a rule, these predictions are very good. However, particularly in the case of large initial deviations or very pronounced dynamic effects, a re-correction is often necessary, owing to the exclusively linear approach employed in mirroring the deviations.

7.2 Testing Bevel Gear Sets

7.2.1 Fundamentals

Industrial requirements for a test method which can be used to assess the manufactured quality of bevel gear sets include:

- reproducibility of the results,
- short test times,
- good emulation of the installed situation,
- a simple good/bad assessment

Running tests on bevel gear sets are cumulative test processes in which a large number of manufacturing-dependent deviations are captured simultaneously. Pinions and wheels can be tested in pairs or compared to appropriate master gears. In most cases, bevel gears are pair-tested and the grading applies to the gear set.

Many running test methods have been developed in the past. They may be categorized as subjective and objective running test methods [EDER01]. Subjective methods include noise tests made by a tester and visual contact pattern testing. Objective methods are single-flank and double-flank tests or the structure-borne noise test. Assessment criteria are the form and position of the contact pattern, the perceived gear noise and standardized or non-standardized parameters of the various objective running test methods. On this basis, it is possible not only to judge running behavior in a certain installed position but also to determine the optimum position in which to install the gear set. To this end, the pinion mounting distance is gradually changed during the test, and an assessment is made for each position. When compared to flank form measurement, the running test has the advantage of short measuring times and conclusiveness on the running behavior of the gear set, coupled to a simple, low-cost test set-up.

7.2.2 Contact Pattern Test

The contact pattern on a gear tooth is the area of a tooth flank which participates in the transfer of force and motion. A contact pattern test on bevel gears usually characterizes the final state of a gear set, and is an integrating variable. Even today, the contact pattern test remains the simplest and most frequently used test, performed either on a running test rig (tester), or on a mounted gear set in the gear housing. Marking compound or contact pattern spray is applied evenly on the teeth of the wheel to make the contact pattern visible. A single or repeated rolling of the pinion and wheel tooth flanks displaces the paste from the contact points. The contact pattern is therefore easily seen on the painted tooth flank. If a spray is used, it is removed in a short test run, rather than being displaced like the marking compound.

Contact patterns are distorted if the compound is applied unevenly or too thickly, or rolled too often, creating a cumulative contact pattern which is shifted by wobble or run-out at each tooth contact, and cumulatively enlarged. The position and size of the contact pattern are dependent on the test load, making it necessary to distinguish between no-load, part-load and full-load contact patterns. However, no conclusion may be drawn from a contact pattern as to the exact load level which has been applied.

The advantage of the contact pattern test is that it can be performed quickly with very little effort. The loaded contact pattern can even be tested with the gear set mounted under the prevailing installation and operating conditions. However, contact pattern assessment is left to the subjective impressions and experience of the particular tester. It is a disadvantage that a contact pattern can be documented only if a copy is made with a strip of adhesive foil or if the contact pattern is photographed. Results and interpretation may change if the contact pattern test is conducted with an infra-red camera (see Sect. 4.4.2.1).

7.2.3 Single-Flank Test

In this method, the pinion and wheel are meshed in their installed positions in the gearbox, only one tooth flank being in contact at any time. The measuring principle of the single flank test is shown in Fig. 7.9. The base of all analyses is the measured fluctuation in the transmission ratio between the pinion and the wheel, referred to as the transmission error or single flank composite deviation. High-resolution incremental angular encoders are connected to the pinion and wheel spindles. The pinion spindle drives the gear spindle to which a braking torque is applied. A control unit transforms the sinusoidal signals from the incremental angular

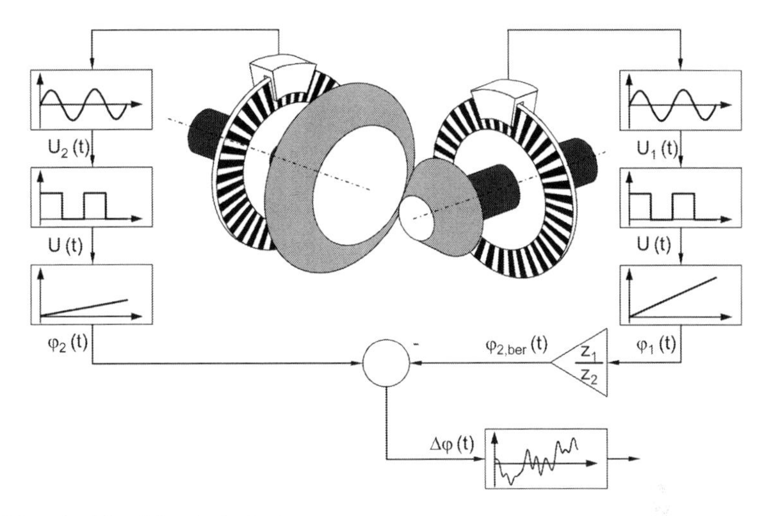

Fig. 7.9 Principle of the single-flank test

encoders such that they can be analyzed by computer software. The software calculates the difference between the measured rotation angle of the wheel and its theoretical rotation angle resulting from the measured pinion rotation angle and the transmission ratio.

The pinion speed and the torque applied to the wheel in the single-flank test must be adjusted so that dynamic effects are largely excluded. The test must be performed under defined flank contact in quasi-static conditions. For this reason, VDI code of practice 2608 suggests pinion speeds of 5–30 min^{-1} at a braking torque of 1–5 Nm. In practice, automotive gear sets are tested at higher speeds of about 60 min^{-1} and braking torques of 20 Nm. To reduce cycle time, the single-flank test is sometimes performed at even higher speeds, but the information conveyed by higher frequency signal components diminishes with rising test speeds [SCHA98].

Since the single-flank test is a cumulative test process, individual deviations are superposed during measurement. This superposition can, on the one hand, lead to a reciprocal reinforcement, but on the other hand to a reciprocal cancellation of the individual deviations.

Measurement during one complete over roll is recommended; this means that, if the gear ratio is not an integer, each tooth of the pinion will mesh once with each tooth of the wheel during the measurement. This ensures that all geometry deviations affecting transmission will be detected. In practice, to shorten the test, measurements over only a few revolutions of the wheel are also usual [EDER05].

Measurement results are presented in the time range as a transmission error curve versus the measured angular position and in the frequency/order range in the form of spectra. Averaging over one revolution of the pinion or the wheel serves to identify transmission error components which are caused by the wheel or the

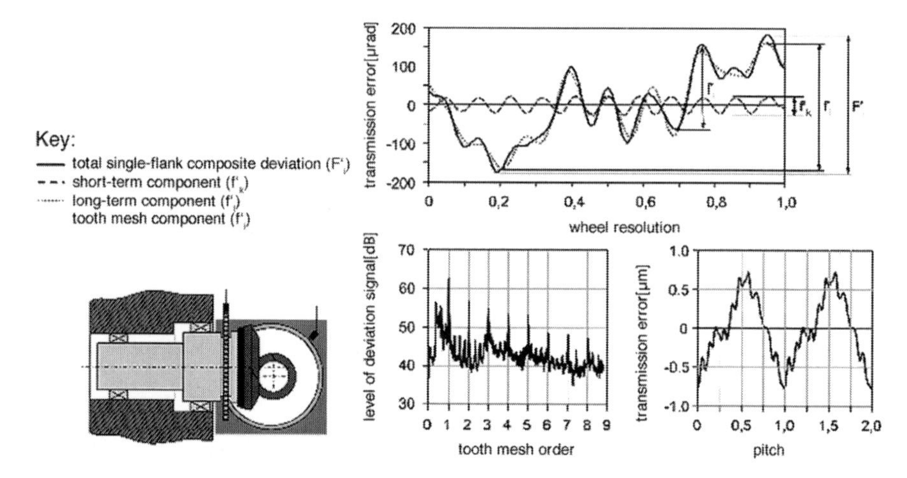

Fig. 7.10 Analyzed variables in the single-flank test according to DIN 3960 or ISO 17485

pinion. The top half of Fig. 7.10 shows an example of a transmission error curve averaged for one revolution of the wheel.

In ISO 17485:2006, four objective parameters are defined for the single-flank test. The total single-flank composite deviation F'_i represents the difference between the maximum and minimum transmission error. A low pass filter with a window of three tooth contacts provides a centering linear curve of the transmission error signal. This long-term component of the transmission error supplies the corresponding parameter f'_l. The difference between the complete signal and the long-term component yields the short-term component of the single-flank deviation. The maximum deviation within a tooth contact is the tooth mesh component f'_i.

In many cases a different analytical method is used. As in acoustics, the transmission error signals are subjected to a Fast Fourier analysis and related to the rotational speed of the wheel. A strong dominance of the tooth mesh order and its harmonics is characteristic of the resulting order spectrum [MARQ95, VDI01].

The transmission error can be indicated as an angle. Alternatively, it can be given as an arc length, the gear diameter of the pitch cone at the calculated point being given as the reference diameter [VDI01]. The advantage of this form of representation is that the transmission error curves are simpler to compare with the ease-off. Another option is to level the transmission error on a reference value. As in acoustics, where measured sound pressure or structure-borne noise signals are shown in leveled form, this facilitates the comparison of signal curves.

Averaging for a single tooth mesh eliminates virtually all long-term influences, allowing consideration of the influences which topography deviations exert on the transmission error curve per tooth mesh. A large number of studies and publications have been made on the effects of gear tooth deviations [FAUL68], [SMIT84], [LAND03]. Section 5.3 examines in detail manufacturing-dependent transmission errors and their effects on noise.

7.2.4 Double-Flank Test

In a double-flank test, cylindrical gear sets are rolled together at a varying center distance while maintaining permanent double-flank contact. The measured variable is the variation in the double-flank center distance $\Delta a''$. The principle of the double-flank test is shown in the left half of Fig. 7.11. During the rolling process, a force in the direction of the center distance prevents separation of the meshing tooth flanks.

Measurement and analysis of the double-flank test curves are analogous to those of the single-flank test. Likewise, measurement of one complete over-roll is time consuming. Therefore, in order to reduce test time, only one revolution of the wheel is usually measured and analyzed.

The double-flank radial composite deviation F''_i is the difference between the maximum and minimum values of the center distance of cylindrical gears. A centering linear curve yields the double-flank run-out F''_r, which is plotted in the top half of Fig. 7.11.

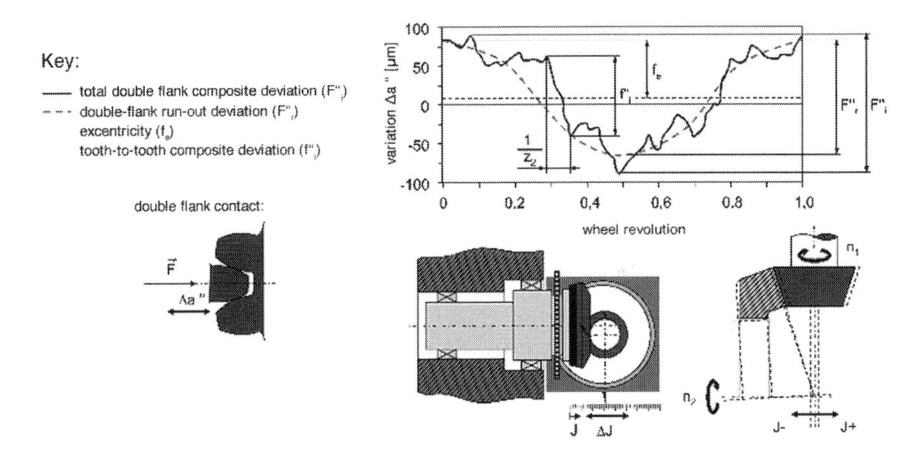

Fig. 7.11 Variables in the double-flank test according to DIN 3960 or VDI 2608

The tooth-to-tooth double-flank composite deviation f''_i indicates the maximum change in the center distance during a tooth contact. Unlike the single-flank test, the double-flank test does not include analysis of the short-term signal component. Instead, VDI code of practice 2608 defines the eccentricity f_e which represents the amplitude of the almost sinusoidal double-flank run-out.

The double-flank test is used with bevel gears as a rapid pre-testing method on lapping and testing machines. The measured parameter is the variation in mounting distance when meshing the pinion and wheel without backlash as shown in the bottom half of Fig. 7.11. The primary objective of the double-flank is to rapidly capture the total run-out and any damage on the pinion or wheel. Because of the double-flank contact, no conclusion as to running behavior can be made on the basis of this method.

7.2.5 Structure-Borne Noise Test

Another objective method in judging the running quality of a gear set in single-flank contact is the structure-borne noise test. Unlike the test methods presented above, this analysis of bevel gear set quality relies on the measurement of acceleration forces rather than on the determination of geometry deviations. For this reason, test rotation speeds in structure-borne noise measurement are significantly higher than those of quasi-static methods, thereby shortening test times. However, when interpreting test results, it is necessary to account for the dynamic behavior of the test machine since measurement results are affected by the entire oscillatory system, i.e. the meshing gear set and the machine. The principle of the structure-borne noise test is shown in Fig. 7.12.

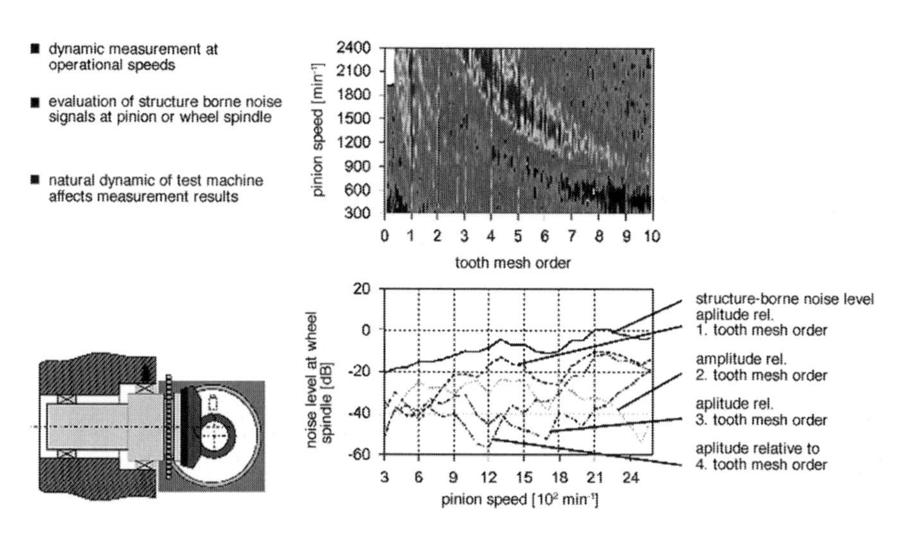

Fig. 7.12 Structure-borne noise test at operationally relevant rotation speeds

In most applications, the radial acceleration in the housing area near the wheel spindle is recorded. Certain test centers also have acceleration sensors on the pinion spindle.

In dynamic test methods like structure-borne noise measurement, the test speed has a substantial influence on the measured parameters which usually are the amplitudes of tooth mesh orders and their harmonics. An exact knowledge of the resonant frequencies of the test machine is crucial when interpreting the amplitudes of the various orders. If a tooth mesh order or one of its harmonics coincides with a resonant frequency in the test machine, the amplitudes are reinforced significantly. The measurement usually takes place at a fixed speed. It is possible to interpret single orders of the structure-borne noise signal in the resonance range of the test machine. The advantage of a measurement in the resonance range is that differences in the running and noise behavior of different gear sets are clearly visible. A disadvantage is the relatively poor reproducibility of the measured results at any operating point because of the high amplitude gradient. Alternatively, the test can be conducted at an operating point at which the dominant excitation frequency and the machine resonant frequencies differ. This stabilizes the results and makes them easier to reproduce.

The optimum operating range of the bevel gear tester is usually determined in preliminary tests, in order to establish the critical frequency ranges for structure-borne noise in the installation.

7.2.6 Comparison of Rolling Test Methods

There are specialized single-purpose devices for the test methods described in Sect. 7.2. Modern bevel gear testers are equipped with measuring systems for the single-flank and double-flank tests and with sensors for structure-borne noise testing. In the automotive industry, test parameters are usually decided according to the application; all the bevel gear test methods presented above may be used [STAH04]. Table 7.4 provides a brief overview of the advantages and disadvantages of the various methods.

Table 7.4 Comparison of three rolling test methods

Parameter	Single flank test	Double flank test	Structure-borne noise test
Measuring time	–	+	+
Equipment cost	–	+	–
Reproducibility	+	+	–
Noise prediction	0	–	0
Combination with contact pattern test	+	–	+

The measuring time for the single-flank test is longer than that of the dynamic structure borne noise test because of restrictions in rotation speeds. Equipment costs for the two methods are roughly the same.

In general, the reproducibility of the structure-borne noise test is unsatisfactory [EDER05] since results depend on the resonant dynamics of the bevel gear tester. Results obtained on different testers are more difficult to compare than those from quasi-static measurements. This means that thresholds have to be established individually for each tester.

A final evaluation criterion of the three above rolling tests is the ability to combine the contact pattern test which continues to be highly informative about the manufacturing quality of a bevel gear set. The contact pattern test can be combined only with the single-flank test and the structure-borne noise test in terms of a correct operating position. In many cases, only these two tests permit the determination of the optimum operating position.

References

[DIN3960] Begriffe und Bestimmungsgrößen für Stirnräder (Zylinderräder) und Stirnradpaare (Zylinderpaare) mit Evolventenverzahnung. Beuth Verlag, Berlin (1987)

[DIN3961] Toleranzen für Stirnradverzahnungen – Grundlagen, Deutsche Norm, August 1978

[DIN3962] Toleranzen für Stirnradverzahnungen, Deutsche Norm, August 1978

[DIN3965] Toleranzen für Kegelradverzahnungen, Deutsche Norm, August 1986

[EDER01] Eder, H.: Messtechnik bogenverzahnter Kegelräder in der Automobilindustrie. In: Tagungsband zum 3. Seminar „Innovation rund ums Kegelrad". Aachen 15.-16. März 2001. Aachen: Eigendruck der Aditec gGmbH (2001)

[EDER05] Eder, H.: In Roll Testing Technology of Spiral Bevel and Hypoid Gear Sets. In: Gear Technology, 2005; Ausgabe Mai/Juni 2005

[FAUL68] Faulstich, H.I.: Zusammenhänge zwischen Einzelfehlern, kinematischem Einflanken-Wälzfehler und Tragbildlage evolventenverzahnter Stirnräder. Diss. RWTH Aachen (1968)

[ISO1328] Cylindrical Gears – ISO system of accuracy (1995)

[ISO17485] Bevel Gears – ISO system of accuracy; First edition 2006-06-15

[LAND03] Landvogt, A.: Einfluss der Hartfeinbearbeitung und der Flankentopographieauslegung auf das Lauf- und Geräuschverhalten von Hypoidverzahnungen mit bogenförmiger Flankenlinie. Diss. RWTH Aachen (2003)

[LAND04] Landvogt, A.: Qualitätsregelkreise in Fertigung und Montage. In: Tagungsband zum 4. Seminar „Innovation rund ums Kegelrad". Aachen 2.-3. März 2004. Aachen: Eigendruck des WZLforum gGmbH (2004)

[MARQ95] Marquardt, R.: Einflankenwälzprüfung – Ein Weg zur Lösung von Geräuschproblemen bei Fahrzeuggetrieben. In: wt-Produktion und Management 85 (1995)

[SCHA98] Schaber, G.: Einflankenwälzfehlermessung und schnelle Wälzprüfung in der Serienfertigung. In: Tagungsband zum 1. Seminar „Innovation rund ums Kegelrad". Aachen 25.-26. März 1998. Aachen: Eigendruck der Aditec gGmbH (1998)

[SMIT84] Smith, R.E.: What Single Flank Measurement Can Do For You. In: Fall Technical Meeting Washington, DC, 15.-17. Oktober 1984

[STAH04] Stahl, K.: Flexibilisierung in der Radsatzfertigung für PKW-Achsgetriebe. In: Tagungsband zum 4. Seminar „Innovation rund ums Kegelrad". Aachen 2.-3. März 2004. Aachen: Eigendruck des WZLforum gGmbH (2004)

[TRAP02] Trapp, H.-J.: Mess- und Korrekturstrategien für die CLOSED LOOP-Zahnradfertigung, VDI-Bericht 1673 (2002)

[VDI01] Richtlinie VDI/VDE 2608 Einflanken- und Zweiflanken-Wälzprüfung an Zylinderrädern, Kegelrädern, Schnecken und Schneckenrädern. Beuth Verlag (2001)

[VOGE07] Vogel, O.: Gear-Tooth-Flank and Gear-Tooth-Contact Analysis for Hypoid Gears, Diss. Humboldt-Universität Berlin, Shaker Verlag, Aachen (2007)

Chapter 8
Dynamics of Machine Tools

8.1 Introduction

The static and dynamic compliance behavior of machine tools frequently limit work piece quality as cutting power increases. Work accuracy is determined by deviations from specified machining motions between tool and work piece. These geometric and kinematic deviations are caused by static and dynamic forces which deflect components like slides, machine bed, carriages, spindles, etc. lying in the force flow from tool to work piece [QUEI05].

Using modern computing methods, it is possible to calculate the static stiffness of machine components in the force flow at the concept phase. However, when predicting machine dynamic properties, the damping characteristics of the joining and coupling points between the individual structural elements are highly uncertain. In industrial practice, it is still necessary to rely on metrological tests in order to judge the dynamic compliance behavior of machine tools [BREC06.1].

Optimization procedures can sometimes be derived directly from the measured results. Therefore, it is necessary that research and development predict the structural properties of a machine as far as possible during development of the machine such that optimization can be introduced early in the process, and time- and cost-intensive design adaptations can be avoided [BREC05].

This chapter is intended to provide an overview of the fundamental characteristics of static and dynamic machine behavior, to present state of the art methods in metrological analysis, and to discuss special features involved in examining gear cutting machines.

J. Klingelnberg (ed.), *Bevel Gear*, DOI 10.1007/978-3-662-43893-0_8

8.2 Static Machine Behavior

The size and form accuracy of a work piece is partly determined by the static compliance behavior of a machine. Metrological investigations analyze the relative displacements at the cutting point in response to simulated static process loads. For example, in the case of a milling machine, a steadily increasing force is applied between the machine table and the spindle. A sensor is used to record the resulting displacement between the spindle and the table, subsequently allowing load-deflection behavior to be represented in chart form as a static curve (Fig. 8.1) [BREC06.1].

Once the clearances at the bearings, guides and joining points have been overcome, the compliance of the system generally declines as the load increases. This is caused by non-linear conditions at these contact points. When the force is relieved, hysteresis is often present as a result of changed contact parameters, such that the relief load-deflection curve and the force load-deflection curve do not coincide. In order to characterize three-dimensional compliance behavior, static compliance is determined successively for each of the three coordinate axes.

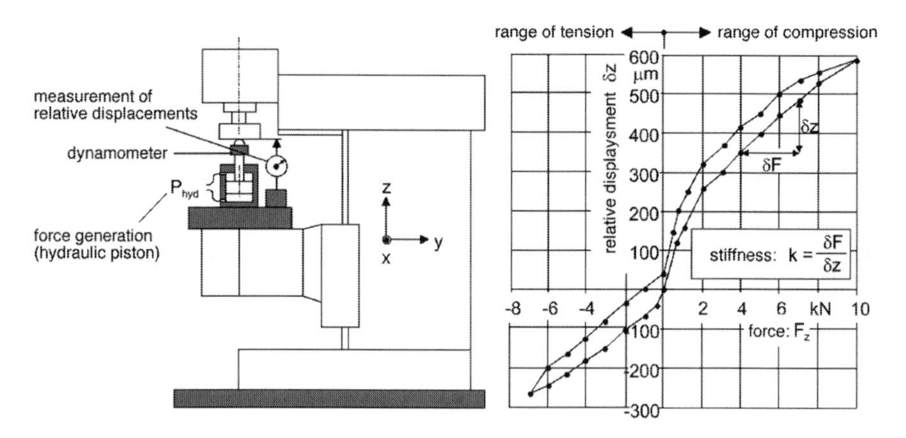

Fig. 8.1 Measurement of the static compliance behavior

A drawback in measuring static characteristic curves is that it is only possible to determine summarily the relative compliances or stiffness between the tool and the work piece at the point where the force is applied. It is not possible to say anything about the shares of the total deflection which befall each of the components and joints lying in the force flow. To answer this question, it is necessary to measure the relative or absolute motion of individual components at many points in the machine structure. The effort required to perform a static load-deflection analysis of this kind is quite substantial. The measuring set-up, in which a large number of sensors are arranged around the machine on a gauging rack, is also susceptible to the influences of temperature fluctuations and vibrations in the immediate environment. The static numerical load-deflection analysis of complete machines is therefore usually preferred to experimental analysis.

8.3 Dynamic Machine Behavior

8.3.1 Simulation Methods

Unbalanced dynamic properties in a machine can lead to process instabilities, i.e. to vibratory phenomena whose consequences include not only poor work piece surface quality and enhanced machine and tool wear, but tool breakage and damage to the work piece and the machine tool. Therefore, the dynamic compliance behavior of a machine tool in response to alternating loads must be seen as a performance criterion.

Machine tools are made up of individual machine components. In terms of dynamic behavior, these components represent a multiple degree of freedom (MDOF) system. In many cases, machine behavior under dynamic load can be described by a system of decoupled single degree of freedom (SDOF) systems. Therefore, it is useful to characterize dynamic machine properties for the example of a SDOF system [HEYL03]. The SDOF system is shown in Fig. 8.2 [BREC06.1]. The equations of motion for the SDOF system are listed in Table 8.1.

Transforming this differential equation into the frequency domain produces a description of the dynamic behavior in the form of the compliance frequency response G(jω) (Formula 8.3) which is caused by the complex ratios between dynamic displacement x(jω) and dynamic force F(jω). The compliance frequency response may be separated into amplitude and phase responses (Fig. 8.3). The amplitude response represents compliance as a function of frequency. The static compliance amplitude, i.e. the reciprocal of the static stiffness l/k, can be read off at a frequency f = 0 Hz. The system possesses its maximum compliance at resonance frequency f_R. The phase response describes the time lag between the excitation force and the resulting displacement.

Table 8.1 Differential equations in the frequency domain

Designation	Formula	No.
Time domain	$m\ddot{x} + c\dot{x} + k\left(x_{dyn} + x_{stat}\right) = F_{stat} + F_{dyn}$	(8.1)
Frequency domain	$\left[m\hat{x}\,(j\omega)^2 + c\hat{x}\,(j\omega) + k\hat{x}\right] e^{j(\omega t+\varphi)} = \hat{F}e^{j\omega t}$	(8.2)
Frequency response	$G(j\omega) = \dfrac{x(j\omega)}{F(j\omega)} = \dfrac{1}{m(j\omega)^2 + c(j\omega) + k}$	(8.3)

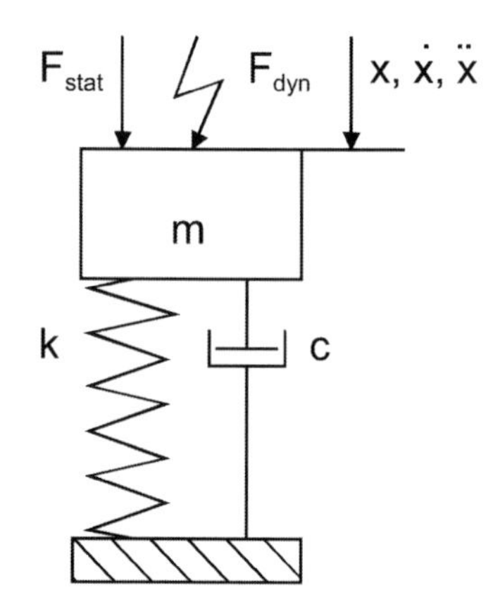

Fig. 8.2 Compliance response principle

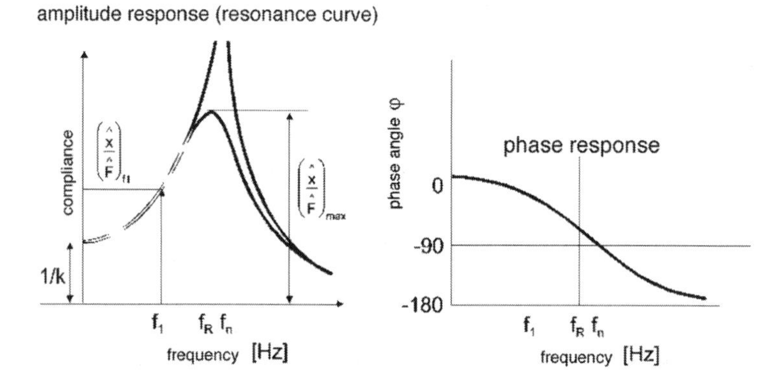

Fig. 8.3 Amplitude and phase response
f_1 frequency at phase angle φ_1
f_R resonance frequency of the damped system
f_n resonance frequency of the undamped system at $\varphi = -90°$
D damping factor at $D = c/2m\omega_n$

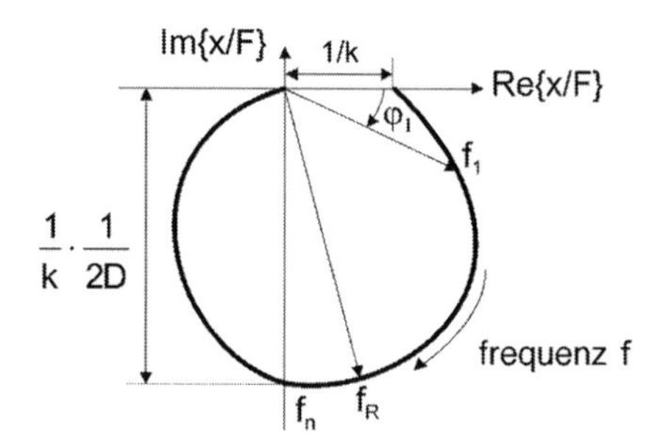

Fig. 8.4 Nyquist plot

A form of representation, equivalent to the amplitude and phase responses, is the frequency response locus, or Nyquist plot, shown in Fig. 8.4. The distance from the origin of a point on the Nyquist plot is the amount of compliance, while the rotation of this position vector in relation to the positive horizontal axis represents the phase. Because of passive system characteristics, there is always a time lag between the force and the displacement which requires negative phase values in the phase response and a frequency parameterization of the frequency response locus in the clockwise direction, i.e. in the mathematically negative direction of rotation.

A fundamental distinction is drawn between two types of machine vibrations: separately excited and self-excited vibrations. Separately excited vibrations are caused by disturbance forces arising inside or outside the machine. These vibrations are, for example, those which

- are introduced via the bed,
- are due to sub-assemblies of the machine (e.g. from imbalances),
- are caused by a discontinuous cutting process.

Resonant frequencies of the machine in the range of separately excited vibrations increase the risk of resonance. The higher the compliance of the machine at the relevant excitation frequency, the greater is the effect of these vibrations.

Unlike separately excited vibrations, self-excited vibrations are initiated not by external energy, but from within the process itself. The occurrence of self-excited vibrations is closely correlated with the dynamic compliance behavior of the machine tool system.

Probably the most frequent cause of self-excited vibrations is the regenerative effect which can occur in cutting processes as well as in grinding processes. The regenerative effect stems from the excitation of vibrations in the machine structure in the region of pronounced resonant points. These are manifested in the form of waviness on the work piece surface. When the tool re-enters the waviness which has been cut previously, the excitation of the machine structure is repeated, possibly leading to unstable process behavior if the system is not adequately damped. Even minimal excitation forces, like cutting force noise which is always present, are enough to initiate the regenerative effect. Apart from system damping or the dynamic compliance behavior of the machine, process stability is also determined by the chosen cutting parameters. The regenerative effect in cutting machines occurs only on the work piece side whereas the regenerative effect in grinding machines can also be induced by wear on the grinding wheel.

A knowledge of what causes vibrations is therefore of crucial importance. Machining tests indicate whether separately excited or self-excited vibrations predominate in the process under investigation. The spectral composition of the vibrations which occur during the cutting operation is analyzed for this purpose.

In the case of separately excited vibrations, the machine tool vibrates at tool mesh frequency or one of its harmonics, while the spectrum for self-excited vibrations primarily reveals vibration components close to a machine resonant frequency. In practice, an important distinction is that, in the case of self-excited vibrations, only slight changes in the frequency of the machine vibrations appear when the process excitation frequency is altered—for example by varying the rotation speed of a milling cutter—while in the case of separately excited vibrations the dominant machine frequency changes in proportion to the speed of rotation. With separately excited vibrations, changing the tool rotational frequency is often enough to stabilize the process.

A simulation at the design phase can determine the frequency and amplitude of machine vibrations before a new machine is built. Subsequent testing of dynamic machine behavior, in the form of compliance frequency responses, can be used to improve modeling for future designs. For example, an actuator can be used to apply an alternating force to the machine structure. Absolute or relative actuators can be used. They are located between the tool and work piece with a static pre-load to simulate the static cutting force component in the process and eliminate the influence of clearances on measurement results (Fig. 8.5).

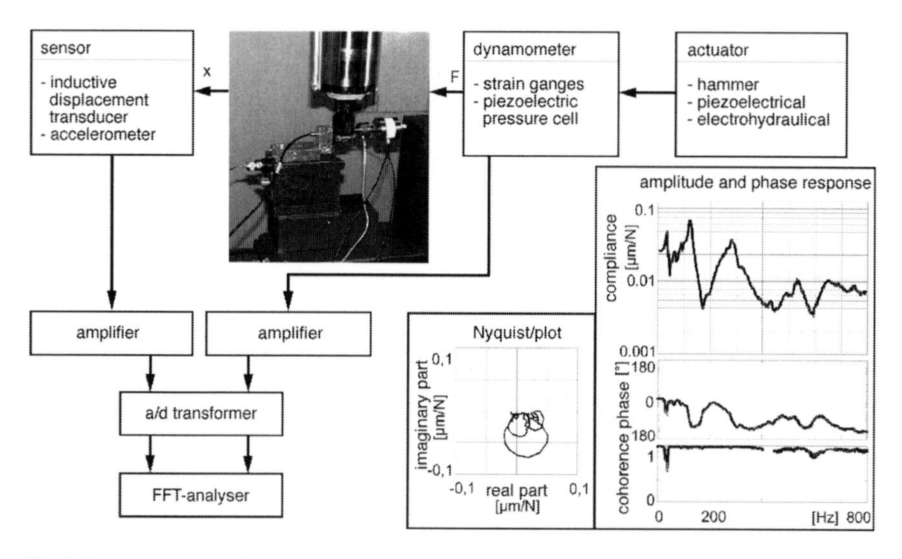

Fig. 8.5 Measurement of compliance frequency responses in machine tools

At the same time, the relative displacement is captured by an inductive displacement transducer and the acceleration signals of both the work piece and the tool are measured by accelerometers.

The sensors and the actuator are often attached separately to the tool and to the work piece. The measuring set-up, which is repeated for each of the three orthogonal directions, is relatively time-consuming.

The measurement system shown in Fig. 8.6, with all the sensors and the actuator integrated in a single housing, is an exception in this respect. Apart from the advantage of a cleanly arranged and time-saving set-up, the type of fitting and the sensor positions are always the same, irrespective of the machine being tested, and this significantly increases measurement repeatability [BREC06.2, HANN06].

In the interest of a stable set-up, on gear cutting machines in particular, it is often advisable to substitute dummy elements for the tool or the work piece, as this allows reliable support of the excitation system. Figure 8.6 shows an example of a gear grinding machine where the grinding wheel has been replaced by a substitute tool of comparable mass.

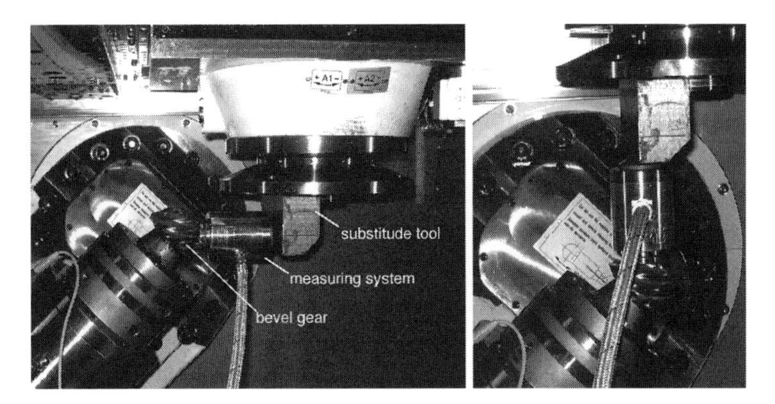

Fig. 8.6 Sample set-up on a gear grinding machine

Different types of signals are used to trigger actuators in frequency response measurements [EWIN03]. The objective in every case is to attain the most evenly distributed vibration energy over the relevant frequency range, i.e. to excite the machine sufficiently in all the frequencies of interest. A periodic excitation using a variable-frequency sinusoidal signal, or a noise signal with stochastic distribution of the excitation frequencies, can be employed. A periodic signal is superimposed on the stochastic signal to systematically excite certain frequency ranges in a machine.

Pulsatile excitation with what is known as an impulse hammer is another possibility. The desired excitation frequency spectrum needs to be taken into account when choosing the material of the striking surface of the hammer, as the duration of the pulse is longer with a relatively soft material than with a hard material. The maximum excitation frequency is inversely proportional to pulse length, such that lower frequencies can be obtained, for example, with a rubber rather than a steel striking surface. Despite the relative technical simplicity of the set-up and the ability to measure delicate components like slender tools and work pieces, it is a drawback that a static pre-load cannot be applied as in the case of a relative actuator.

The measured acceleration signals are converted into displacements by double integration in order to obtain the compliance frequency response $G(j\omega)$. The recorded force and displacement signals are then translated into the frequency domain by means of Fast Fourier Transform. The relative extremes of the compliance curve, at which the amplitude can often measure many times the static compliance, are the resonance points of the machine. Resonance of the machine structure, where there is a large phase offset between the exciting force and the resulting displacement, may be critical for process stability. An unfavorable phase position with values below -90° can result in self-excited vibrations in machine tools used for cutting.

Coherence is an important criterion for measurement quality; it indicates the extent to which measured displacements correlate with the applied force signals (ideal correlation: coherence = 1; no correlation: coherence = 0). Reliable measurements should therefore possess coherence values greater than 0.8 across the entire test frequency spectrum. If not, the measurement set-up should be checked. Figure 8.7 shows a dynamic compliance measurement set-up on the dressing spindle of a gear grinding machine.

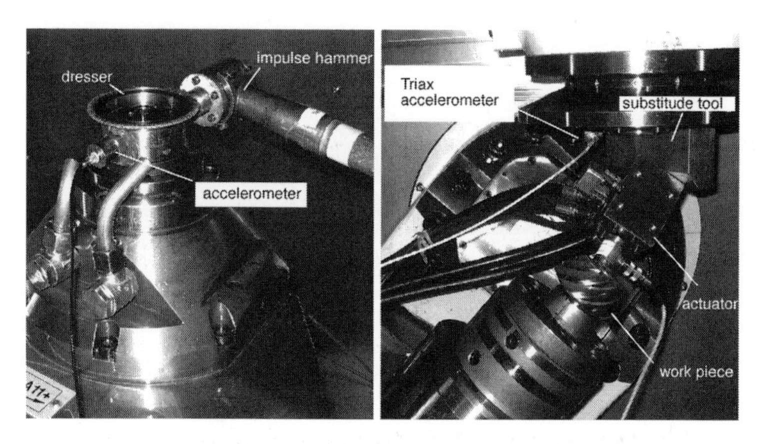

Fig. 8.7 Compliance measurement on a dressing spindle (*left*) and three-dimensional measuring set-up for modal analysis

8.3.2 Modal Analysis

The first step in modal analysis is to approximate the geometry of the machine in a mesh structure. The nodes of the mesh represent measuring points at which displacements of the individual machine components will subsequently be detected in three orthogonal coordinate directions. A three-dimensional application of force to the machine, relative to the tool and work piece, is preferable in order to include all the resonant frequency modes of the structure in the analysis. As in the frequency response measurement, stochastic excitation signals ranging from 0 Hz to 1,000 Hz are generally employed. The displacements in the machine are measured by accelerometers at each of the measuring points, preferably in the coordinate directions, and are related to the applied forces. The transfer functions determined in this manner are then interpreted in terms of amplitude peaks and phase positions in the range of dominant resonant frequencies extracted from the compliance frequency responses [EWIN03].

Determination of the transfer functions for each point in the structure is followed by "curve fitting" [KIRC89] in which the measured transfer functions are described mathematically. In this approximation of the transfer functions through complex analytical equations, the frequency and the damping are required for each important resonant frequency mode, each point in the structure and each direction of the complex compliance vector. There are many different curve fit processes, generally distinguished into two categories, [NATK92]:

- single degree of freedom (SDOF) processes and
- multiple degree of freedom (MDOF) processes

In the SDOF process, the structure is described by a system of decoupled single mass oscillators with a single degree of freedom, i.e. the mode is determined as if this were the only resonant frequency of the structure. The cumulative compliance frequency response of the structure can then be described as the sum of all individual frequency responses. It should be pointed out that the individual vibration modes of the tested structure must be clearly decoupled if the SDOF method is used. This is manifested in the compliance frequency responses by individual resonance points which are well separated in terms of their frequencies, and hence have only a slight influence on each other.

SDOF processes produce quick results and require only limited computing power. They are usually much less accurate than MDOF processes, since the requirement for decoupled systems is met only in very rare cases. MDOF methods simultaneously determine the resonance points, damping and vibration amplitudes for several vibration modes. They can provide a more precise description of a dynamic behavior through their ability to detect decoupled and heavily damped vibration modes [LMS00].

By assigning these values to the individual measuring points in the geometry mesh, it is possible to display the vibration modes of the machine in animated form. Coupled with the compliance frequency responses, which allow quantification of vibration amplitudes, modal analysis can then become an important tool in the analysis of dynamic machine behavior.

References

[BREC05] Brecher, C., Weck, M.: Werkzeugmaschinen – Konstruktion und Berechnung, Band 2. Springer, Berlin (2005)

[BREC06.1] Brecher, C., Weck, M.: Werkzeugmaschinen – Messtechnische Untersuchung und Beurteilung, Band 5. Springer, Berlin (2006)

[BREC06.2] Brecher, C., Hannig, S.: Entwicklung eines Systems zur Stabilitätsanalyse und Prozessauslegung von Schleifprozessen, Abschlussbericht FWF Forschungsvorhaben Nr. AiF 14185N (2006)

[EWIN03] Ewins, J.D.: Modal Testing: Theory, Practice and Application. 2nd edn. Research Studies Press. ISBN 0863802184 (2003)

[HANN06] Hannig, S.: Analysis and Modelling of the Dynamic Behavior of Grinding Processes, Fortschritt-Berichte VDI Nr. 660, Bremen (2006)

[HEYL03] Heylen, W., Lammens, S., Sas, P.: Modal Analysis Theory and Testing, KU Leuven. ISBN 90-73802-61-X (2003)

[KIRC89] Kirchknopf, P.: Ermittlung modaler Parameter aus Übertragungsfrequenzgängen; Diss. TU München (1989)

[LMS00] Theory and Background, Firmenschrift, LMS International. Leuven, Belgien (2000)

[NATK92] Natke, H.G.: Einführung in Theorie und Praxis der Zeitreihen- und Modalanalyse. Vieweg, Wiesbaden (1992)

[QUEI05] Queins, M.: Simulation des dynamischen Verhaltens von Werkzeugmaschinen mit Hilfe flexibler Mehrkörpermodelle, Diss. RWTH Aachen (2005)

Erratum to: Bevel Gear

J. Klingelnberg (ed.), *Bevel Gear*,
DOI 10.1007/978-3-662-43893-0

DOI 10.1007/978-3-662-43893-0
DOI 10.1007/978-3-662-43893-0_2
DOI 10.1007/978-3-662-43893-0_3
DOI 10.1007/978-3-662-43893-0_4

Frontmatter
Page VII, line 9 from the bottom
B was replaced by b

Page VII, line 11 from the bottom
b_M was changed B_M

Page XXVI
Garching was replaced by Eching-Dietersheim

The updated original online version for this book can be found at
DOI 10.1007/978-3-662-43893-0
DOI 10.1007/978-3-662-43893-0_2
DOI 10.1007/978-3-662-43893-0_3
DOI 10.1007/978-3-662-43893-0_4

J. Klingelnberg (ed.), *Bevel Gear*, DOI 10.1007/978-3-662-43893-0_9

Chapter 2

Page 19, Table 2.1 (last column)

"Radius difference" was replaced by "Cutter tilt" 4 times

Page 36, Table 2.7

σ was replaced by Σ in the whole table

Page 42, Table 2.12, equ. 2.79

D was replaced by C

Page 45, Table 2.16

"the face hobbed pinion" was replaced by "all pinions"

Chapter 3

Page 67, Figure caption 3.6

"γ" was moved to the next line

Chapter 4

Page 128, Table 4.16

Equation 4.115 was changed to:

Lengthwise curvature factor	$K_{F0} = 0{,}211 \left(\frac{r_{c0}}{R_{m2}}\right)^q + 0{,}789$	For spiral bevel gears (face milled)	(4.115)

Page 128, Table 4.18, Equation 4.122

Y_ε was replaced by Y_{LS}

Page 184

Sentence "where: $\xi_V{}^* = \xi_V / m_{mn}$ and $\xi_R{}^* = \xi_R / m_{mn}$ (see Fig. 4.35)" was moved to right

Page 189

Layout of equation 4.349 was changed

Trademarks

Coniflex	Registered trademark of The Gleason Works, Rochester/NY (USA)
FORMATE	Registered trademark of The Gleason Works, Rochester/NY (USA)
Palloid	Registered trademark of Klingelnberg GmbH, Hückeswagen (D)
PENTAC	Registered trademark of The Gleason Works, Rochester/NY (USA)
Phoenix	Registered trademark of The Gleason Works, Rochester/NY (USA)
Revacycle	Registered trademark of The Gleason Works, Rochester/NY (USA)
RIDG-AC	Registered trademark of The Gleason Works, Rochester/NY (USA)
SINGLE CYCLE	Registered trademark of The Gleason Works, Rochester/NY (USA)
Spirac	Registered trademark of Klingelnberg AG, Zürich (CH)
TRI-AC	Registered trademark of The Gleason Works, Rochester/NY (USA)
TwinBlade by Klingelnberg	Registered trademark of Klingelnberg GmbH, Hückeswagen (D)
WEDGE-AC	Registered trademark of The Gleason Works, Rochester/NY (USA)

J. Klingelnberg (ed.), *Bevel Gear*, DOI 10.1007/978-3-662-43893-0

Zerol	Registered trademark of The Gleason Works, Rochester/NY (USA)
Zyklomet	Registered trademark of Klingelnberg GmbH, Hückeswagen (D)
Zyklo-Palloid	Registered trademark of Klingelnberg GmbH, Hückeswagen (D)

Index

J. Klingelnberg (ed.), *Bevel Gear*, DOI 10.1007/978-3-662-43893-0

D

E

U

V

W

Z